The
Sacred Mushroom Seeker

Gordon Wasson in Mexico (1955). *Allan B. Richardson. Courtesy Wasson Collection.*

The Sacred Mushroom Seeker

Tributes to R. Gordon Wasson

*Edited by
Thomas J. Riedlinger*

*by Terence McKenna, Joan Halifax,
Peter T. Furst, Albert Hofmann,
Richard Evans Schultes, and Others*

Park Street Press
Rochester, Vermont

Park Street Press
One Park Street
Rochester, Vermont 05767
www.gotoit.com

Copyright © 1990, 1997 by Timber Press, Inc.
This edition published by Park Street Press

All rights reserved. No part of this book may be reproduced or utilized in any form or by any means, electronic or mechanical, including photocopying, recording, or by any information storage and retrieval system, without permission in writing from the publisher.

LIBRARY OF CONGRESS CATALOGING-IN-PUBLICATION DATA

McKenna, Terrence K., 1946–
    The sacred mushroom seeker : tributes to R. Gordon Wasson / by Terrence McKenna . . . [et al.] ; edited by Thomas J. Riedlinger.
      p. cm.
    Originally published: Portland, Or. : Dioscorides Press, c1990, in series: Historical ethno- & economic botany series ; v. 4, and Ethnomythological series ; no. 11.
    Includes bibliographical references (p.   ) and index.
    ISBN 0-89281-338-5
    1. Wasson, R. Gordon (Robert Gordon), 1898– .
    2. Ethnobotanists—United States—Biography.   3. Mycology—Mexico.
    4. Mushrooms, Hallucinogenic—Religious aspects—Mexico.   5. Indians of Mexico—Ethnobotany.   6. Indians of Mexico—Rites and ceremonies.
    7. Hallucinogenic drugs and religious experience—Mexico. I. Wasson, R. Gordon (Robert Gordon), 1898–   . II. Riedlinger, Thomas J.
    [GN21.W38M35 1997]
    394.1'4—dc21   96–54885
                                               CIP

Printed and bound in China

10 9 8 7 6 5 4 3 2 1

Park Street Press is a division of Inner Traditions International

Distributed to the book trade in Canada by Publishers Group West (PGW), Toronto, Ontario

Distributed to the book trade in the United Kingdom by Deep Books, London

Distributed to the book trade in Australia by Millennium Books, Newtown, N.S.W.

Distributed to the book trade in New Zealand by Tandem Press, Auckland

Distributed to the book trade in South Africa by Alternative Books, Randburg

Contents

Preface
THOMAS J. RIEDLINGER 7

Foreword
RICHARD EVANS SCHULTES 13

R. Gordon Wasson:
Brief Biography and Personal Appreciation
J. CHRISTOPHER BROWN 19

Wasson's Help to a Student of Drug History
MICHAEL R. ALDRICH 25

My Life with Gordon Wasson
MASHA WASSON BRITTEN 31

A Vote for Gordon Wasson
MICHAEL D. COE 43

A Bibliophile's View of
Gordon Wasson's Books and Bookplates
ROBERT DEMAREST 47

'Somatic' Memories of R. Gordon Wasson
WENDY DONIGER 55

R. Gordon Wasson: The Poet of Ethnomycology
WILLIAM A. EMBODEN 61

'Vistas Beyond the Horizons of this Life':
Encounters with R. Gordon Wasson
PETER T. FURST 67

Wasson and the Development of Mycology in Mexico
GASTÓN GUZMÁN 83

The Mushroom Conspiracy
JOAN HALIFAX 111

Ride Through the Sierra Mazateca in Search
of the Magic Plant 'Ska María Pastora'
ALBERT HOFMANN 115

Collecting Wasson
MICHAEL HOROWITZ 129

Remembrances of Things Past
IRMGARD WEITLANER JOHNSON 135

The People of Miniss Kitigan Who Were and Are
Honor the Spirit of WaussungNaabe Who Was and Is
KEEWAYDINOQUAY ET AL. 141

My Friend Gordon
WESTON LA BARRE 147

Mixe Concepts and Uses of Entheogenic Mushrooms
FRANK J. LIPP 151

'The Banquet of his Interests'
BERNARD LOWY 161

Wasson's Literary Precursors
TERENCE McKENNA 165

A Posthumous 'Encounter' with R. Gordon Wasson
CLAUDIO NARANJO 177

A Twentieth Century Darwin
JONATHAN OTT 183

Recollections of R. Gordon Wasson's
'Friend and Photographer'
ALLAN B. RICHARDSON 193

A Latecomer's View of R. Gordon Wasson
THOMAS J. RIEDLINGER 205

Mr. Wasson and the Greeks
CARL A. P. RUCK 221

Celebrating Gordon Wasson
ALEXANDER T. SHULGIN 227

Travels with R. Gordon Wasson in Mexico, 1956–1962
GUY STRESSER-PÉAN 231

Appendix I. Gordon Wasson's Account of his Childhood
R. GORDON WASSON 239

Appendix II. Bibliography:
R. Gordon Wasson and Valentina Pavlovna Wasson
J. CHRISTOPHER BROWN 257

Appendix III. The R. Gordon Wasson
Ethnomycological Studies Series. 265

Index 267

Contributors 281

Preface

THOMAS J. RIEDLINGER

On May 13, 1957, *Life* magazine published R. Gordon Wasson's article "Seeking the Magic Mushroom" as part of its Great Adventures series. *Mushrooms, Russia and History,* a book by him and his wife Valentina, was issued simultaneously. Both publications told how he, a vice president of the Wall Street banking firm J. P. Morgan & Co., and she, a pediatrician, had sought and found hallucinogenic "magic" mushrooms being used by religious cults among the Mexican Indians, and how Gordon had eaten the mushrooms in their sacred rituals. This confirmed earlier reports about the Indians by other explorers who published their findings in scholarly journals with limited circulations. However, until Gordon's *Life* article and the Wassons' book, the general public was completely unaware that such a thing as mind-altering mushrooms existed. No "white outsider" before Gordon had partaken of the "mushrooms that cause strange visions" (as the cover of *Life* referred to them). He was the first to whom the Indians fully conveyed their ancient secret, handed down from the Mayas and Aztecs, by inviting his direct participation. And he, by announcing it publicly, shared it with everyone.

After Valentina's death in 1958 Gordon, then 60 years old, continued seeking evidence of sacred mushroom use by indigenous cultures around the world. He had no formal education past the level of a Bachelor's degree in Literature, yet during the last three decades of his life he wrote many scholarly books and articles on hallucinogenic mushrooms that earned the respect of professional scientists. In these publications Gordon argued, for example, that Soma, a previously unidentified sacred substance often mentioned in ancient Hindu writings, was the psychoactive mushroom *Amanita muscaria,* and that the ancient Greek Mystery Rites of Eleusis were secretly based on a sacramental drink prepared with *Claviceps purpurea,* a fungus containing what might be described as a primitive form of LSD.

Bibliophiles also were impressed. Gordon arranged for his books to be privately printed in limited editions of extraordinary quality. Their monetary value as collectibles grew rapidly: today, almost all of them cost between $300 and $750 if purchased from rare book dealers, with copies of the two-volume set *Mushrooms, Russia and History* having sold for well over $2,000.

Gordon took great pride in these important works and in his role as a pioneer "ethnomycologist"—one who studies the cultural uses of mushrooms. In fact, it would not be inappropriate to call him and Valentina the Father and Mother of Ethnomycology, for they were the first to develop this field as a systematic discipline and coined its name.

But Gordon's writings had other effects that were more controversial. Many contend his *Life* article launched, or at least helped trigger, the "psychedelic revolution" of the 1960s. (Timothy Leary, for example, was influenced by it to eat magic mushrooms in Mexico before trying LSD or any other hallucinogen.) Gordon deplored this development because he felt the mushrooms were being used casually, for recreation, rather than respectfully for spiritual enlightenment. This brought a harsh reaction from liberal critics such as Andrew Weil, who complained, in *The Journal of Psychoactive Drugs* [October–December 1988; 20(4):489–490]:

> Wasson was a snob and an elitist about psychedelics, relegating most of those who have experimented with sacred substances to the category of "the Tim Learys and their ilk." Who is he to judge whether others' uses of psychedelics are or are not religious? A great many people in this century have experienced the joy, terror, and mystery of existence through these substances, and there may not be a clear boundary dividing recreation from religion. Wasson was partly responsible for bringing knowledge of sacramental plants and molecules to the masses, yet he was never comfortable with his role as a popularizer and founder of the psychedelic movement. Perhaps the elitist tone of his writing was rooted in that discomfort.

On the other hand, as evidenced by certain of his writings quoted in the present volume, Gordon opposed legislation to ban hallucinogens. In 1978 he told one correspondent:

> You ask for my opinion on legislation to control the use of hallucinogenic mushrooms. I think such legislation is futile and counterproductive. These growths are everywhere. They do no harm, or at any rate minimal harm to the few who abuse them. The harm that they do cannot begin to compare to the harm done by tobacco or alcohol.

He later added, in a draft of the foreword he wrote for a book seller's catalog: "If I had my way, I would make the psychoactive drugs (except alcoholic beverages) as cheap as possible. I would make them available in every drug store without prescription to anyone." This, he believed, would destroy overnight the financial incentives for drug pushers. But he would also "make it clear to those who will succumb to the drugs that not one cent of public money would be spent on their rehabilitation."

Others accused him of violating what they felt to be a sacred trust. The magic mushrooms were profaned, they said, when Gordon shared the secret of their existence with all humankind. He agonized over this charge, which was leveled by some of his Indian contacts as well as less qualified critics, but eventually concluded he had done the right thing. "I arrived [among the Mexican Indians] in the same decade with the highway, the airplane, the alphabet," Gordon wrote in *The Wondrous Mushroom* (1980). "The Old Order was in danger of passing with no one to record its passing." His work thus preserves information about the mushroom cults that otherwise might have been lost forever.

Elsewhere in the same book, Gordon reported that he was repeatedly asked what the future might hold for the mushrooms and other hallucinogens. On this he refused to speculate, claiming he was interested only in past uses of these substances by primitive peoples. He was, however, willing to comment that "the awe and reverance that these plants once evoked" seemed to him gone forever. Yet Gordon's writings paradoxically belie such assertions. He kindles in readers a sense of awe and reverence about the sacred mushrooms that is vital and immediate rather than retrospective. Some perceive it as a subtle invitation to experience the sacrament themselves.

Was Gordon acknowledging "between the lines" a viable role for the mushrooms today and in the future? If so, it almost certainly was tied to his most controversial assertion about them, succinctly expressed in his book *The Road to Eleusis* (1978):

> As man emerged from his brutish past, thousands of years ago, there was a stage in the evolution of his awareness when the discovery of a mushroom (or was it a higher plant?) with miraculous properties was a revelation to him, a veritable detonator to his soul, arousing in him sentiments of awe and reverence, and gentleness and love, to the highest pitch of which mankind is capable, all those sentiments and virtues that mankind has ever since regarded as the highest attributes of his kind. It made him see what this perishing mortal eye cannot see. . . . What today is resolved into a mere drug . . . was for him a prodigious miracle, inspiring in him poetry and philosophy and religion. Perhaps with all our modern knowledge we do not need the divine mushrooms any more. Or do we need them more than ever?

This passage articulates Gordon's belief that hallucinogens played an instrumental role in the origin of religion during prehistoric times, a belief he often discussed in his books. The final sentence is atypical. It seems to make explicit something many find implicit in his writings: the seeds—or more appropriately, spores—of religious renewal. Indeed, some believe the timing of Gordon's discoveries may have been purposeful; that sacred mushrooms are a gift bestowed on suffering humanity by God or extraterrestrials. Such notions may sound fanciful, but Gordon unintentionally fueled them by campaigning to call hallucinogenic substances "entheogens," a term with strong religious connotations meaning "god generated within." His friend and colleague Wendy Doniger even observed, in her review of his last book for *The Times Literary Supplement* (December 23–29, 1988:1412):

> *Persephone's Quest* is the pious meditation of an inspired devotee, a religious book in the deepest sense, the credo of a passionate initiate, indeed a convert. It is, not an academic book at all, but an original religious document; it is the work of a shaman, not of a scholar (even an amateur scholar, which is what Wasson rightly claimed to be).

An alternative view held by some is that Gordon respected religion but was not himself religious. Like William James in *The Varieties of Religious Experience*, he endorsed the psychological validity of strong religious feelings regardless of whether or not they involve an authentic communion with actual deities. From this perspective, the metaphysical significance of drug-induced epiphanies remains an open question.

In any case, the present volume does not seek to resolve speculations about sacred mushrooms per se, but is focused instead on the man who made modern society aware of their existence. Why Gordon is worth our attention should be clear from the foregoing paragraphs. His apparent contradictions were the outward indications of an enigmatic, complex personality. He was both a respectable banker and, like it or not, a "founder" of the psychedelic movement; an elitist about sacred mushrooms but also, through his article in *Life,* their popularizer; a level-headed scientist whose scholarly writings, while grounded in fact, yet inspire many readers to regard the sacred mushrooms with religious awe and reverence; the Father of Ethnomycology but also, to many, a kind of New Age patriarch.

These and other fascinating angles are explored in the following essays. Written with candor by Gordon's close friends, valued colleagues and family members, they cast light on his personal life and ambitions while also contributing new information on his scientific field work and academic research. Even those who knew him well will find at least a few surprises.

The format is convivial. Imagine that the authors have gathered for an afternoon party on the grounds of Gordon's lovely home in Danbury, Connecticut. There, in the context of his famous hospitality, they mingle freely in an atmosphere of pleasant relaxation. Conversations include recollections of working with Gordon and traveling with him. Amusing anecdotes bring smiles and even occasional laughter. Photographs are passed around. Some of the guests are discussing Gordon's scientific work, and others their own in the same or a related field. One or two are engaged in polite disputation with Gordon himself: arguing, perhaps, that Soma was not *Amanita muscaria,* or that those images in ancient Maya codices were water lilies rather than mushrooms; and Gordon, unperturbed and unpersuaded, explains patiently why he believes the disputant is wrong.

In short, this is a tribute to Gordon that readers should find as engaging and as painlessly informative as just such a gathering.

The essays, including a highly revealing autobiographical account of Gordon's childhood, need not be read in sequence. Each stands on its own, with some information repeated in several of them to make this possible. The redundancy also allows the same events to be described or assessed from different perspectives.

Acknowledgments

My debt of gratitude to many who helped in preparing this book is substantial. To acknowledge them here with my deep appreciation is the least I can do to repay them.

The list begins with Michael Horowitz, who initially suggested the idea for this book and convinced me I should undertake the task of pulling it together. Gordon's daughter Masha Wasson Britten and his friend, the redoubtable Richard Evans Schultes, both endorsed my decision to do so, then provided expert guidance and facilitated access to important information that could not have been obtained from other sources.

Christopher Brown freely gave of his time to familiarize me with the arrangement of the Wasson Collection at Harvard. He also allowed me to publish in this book, for the first time anywhere, a comprehensive bibliography of works by Gordon and Valentina Wasson that he compiled with Gordon's assistance.

Allan Richardson's firsthand accounts of his Mexican forays with Gordon both charmed and enlightened me. His dozens of wonderful photographs appearing in this book include several not published previously, along with a number that by now are famous. I am certain many join me in applauding him, with gratitude, for having made so great a contribution to the scientific record.

My friend Stuart Abelson gave me advice and encouragement, including financial assistance, at crucial points in the book's long gestation. I consider him its godfather.

George Tarbay processed many of the photographs published herein. His magic touch made presentable several old prints and faded transparencies that otherwise could not have been included.

Rick Mason, my research assistant, was efficient, insightful and diligent during our highly productive working visit to the Wasson Collection.

For critical comments and useful suggestions regarding those parts of this book that were written by me, I thank my daughter Jennifer and my close friends Catherine Ami, Stan Johnson and Thomas Roberts.

My publisher, Richard Abel, deserves special mention. He helped launch this book, my first, by expressing strong enthusiasm for it from the start. As it progressed, his expert guidance kept it steadily on course. Thus did I safely negotiate doldrums, stormy seas and shallow waters. I was lucky indeed to have such a seasoned navigator with me on my maiden voyage.

Periodically while working on this project I failed to keep in touch with friends and family, often for weeks at a time. For this I apologize. Yet they were always understanding and supportive, for which I am grateful. I return to them now, like the prodigal son, travel-weary and glad to be home—at least until the next book beckons.

Finally, I thank my dear wife June. Her support was unflagging; her research skills invaluable; her scientific knowledge a convenient and dependable resource. She made numerous suggestions for deletions, corrections, expansions and other such changes that were frequently astute. This book as a whole is a tribute to Gordon; but my labors in preparing it I dedicate to her.

Gordon relaxing at his home in Danbury, Connecticut (c. 1970s).
Courtesy Richard Evans Schultes.

Foreword

RICHARD EVANS SCHULTES

It was in 1953 that my long association and collaboration with a remarkable gentleman and scholar, R. Gordon Wasson, began with a telephone call.

For many years previously, he and his wife Valentina had nurtured a deep and penetrating interest in the influence of fungi, especially mushrooms, on the historical, cultural, social and religious development of civilizations. In fact, they—and particularly Gordon, after Tina's death in 1958—have rightly been called the founders of this new branch of ethnobotany, for which they coined the appropriate term *ethnomycology.*

Of the many diverse facets of Gordon's research, perhaps the most outstandingly original and revolutionary was his interdisciplinary study of the magico-religious role of inebriating mushrooms in Mexico and Mesoamerica and in ancient India.

Although the early European conquerors of Mexico had written profusely about the powerful hold that intoxicating mushrooms had on the natives—toadstools known to the Aztecs as *teonanácatl* ("flesh of the gods")—their ceremonial use in modern times had not been seen. A highly improbable proposal had been offered: that the Spaniards had confused mushrooms with another hallucinogen, the dried tops of the peyote cactus, and that consequently *teonanácatl* and *peyote* were merely two names for the cactus. According to this theory, which was widely accepted in the literature, intoxicating mushrooms had never been used.

Dr. Blas Pablo Reko, an Austrian physician and amateur botanist who had practiced medicine in Oaxaca, stoutly maintained, however, that psychoactive mushrooms were indeed still used by Indians in the hills of southern Mexico. With Dr. Reko, I collected identifiable specimens amongst the Mazatec Indians of Oaxaca; and, in 1941, I published an article in the Harvard *Botanical Museum Leaflets*—a journal of extremely limited circulation—proposing the first botanical identification based on voucher botanical material of a "sacred" mushroom of Mexico: a species of *Panaeolus.*

Many years later, this publication fell into Gordon's hands. It was a turning point in his ethnomycological investigations. As Gordon has acknowledged, it opened for him a wholly new vista in field and literature research and led to the eventual publication of sundry papers and books on the role of hallucinogenic mushrooms, not only in still-extant cultures in Mexico but also in pre-Conquest Mexico and Guatemala, and eventually to his penetrating exploratory ethno-mycological studies in Europe and Asia.

After reading my article in 1953, Gordon, then an official of a prestigious New York bank, was so fired up with enthusiasm for this new and challenging aspect of ethnomycology that he decided to visit the Mazatecs. He wanted to see for himself and understand fully this still-living but ancient fungal role in a magico-religious context. For this reason, he telephoned me at Cambridge during one of my infrequent vacations from field work in the Amazon of Colombia, seeking information for the preparation of his first visit to Oaxaca.

Many years had elapsed since my work in Oaxaca; I had subsequently trans-ferred my attention to the Amazon in 1941. I therefore referred Gordon to Dr. Reko in Mexico City. He contacted Reko, who supplied him with up-to-date information. Unfortunately, Reko died shortly thereafter and was never able to appreciate the extraordinary scientific programme that Gordon and his collaborators eventually carried out and published.

Gordon's first visit to the Mazatecs was highly promising. He established an immediate and long-lasting friendship and confidence with one of the principal shamans, María Sabina, a woman of advanced age. The initial trip was followed by a series of well-organized expeditions, on each of which he took an expert in a field pertinent to the interdisciplinary nature of the research that he was carrying out: specialists in mycology, photography, chemistry, linguistics and other fields. On one of these visits, he and his group were the first "outsiders" fully to participate in the all-night ceremony, or *agapé*, in which the sacred mushrooms were used. The whole ritual eventually was taped and, with the help of two linguists familiar with the Mazatec language, George and Florence Cowan, was published in Mazatec with English and Spanish translations. Four long-playing records that accompany this volume, *María Sabina and her Mazatec Mushroom Velada* (1974), preserve all the chants of the night-long ritual.

Amongst the numerous collaborators with whom Gordon worked very closely in his research was the late Dr. Roger Heim, the leading French mycolo-gist, who identified the mushrooms employed by Mexican natives and described numerous species new to science. As a result of their joint efforts and those of later Mexican and other mycologists who followed, more than 20 species of psychoac-tive mushrooms are now known to be part of the shaman's "stock in trade" in Oaxaca.

Another colleague who worked closely with Gordon was the Swiss chemist Dr. Albert Hofmann, who isolated, identified and synthesized the active prin-ciples of the mushrooms: two new indole alkaloids, psilocin and psilocybin. As a result of his initial chemical studies which, in a way, were engendered by Gordon's research programme, psilocybin is available today for use in psychiatry and also is the basis for semi-synthetic analogues employed medicinally for certain cardiac problems.

Following his assiduous field studies of the contemporary role of mushrooms in Mexico, Gordon delved into a critical evaluation of their use in pre-Conquest Mexico and Guatemala as recorded in the 17th Century Spanish records and earlier archeological artifacts, wall paintings, idols and monuments of the Aztec empire, as well as the even more ancient "mushroom stones" of the highland Maya of Guatemala.

The enthusiasm engendered by his investigations of the Mexican mushrooms led to his later interest in the antiquity and current use of other Mexican hallucinogens, especially *ololiuqui (Turbina corymbosa)*, the seeds of a morning glory. It was responsible for his research into the role of *badoh (Ipomoea violacea)*, a second species of morning glory, the seeds of which the Zapotecs of Oaxaca employ as a psychoactive agent; and for his identification of a new species of mint, *Salvia divinorum*, cultivated almost secretly by the Mazatecs as a sacred hallucinogen, *hierba de la Pastora*, the leaves of which are used when the mushrooms are not available.

Not resting on his laurels with these studies in Mexico, Gordon turned his attention to Asia, where his interdisciplinary research centered on the identification of Soma of ancient India, an hallucinogenic plant so sacred that it was considered itself to be a god. As its use died out many centuries ago, the identity of the plant was no longer known. He attacked the problem with his wonted inclusive and meticulous thoroughness, analyzing the many references to the plant and its uses that survive in the Indian Vedas. Typically, he sought the collaboration of specialists, particularly that of a now-renowned Indic scholar, Dr. Wendy Doniger O'Flaherty. His investigations led to the identification of Soma as the widespread fly-agaric mushroom, *Amanita muscaria*, an inebriant still ritualistically employed by the primitive tribesmen in Siberia. Subsequently, Gordon learned of the ceremonial use of the fly-agaric in North America amongst the Ojibway Indians— a discovery made with the help of a woman shaman of the tribe, Keewaydinoquay.

Still another ethnomycological curiosity that Gordon investigated with Heim concerned reports that the natives in the hills of Papua, New Guinea, occasionally went berserk in wild orgies after ingesting certain mushrooms.

Another and novel excursion into the fungal world was written up in *The Road to Eleusis: Unveiling the Secret of the Mysteries* (1978), a book which he co-authored with Hofmann and Carl A. P. Ruck, a professor of the classics at Boston University. Its theme offered the suggestion, based on historical analysis of Greek literature and botanical and chemical evidence, that the psychoactive agent employed in the Eleusinian Mysteries was a species of ergot growing on a wild grass common in Greece—a different ergot than the one infecting rye and which has a long history as a medicine and poison in more northern parts of Europe.

Gordon's last book, with several chapters from his own pen and contributions from three colleagues—Stella Kramrisch, Jonathan Ott and Ruck—was published in December 1986. Entitled *Persephone's Quest: Entheogens and the Origins of Religion*, it summarized his theme, expressed throughout earlier publications, that the mysterious, unworldly, psychic effects of hallucinogens were responsible for the beginnings of man's beliefs in the supernatural in the infancy of the human race. Although he and I had read galley proofs, Gordon did not live to see the book in its beautiful format and binding, as it appeared in Europe only several days before his death 23 December 1986.

Two very original proposals suggested by Gordon stand out as contributions to ethnobotany. He and Tina proposed the division of peoples into two classes for which they coined the terms *mycophiles,* those who love and know intimately their mushrooms, and *mycophobes,* those who fear, abhor and do not know their mushrooms.

One of the last contributions made by Gordon and several colleagues arose from his dissatisfaction with any of the numerous words applied to plants with mind-altering properties—terms such as hallucinogen, psychotomimetic, schizogen and others. None of these words, he believed, have true specificity in their meaning, nor do they represent the sacredness which native peoples so often grant their psychoactive plants. As an alternative, Gordon and his colleagues—Jeremy Bigwood, Danny Staples, Ott and Ruck—proposed the term *entheogen,* interpretable as "generating the idea of god." This reflects the aforementioned theory that, very early in the evolution of the human animal, the weird, unearthly effects of hallucinogenic plants introduced man to belief in the supernatural, giving rise to prototypical ideas of the religious experience.

Numerous honours recognized Gordon's scholarship and contributions. In 1972, he was selected as president of the Society for Economic Botany and was given the Distinguished Economic Botany award. The following year marked his election to the prestigious Linnean Society of London. The University of Bridgeport conferred on him an honorary Sc.D. degree and Yale University presented him with the Addison Emery Verrill Medal for his research in the natural sciences. The citation that accompanied the latter appropriately described his accomplishments in these words: "You have combined aspects of anthropology, archeology and botany into an organic whole. . . ."

During his years of retirement, which also saw his most intensive ethnobotanical research and writing, Gordon quietly cherished and often described as a most appreciated honour his membership on the staff of the Botanical Museum of Harvard University as Research Associate, an appointment he fulfilled with distinction for 23 years until his death. It was Professor Paul C. Mangelsdorf, my predecessor as Director of the Museum, who, in 1963, recognized in Gordon the potentialities of a pioneer investigator in a long-neglected field of economic botany, and who offered him the appointment. It was not, however, Gordon's first connection with Harvard, for during the 1950s he had served the University as a member of the Visiting Committee to the Slavic Department.

In 1983, he took a step characteristic of his selfless interest in the field of research that he had so firmly established. He donated the Tina and Gordon Wasson Ethnomycological Collection to the Botanical Museum. It comprises more than 4,000 books, pamphlets and reprints, as well as original water colours, photographs, charts, slides, art objects and archeological artifacts. Perhaps most significant of all, it also includes his collection of personal manuscripts and correspondence. The only collection of its kind in the world, it is now available to scholars and students in the field. He furthermore created a fund to support the collection and ethnomycological publications.

Having discussed his distinguished accomplishments, I turn as a friend to the question: what kind of person was Gordon Wasson?

He was most certainly a "gentleman of the old school." Whether in his New York or London clubs or in the hills of Oaxaca, he knew how to respect those with whom he came in contact. But his strict meticulousness in every aspect of his activities knew no bounds and even occasionally irritated a few scholars who did not understand this basic characteristic of his being. He could not bear sloppiness, especially in writing, and showed no patience with mediocrity. I have never known a man more meticulous in his bearing, his speech, his writing and his thinking. To his closest friends and colleagues, his seriousness and apparent exclusivity of purpose did not obscure his extraordinary breadth of interest and knowledge. Above all, he was intrinsically gentle and humble, a part of his nature that became ever more obvious the longer one knew and worked with him. Always self-effacing, he shunned publicity—a characteristic which, unfortunately, some who did not know him personally interpreted as snobbish aloofness, a trait completely foreign to his personality.

Gordon was a true ethnobotanist in the tremendous breadth of his interests, for ethnobotany *is* an unusually interdisciplinary field of research. This can be seen in the multi-faceted methods he used in attacking the many projects that occupied his attention for so many years.

But there is one anecdote which, to my mind, illustrates better than anything else his willingness to use any road to unravel difficult questions—and in a way it is truly humorous! In his research on *Amanita muscaria,* the fly-agaric, which he identified—correctly, I am certain—as the ancient Soma of India, he read that reindeer in Lapland ate this mushroom. Gordon wrote to a scientific colleague in Sweden requesting him to carry out an experiment: feed *Amanita muscaria* to a reindeer in the Stockholm zoo, collect urine from the animal and examine it for muscimole to see if, as in humans, it passes through the deer's body and is excreted unmetabolised. The colleague was not so enthusiastic as Gordon, and the experiment was never done!

A perfectionist, his insistence on the highest quality in publication of his books is legendary. It constantly reminded me of my former professor at Harvard, Oakes Ames, who, when criticized for setting up a press in the Botanical Museum and spending so much money on publications of the highest quality, defended the practice by saying, resolutely: "A scientist's research is a jewel worthy of a proper setting." Gordon most certainly held an identical view: his books contain not only jewels of research but will long be admired as models of the very finest specimens of the printer's art.

Gordon could appreciate a good joke, especially when it was at his own expense; and he was excellent company. I recall his pleasure and even his jocularity when, on our visit to Castle Howard in Yorkshire, we stopped at a tiny inn and were invited by several English youths in the pub to play darts with them. To describe our performance as tragically bad would be a gross understatement. I shall never forget that evening: it was a jolly affair, mostly because of Gordon's ability to laugh at his own and my awkwardness and to show his amazement at the skill of our young friends.

During the last year or two of his life, Gordon noticed his failing health, yet he carried on his research and writing and even planned future trips abroad in connection with his work. He died whilst visiting his daughter in northern New York state two days before Christmas, 1986. His host of friends and acquaintances in the United States and around the world, as well as the many admirers of his trail-blazing scientific contributions who were not fortunate enough to have known him personally, will sorely miss this great man who contributed much to scientific advancement and firmly established a new field of ethnobotanical endeavor.

R. Gordon Wasson:
Brief Biography and Personal Appreciation

J. CHRISTOPHER BROWN[1]

Dr. R. Gordon Wasson, outstanding amateur scientist, explorer and traveller, died 23 December 1986. Like many of his kind who in the Victorian era contributed so much, he was an amateur scientist in the original sense of the word *amateur*, being a lover of the relentless pursuit of knowledge in his chosen field of ethnomycology.

It was my privilege to have known Gordon Wasson, and to have worked closely with him during my association with the Botanical Museum of Harvard University, first as a student assistant and later as a staff member, from 1982 through 1985.

A man of august bearing, insatiable curiosity and insistence on meticulous accuracy, Wasson was aptly described by a British colleague as "the perfect old-fashioned gentleman, with a twinkle in his eye." Although best known as an ethnomycologist, Wasson was at home in numerous apparently unrelated fields.

Wasson was born 22 September 1898 in Great Falls, Montana, where his father served as an Episcopalian clergyman, and spent his childhood in Newark, New Jersey, where he attended the public schools. He was sent by his parents at age 16 to France and Spain, where he spent more than a year learning the languages and vagabonding. Upon his return he began college studies at the Columbia University School of Journalism in New York. On 25 June 1917 he withdrew to enlist as a private in the U.S. Army.[2] In the First World War he served in the American Expeditionary Forces, and saw 14 months of service in France as a radio operator. Returning to Columbia University after the war, he graduated with honors in 1920, earning a Bachelor's degree in Literature, and was awarded the first Pulitzer Traveling Scholarship ever granted. He then studied at the London School of Economics and spent the spring of 1921 walking through the Pelopennesus with Axel Boethius, who later became famous as a Roman and Etruscan archaeologist, founder and head of the Swedish Academy in Rome. On his return in the fall of 1921 he taught in the English Department of Columbia University for a year, where his students included the poet Langston Hughes.

Wasson then worked as a journalist for several years. He started, with the *New Haven Register*, as an editorial writer and State political correspondent covering the legislature in Hartford. In 1925 he moved to New York to take the job of associate editor at *Current Opinion,* a monthly magazine that has since disappeared. He then joined the *Herald Tribune* in the financial news department, where he wrote a daily signed column.

In 1926 Wasson married Valentina Pavlovna Guercken, a White Russian and a physician, who died in 1958. They had two children, Peter and Mary (Masha), who survive.

Wasson entered the banking profession with the Guaranty Company of New York in 1928, and was sent shortly afterward to Argentina and then to London for long periods. In 1934 he joined the staff of J. P. Morgan & Company, where he remained until 1963, from 1943 on as a vice president. He even served a term as a member of Harvard University's Visiting Committee to the Slavic Department.

Wasson and his late wife were pioneers in ethnomycology, and promoted it as a distinct and promising branch of ethnobotany. They wove mycological information with data from many fields—history, linguistics, comparative religion, mythology, art, archaeology and others—to analyze and explain the role of fungi in the cultural development of numerous peoples. They explored the philological aspects of the vernacular names of mushrooms, the uses to which mushrooms were put in early times and the aura of mystery that surrounds them; they hypothesized that the peoples of the world could be separated into "mycophiles" and "mycophobes." In 1952 the poet Robert Graves sent the Wassons an article which mentioned the discovery in 1938, by Dr. Richard Evans Schultes, of the survival of the use of intoxicating mushrooms among the Indians in Mexico. Immediately Gordon Wasson telephoned Schultes at Harvard, and the confirmation and encouragement he received from Schultes focused his attention on Mexico. Beginning in 1953, Tina and Gordon Wasson organized yearly research expeditions to the remote mountain villages of the monolingual Mazatec Indians of Oaxaca, Mexico, and in 1955 were the first outsiders in modern times to participate in the night-long rites of the cult of the sacred mushroom. They announced their discovery in 1957 in their jointly written book *Mushrooms, Russia and History.* Long unavailable, there is now such a demand for this book that it may be republished.

Following Valentina's death, Gordon continued his research, working closely with Dr. Roger Heim, the world-famous French mycologist and Director of the Muséum National d'Histoire Naturelle, who had accompanied the Wassons on several expeditions to Mexico. In 1958 they published *Les Champignons Hallucinogènes du Mexique;* this was followed by a second volume, *Nouvelles Investigations sur les Champignons Hallucinogènes,* in 1967. Many of the mushrooms used in the Oaxacan mushroom rite were previously unknown to science and were described by Heim. Roger Heim even named a new species of *Psilocybe* in honor of Wasson: *Psilocybe wassonii* Heim. The Mexican mycologist Gastón Guzmán named a species of *Psilocybe* after both the Wassons: *Psilocybe wassoniorum* Guzmán.

On his retirement from the banking business in 1963 Wasson went to the Far East, where he gathered material that led him to publish, in 1968, *Soma: Divine Mushroom of Immortality,* written in collaboration with the Indologist and Vedist Wendy Doniger O'Flaherty, in which he dealt with the Soma enigma. The ancient

Gordon with mushroom stone at the Harvard Botanical Museum in the early 1980s. *Jane Reed. Courtesy Richard Evans Schultes.*

Aryans had worshipped a plant of that name, one of the very few plants that has ever been made a god, but the identity of the plant had been lost. Wasson undertook to prove that it was a mushroom, *Amanita muscaria* L. His book stirred up violent controversy among scholars who specialize in ancient Hindu culture and religion, the reception on balance being favorable to Wasson's theory.

On his return from the Orient in 1965 Wasson moved to a beautiful country home in Danbury, Connecticut, where, until his death, he devoted his energies wholly to ethnomycology, traveling widely in Mexico, Europe, the Far East and Oceania, and to prolific writing. From his penetrating research and correspondence with eminent scholars in diverse fields around the world have come further trail-blazing volumes and technical papers, in which are set forth discovery after discovery.

As a result of his ethnomycological research in the field, in museums and in libraries, he made contributions to other aspects of ethnobotany, including investigations of hallucinogens from the higher plant families and the discovery of a new psychoactive *Salvia* from Oaxaca.

In 1974 Wasson, working closely with George and Florence Cowan, experts in the Mazatec language, and Willard Rhodes, a musicologist, published *María Sabina and her Mazatec Mushroom Velada.* This book, 15 years in preparation, gives the complete transcript of a sacred mushroom ceremony, held to diagnose a serious illness, that Wasson recorded in the field in 1958. The entire proceedings are printed in Mazatec, along with translations into Spanish and English, and are accompanied by linguistic and musicological analysis, color photographs and four LP records. In spite of the slight scholarly notice given to this work, Gordon often said that he was prouder of it than of any of his other books.

In 1978 he published with Albert Hofmann, the discoverer of LSD, and Carl A. P. Ruck, a Classical Greek scholar, *The Road to Eleusis: Unveiling the Secret of the Mysteries,* with a new translation of the *Homeric Hymn to Demeter* by Danny Staples, in which he set out to show that the Mysteries of Eleusis were based on ergot of rye.

Wasson, Ruck and Staples collaborated with Jeremy Bigwood and Jonathan Ott in inventing a new word, *entheogen*—a plant substance that so gripped the imagination of our ancestors in prehistory as to lead to profound religious feelings.

Late in life Wasson wrote *The Wondrous Mushroom: Mycolatry in Mesoamerica,* published in 1980, in which he elucidated the details of the religion of the Indians who held the mushroom sacred, a subject which had been largely ignored by scholars until then.

His gift to the Harvard Botanical Museum, in February 1983, of the Tina and Gordon Wasson Ethnomycological Collection, established the only ethnomycological library in the world. A research facility, supported by a fund given by Wasson, this library contains more than 4,000 items, not only books but also photographs, films, videotapes, stereo slides, prints and watercolors, as well as an extensive collection of art and archaeological objects, and his entire scientific correspondence. When his books were transferred to the Botanical Museum, one of the most gratifying assignments I have ever had was the organizing of this library and the related collections.

Just prior to his death Wasson completed his final book, *Persephone's Quest: Entheogens and the Origins of Religion,* which added further evidence to support his thesis that the entheogens dominated in prehistory the religion of Eurasia. *Persephone's Quest* includes contributions by Stella Kramrisch, Ott and Ruck. It has been published by Yale University Press, but Gordon did not live to see the final version printed and bound.

Wasson attended and participated as an invited speaker in numerous national and international symposia. He was a member of the Society for Economic Botany, and was President (1972) and Distinguished Economic Botanist. His devotion to scientific research led to his appointment as Research Fellow in Ethnopharmacology in the Botanical Museum of Harvard University, a position he held from 1963 until his death. He was honored in 1973 by election as a Fellow of the Linnean Society of London. Yale University's Peabody Museum awarded him in 1983 the Addison Emery Verrill Medal for distinction in the field of natural history. The citation accompanying the medal read in part:

> All of your many writings have that touch of clear and good writing and of immense erudition that mark the true scholar. You have combined aspects of anthropology, archaeology, and botany into an organic whole; you yourself are the embodiment of the word *amateur,* which in its original sense means "a lover of things." You have obviously loved your subject . . .

The University of Bridgeport conferred on Wasson the degree of Sc.D. on 26 May 1974, in recognition of his record of research and writing, and the New York Botanical Garden, with which he was associated for many years, named him Honorary Research Associate and Honorary Life Manager.

Gordon never sought publicity as the result of his discoveries, and continued working quietly on each successive book. But there is a documentary film in preparation, by Philip Black of London, that deals with Tina and Gordon's lifelong work in the development of ethnomycology. It was made with Gordon's cooperation in the last few years of his life.

In addition to his ethnomycological writings, he also wrote a book analyzing the development of historical myth, *The Hall Carbine Affair,* which went through three editions between 1941 and 1971. This book grew out of his banking career and his interest in the history of J. P. Morgan & Company; it was a definitive refutation of the false story that the Morgan Company owed its early success to a deal in arms on the eve of the Civil War, in which, it was alleged, J. P. Morgan sold a

Christopher Brown peruses the Tina and Gordon Wasson Ethnomycological Collection he helped organize (1987). *Thomas Riedlinger.*

large quantity of defective carbine rifles to the U.S. Army. Wasson not only proved that the story was false, but he also traced the process by which a fictitious tale may be told and retold, embellished at every step, until it acquires the status of a fact among those eager to believe it.

Gordon Wasson also published numerous articles over the years, dealing with finance, foreign policy, linguistics, literary criticism and history, and the nature writer W. H. Hudson. Gordon, like Hudson, loved birds and animals, and he was completely content in his home in the country, observing the deer in the orchard, the hummingbirds in the honeysuckle vines on his porch, or the family of skunks that lived in his garden. He was as gentle and kind as he was intellectually tenacious and erudite, and he will be deeply missed by his many friends. He died two days before Christmas at the home of his daughter in Binghamton, New York. His remains were cremated, and his ashes were interred on 2 January 1987 at a beautiful and simple ceremony in the Bethlehem Chapel of the National Cathedral in Washington, D.C.

Notes

1. A slightly different version of this article first appeared as "R. Gordon Wasson" in *Economic Botany* September–October 1987 41(4):469–73.
2. *Editor's note:* Gordon formally asked the school's director, Dr. Talcott Williams, to allow him to withdraw from his program of studies in order to enlist. His "request" is preserved in the following letter dated June 25, 1917, which the school still has on file:

> My dear Doctor:
> Twice have I made an effort to meet and consult you about what is to me a most important step,-my enlisting in the military service. On appointment with Miss McGill I went from Newark to your office on Friday, only to find that after all you would not be there, and again afterwards I tried to effect an appointment. I am now taking the liberty of going ahead, with my parents' approval; and I trust that I shall have yours also. This morning I expect to become a first-class private in the Signal Enlisted Reserve Corps. My haste in entering the service is due to the circumstance that I can be made linguist of my battalion by enlisting at once,-an opportunity that would be lost by delay, as there is but one linguist to a battalion. My knowledge of French and Spanish, with English, gives me this advantage.
> I have every intention of returning and completing my course in the School of Journalism when the war is over. My experience meanwhile may enable me to get even more out of it. I am grateful for all the School has done for me, and trust that I shall be a better soldier for it.
> I sincerely hope you will approve my course [of action].
> Very truly yours,
> R. Gordon Wasson

Wasson's Help to a Student of Drug History

MICHAEL R. ALDRICH

> The extent to which we have leaned on persons more knowledgeable than we will be apparent to all who know us. The patience of the learned fraternity with our questions has amazed us. Doctors and lawyers and engineers ask fees for their professional advice. But it seems that men of learning, like men of God, stand above these others, bestowing their counsel and blessings freely on all who turn unto them.
> —Valentina Pavlovna Wasson (speaking also for R. Gordon Wasson), from the preface to *Mushrooms, Russia and History* (1957), p. xviii

In the spring of 1968, I was working on a term paper about Euripides' *Bacchae* for a graduate course in Greek at the State University of New York (SUNY), Buffalo. I was interested in the ancient history of psychoactive drugs, and I knew that the *Bacchae* was the quintessential drug myth of Western civilization: the story of Dionysus, the god of wine; his reappearance in Greece after a trip to India; and his reception by Pentheus, the prim and proper tyrant of Thebes who clapped the god in jail and thereafter paid an awful price for this *hubris,* eventually being torn apart by his own mother. What caught my attention was Euripides' account (*Bacchae* lines 13–22) of the route Dionysus took to and from India: through Lydia and Phrygia across Persia to Bactria (Afghanistan), then returning through Medea, Arabia Felix and Asia Minor to Thebes.

Intoxication cults had existed in each of these places, and Euripides was, in effect, outlining a network of psychoactive drug religions in the ancient Middle East. Most of them involved the use of alcohol, but some were more mysterious. What, for instance, was the Soma of ancient India, recorded in the Vedas? It was not alcohol (*surya*) or marijuana (*bhang*), for both are distinguished from Soma in the Vedas. Other candidates were species of *Asclepias* or *Ephedra,* neither of which has the great psychoactive power attributed to Soma in the ancient texts.

I wrote to Richard Evans Schultes at the Harvard Botanical Museum to learn if he could identify Soma. Schultes replied that the only man qualified to help in my search was R. Gordon Wasson, whose address he gave me. Wasson's response to my query was most generous. Although he was then in the process of publishing on the subject, he forwarded two articles outlining his views (Wasson 1967a; 1967b) and asked for a copy of my paper when completed. His discovery

25

that Soma was the fly-agaric mushroom moved Robert Graves (1960, p. 300) to conclude that the worshippers of Dionysus perhaps added *Amanita muscaria* to their wine for the hallucinogenic frenzy Euripides depicts in the *Bacchae*. This made a fine conclusion to my Greek term paper. When it was finished, I sent a copy to Wasson, requesting criticism.

Ask and ye shall receive! On July 15, 1968, he sent back a sharp letter criticizing the grammar and substance of the paper, correctly calling it "a catch-basin with arguments of every quality . . . dashed off at breakneck speed." I was chagrined, and at the same time honored that he took my novice efforts seriously enough to offer a paragraph-by-paragraph critique, both pedantic and profound. Some examples from Wasson's long letter are here reproduced in facsimile, with corrections in his own handwriting:

Dear Mr. Aldrich:

How kind of you to dedicate your paper to me, as well as Robert Graves. I was moved by your striking tribute to Mary Barnard. Do you know her?

I am not learned enough to pass on the merits of your arguments, with their thousand facets -- sahásrapajas in Vedic. You estop criticism by the disclaimers in your preface, but I will criticise nonethe less. It seems to me that you have presented a catch-basin with arguments of every quality, some so flimsy that you almost disown them yourself. Scholarly works cannot be dashed off at breakneck speed. If this is worth doing, it is worth doing well. Your superb enthusiasm is not tempered by caution.

* * *

-- You use 'Zend-Avesta'. It is 30 or 40 years since this term was completely abandoned, with good reason, in scholarly writing. Avesta, simply. (pp. 43, 78, 79, 81, 89)

-- The religion was officialized. ...weird fire-worship. (p. 42) Have you no ear for the music of the language, for the meaning of words? Why do you adopt from the mouths of cheap and ignorant people their abuse of the language?

* * *

-- ...like the .... remnant of the ... Highway underlies the ... Royal Road. 'Like' is a preposition; 'as' is the conjunction you see

* * *

-- laurel, whose narcotic properties qualities when chewed by the Pythian priestesses are well known. (p. 61) Certainly some of the best informed specialists do not believe that they chewed (nor that they inhaled smoke from) laurel leaves. Have you read La Mantique Apollinienne à Delphes: Essai sur le fonction-nement de l'Oracle, by Pierre Amandry, E. de Boccard, publishers, Paris, 1950?

-- On p. 65 you discuss the Hunza Valley. You speak of Burushaski and cite someone who says he was told that it most nearly resembles Welsh. This puts you in the class with those who say Basque sailors stranded in Japan found they could converse with the Japanese, or those others who say that two or three Plains Indians, finding themselves in Tibet, discovered they could converse in Tibetan. There is always a demand for marvelous stories but this is no reason why you should quote them as authorities.

* * *

-- You quote Griffith. He is one of the early translators that no one
uses any more. He and Wilson and Langlois were all worthy men but
they had set for themselves a different task from the recent
translators. They were profoundly impressed with their mission to
the cultivated classes in the West: they were making known the
treasure that they had discovered in the RgVeda. They translated
what they understood of it, and when they did not understand it
they invented, trying to compose an impressive poem embodying their
conception of the Indus Valley before culture before 1000 B.C.
They also bowdler bowdlerized the text. The result was, in Griffith's
Griffith's case, a poem half way between Tennyson's Idylls of the
King and Ossian. Today's translators are translating for other
scholars to read. They make the reader stumble every time they
stumble, and go into the reason why, and what other scholars have
done at that point. Though they write for other scholars,
as it happens they are more helpful to interested outsiders, not
specialists in Vedic, than the old translators were.

* * *

-- You perpetuate the error that bufotenin is in A. muscaria. Wiegand
in an unhappy moment said he thought He detected bufotenin traces
of bufotenin in A. muscaria. He is a distinguished chemist. Though
every chemist since his experiments has looked for those 'traces',
they have not found them. Their results are positively negative.
There is no bufotenin in A. muscaria. There is hardly any muscarine
either. If one consumed 24 kilograms of fresh A. muscaria at a
sitting, one might ingest would ingest enough muscarine to have a
reaction.

* * *

In reading the RgVeda you should keep in mind that the latest of the

hymns were certainly composed after Soma was abandoned. The tone of those

hymns and the vocabulary are easily distinguished from the earlier, purer ones

The late hymns include those in Maṇḍala X from 90 on. All the Brāhmaṇas

were written much later than the body of the RgVeda, and they

are preoccupied with the substitutes for Soma, expressly so called.

* * *

There are two matters in your paper that excited my lively interest.

On p. 73 in Dr. Sampurnananda's quotation he says:

> There were religious injunctions that, even if two
> states were at war, there should be no obstruction
> to the passage of Soma from one to the other.

What is his authority for saying this? Does he give any? There was a like
tradition concerning the conveyance of holy objects in Greece, recorded I

think in Herodotus. I have done no research on this.

I shall try to do some work prompted by your remarks on p; 68

concerning mēros.

Good luck.

                    Sincerely yours,

27

A similar concern for detail and respect for the ideas pursued illuminate all of Wasson's work.

I've always been grateful for his help at the start of my career as a drug historian. Later, when Gordon was appointed a trustee of the Fitz Hugh Ludlow Library, I learned that this kindness was typical of his response to students. As his mycological work became better known, students—and teachers, for that matter—flocked to him wherever he appeared. At the 1978 hallucinogenic conference in San Francisco, he was surrounded by dozens of young enthusiasts, answering their questions with scholarly patience and aplomb. I was taking pictures of our trustees in front of a colorful mural depicting Japanese fishermen (since demolished to make way for a theater complex) when Joan Halifax suddenly noticed that Gordon was tiring of all the adulation. She went to him, took his hand, led him out of the throng—and gave him a great big smooch! I captured with my camera the look of surprise on Gordon's face, though not the smile that replaced it almost instantly. The photo brings back my fondest memories of this wonderful old gentleman, who, at just that time, with Albert Hofmann and Carl A. P. Ruck (Wasson et al., 1978), was unveiling the secrets of the Greek Mysteries to the world—even as he had revealed the "secrets" of Vedic scholarship to me, now 20 years past.

Gordon surprised by a kiss from Joan Halifax (1978). *Michael Aldrich.*

References

Graves, R. 1960. *Food for Centaurs*. Garden City, New York: Doubleday.

Wasson, R. G. 1967a. Fly-agaric and man. In *Ethnopharmacologic Search for Psychoactive Drugs*, edited by D. H. Efron, B. Holmstedt, and N. S. Kline. NIMH Workshop Series No. 2. Dept. of Health, Education and Welfare. Washington, D.C.: U.S. Government Printing Office. Pp. 405–414.

Wasson, R. G. 1967b. Soma: the divine mushroom of immortality. *Discovery* Fall 1967; 3(1):41–48.

Wasson, R. G., A. Hofmann, and C. A. P. Ruck. 1978. *The Road to Eleusis: Unveiling the Secret of the Mysteries*. New York and London: Harcourt Brace Jovanovich.

Wasson, V. P., and R. G. Wasson. 1957. *Mushrooms, Russia and History*. New York: Pantheon Books.

Gordon, Masha and Valentina in Mexico (1955). *Allan B. Richardson. Courtesy Wasson Collection.*

My Life with Gordon Wasson

MASHA WASSON BRITTEN

The invitation to participate in this collection of essays in tribute to R. Gordon Wasson has stirred mixed feelings. As much as I loved and respected my father, the task of writing about him "in memoriam" is painful and difficult. Apart from that, I have always felt strongly that the mushroom search and research conducted by him and my mother "belonged" to them, and that my personal perspective on the matter was really irrelevant. For this reason I have never before written anything about it. However, since I have been asked, I will do so now in honor of both my parents.

First, I wish to take this opportunity to thank Tom Riedlinger for his initiative in preparing this tribute to my father and for all the work entailed. Also, on behalf of my parents and my brother Peter, I wish to thank all the contributors to this volume. Please know that your efforts are much appreciated. I am sure that my father would be very proud, as I am, and equally appreciative.

Life with "RGW" was never dull. I feel privileged, not only to have grown up under his tutelage, but also to have known him. Whether or not he enjoyed the playful antics of children in his household is uncertain. We were definitely challenged to be seen and not heard. And my brother and I felt the pressure of a different challenge growing up: to follow in our father's footsteps, something neither of us felt that we could or even wanted to do. Parents have a tendency to want their children to emulate them and pursue their profession, or at least share their interests. But Peter and I had different interests than those of our mother and father. (My parents, in fact, did not share each other's occupational interests. When my father discussed his banking business, my mother would turn green. And when she, a pediatrician, talked about medicine, my father had the same reaction.) I like music, for example, and my father did not. Neither did my mother; she was much more interested in art. But they encouraged my interest in music while they separately pursued their careers and, together, the mushroom research.

The pressure on my brother Peter to follow in my father's footsteps, including a career in banking, was much greater. But his interests also differed from my father's, as did his temperament. Peter did not accompany us on the early Mexican expeditions; he had little interest in the mushrooms. It was not until many years later, near the end of my father's life, that the relationship between my

31

Gordon Wasson as a young man (date unknown).
Courtesy Masha Wasson Britten.

father and brother became close. My father finally accepted and respected Peter's individuality, talents and interests on their merits.

In any case, my mother was accepting that we children took a different path. I clearly remember her saying to my father, "There is only one of you. And one of you is certainly enough!"

There is no question but that my parents' intellectual pursuits enriched my life. Interesting scholars and other fascinating people, such as the poets Robert Graves and Octavio Paz, dropped by to visit, or we went to visit them. The discussions that ensued kept my mind racing to keep up with and understand them. My father could discourse for hours on any of numerous subjects and in different languages. I shall always remember, from these discussions, his probing curiosity, his vast and varied knowledge, and his enthusiasm and zest for life.

He spent many hours sharing his ideas with me, and of course I learned much from his example. He communicated vividly his reverence and awe for the mushrooms and his interest in the role they played in the cultures of different peoples. His respect for both the mushrooms and the people who used them, and his excitement about his findings and conclusions were contagious. He was meticulous in his research, in his notes and in his attention to detail. He shared with me his feelings of respect for the scholars upon whom he relied so heavily, as well as

Valentina as a young woman (1925).
Courtesy Masha Wasson Britten.

his disdain for those he felt were sloppy and inaccurate in their research or conclusions. And, as is well known, he had little use or patience for those he considered to be hippies.

In short, my father's methods and manner suggested to me something like a pilgrimage in search of the meaning of the mushroom. His willingness and ability to try new and different approaches in pursuit of that goal, and to see patterns and relationships among different disciplines, impressed both my brother and me as remarkable. His scholarship was not of the desk-bound variety. He was fearless in trying new things that might increase his knowledge. One example demonstrates not only this quality but also his sensitivity to the values of others. Once, while wandering about an Asian port town where the ship on which he was traveling had stopped to unload, he entered a place where he found a man with a jar of live beetles. My father, ever curious, inquired about them. The man explained that they were eaten live for some purpose I forget—perhaps longevity. Then he asked my father, "Would you like to try them?," and handed some live beetles to my father which he was expected to swallow whole. And he did. I think he felt he had no choice. He did not want to insult the man.

As much as I respected my father's integrity, I recall that for years I did not believe him when he said his interest in mushrooms began on his honeymoon in

1927. Such an explanation seemed to me like a Hollywood soap opera, something out of character for my father. Eventually, however, I concluded the story was true, for he told it sincerely and consistently.

As for *Mushrooms, Russia and History*: I remember when my parents started writing it during the mid-1940s—as a cookbook! The mushroom search that finally led them to Mexico began as a footnote. They acknowledged this briefly, but explicitly, in their preface to *MR&H*, which includes these lines on page xvii:

> Our essays began as a mere footnote on the gentle art of mushroom-knowing as practiced by the northern Slavs—a footnote in a larger work that was to have dealt with the Russians and their food. The footnote grew, and rose in status to a place in the body of the text, and then mushroomed into a whole chapter, and finally one chapter by fission made five. The manuscript has burst all its seams through successive revisions.

The manuscript they were referring to (or one of what must have been several revisions) survives in the Tina and Gordon Wasson Ethnomycological Collection at the Harvard Botanical Museum. It confirms the account in the preface (and my recollections) and provides some additional insights worth mentioning here.

The book was originally titled *Mushrooms Russia and History: An Introduction to Russia through the Kitchen.* My mother, Valentina Pavlovna Wasson, was originally bylined as the sole author. Its six chapters were: (1) Mushrooms, Russia, and History; (2) Caviar; (3) Vodka and other Russian Drinks; (4) Elena Molokhovets and 19th Century Russia; (5) The Siren, by Anton Pavlovich Chekhov; and (6) . . . *the most copious and elegant language in the world.* My father appended his name only to the introduction, in which he provides some historical background on Western relations with Russia and argues persuasively that Russian cuisine compares favorably with that of the French. It also includes the following passage:

> In the spring of 1945 it occurred to my wife and to our cook, Florence Ada James ("Florrie"), that there might be a place in the English-speaking world for a comprehensive collection of Russian recipes adapted to the markets and tastes of the West. My wife was born and brought up in Russia and Florrie has cooked for Russians for thirty years. They proceeded to assemble upwards of 500 recipes and were ready to organize the text of their cookbook when it dawned on them that, detached from the background, these recipes lost much of their appeal.

In what seems to be a later revision included in the same manuscript, my father returns to this point and elaborates:

> The authors of this book started out with a modest purpose,—to assemble a comprehensive collection of Russian recipes adapted to the markets and tastes of the Western world. It became apparent that, detached from their background, these recipes lost much of their appeal. The authors undertook to supplement the recipes for the kitchen with a commentary for the drawing room. They invite the reader to enter Russia, so to speak, by the kitchen door, and while their commentary for the drawing room attempts no comprehensive picture of Russian life and manners, they have assembled information, directly or indirectly linked to Russian food, that they hope will prove enlightening and entertaining.

Note that by the time he wrote this passage, my father no longer referred to an author, singular, but rather to "the authors." However, he still saw it as "a commentary for the drawing room" linked solely to Russian food, which was to accompany the recipes for the kitchen. These recipes are not included as part of the manuscript. But I have the original recipe cards, some examples of which are reproduced below:

Recipe cards above were handwritten by Valentina Wasson, the card below by the Wasson's cook Florrie. *Courtesy Masha Wasson Britten.*

35

At some point, *Mushrooms, Russia and History* evolved into something much different than my parents had initially conceived. This apparently happened sometime after February 23, 1949, when my father wrote a letter to V. Molohovetz in which the book is still described as "a volume of six essays about Russia," and prior to July 14, 1950, when he wrote his first letter to Giovanni Mardersteig, quoted below in its entirety.

Dear Mr. Mardersteig:

On the advice of a number of friends, I write you to explore the possibilities of your printing and binding a book for my wife and me.

The book deals with mushrooms. I enclose the text of the preface and first three chapters. The preface will give you the idea of the book. In all there will be five or six chapters, bibliography and notes, and index. The complete text is likely to be about twice the length of the preface and three chapters that I am sending you.

The page will be either 25 cm. × 32½ cm. or 22½ cm. × 28 cm., and the paper must be "un papier collé" to accommodate illustrations in color that will be done, by the pochoir process, by Daniel Jacomet of Paris (20 bis, rue Bertrand, Paris 7*eme*). The smaller page would require a slight reduction in the size of two or three of the full page illustrations. I send you herewith a set in black and white of the illustrations that we propose to use. A number of the smaller ones would serve as decorative endings for chapters, etc. The photographs that I send you carry numbers for reference purposes in our correspondence. The whole project will call for the closest collaboration between you and Monsieur Jacomet, as well as with us.

My wife and I want not more than 200 copies of the book, perhaps 100 copies would suffice. We suggest an edition of 500 copies. I have been told that you have facilities for selling a certain number of copies of your books, and that the proceeds can be applied toward the costs of publication. My wife and I know that we cannot make money out of this project of ours, but we should like to keep our own expenses down to a minimum. We feel sure that both our illustrations and our text, being of an original character, will appeal to a certain number of bibliophiles.

Before entering upon a discussion of the details, I should like to get your views regarding this project, and some idea of the expenses that may be involved at your end. Of course I must also learn from Monsieur Jacomet what he will charge.

If you are not interested at all, will you please return the enclosures. If we arrive at an agreement for the printing of this book, the completed manuscript should be ready for you in a few months.

Sincerely yours,
R. Gordon Wasson

Of course, the finished manuscript was not ready until several years later. It was finally published in 1957. The delay was caused by my father's ever-expanding knowledge of and insights concerning mushrooms and their uses, and his deeply held belief that there was more to the mushrooms than met the eye. He and my mother were struck by the fact that peoples of every culture seem either to be mycophiles or mycophobes. To them this was obvious. But they felt it implied something hidden and secret, some forgotten truth of major importance. So when they learned, in 1952, that the Mexican Indians reportedly still used a sacred mushroom in their ceremonies, work on the manuscript was deferred until they were able to investigate in person and incorporate what they discovered there.

Ironically, one of the Mardersteigs helped provide the information that delayed publication of *Mushrooms, Russia and History* for many years. In "The Hallucinogenic Mushrooms," an article by Valentina P. Wasson and R. Gordon Wasson published in the January–February 1958 issue of *The Garden Journal,* my parents reported:

> We had been pursuing our inquiries into the role of mushrooms in the cultural history of the Old World for many years. We had found indications in the etymology of the names of mushrooms and in folklore, as well as in contemporary attitudes toward "toadstools," that at one time mushrooms had played a part in our ancestors' religious beliefs. At this point, in September, 1952, in almost the same mail, we received two communications, one from Robert Graves in Majorca and the other from Hans Mardersteig in Verona, alerting us to the peculiar place of mushrooms in the Meso-American cultures. We had known nothing before then of the indigenous cultures of that region. Quickly we got in touch with Gordon Eckholm of the American Museum of Natural History and Richard Evans Schultes of the Botanical Museum at Harvard.

Dick Schultes referred them to Blas Pablo Reko, whose assistance, together with important information supplied in two earlier papers by Schultes, put them on the track that eventually led to my parents' participation in the mushroom ceremonies.

I accompanied them on those Mexican expeditions. The first time was in 1953, the year I turned 17.[1] I share my father's excitement about traveling and thoroughly enjoyed the opportunity to participate in and observe other cultures.

Since my father has written extensively on the mushrooms of Mexico, I will mainly describe my impressions of the people and sights we encountered on our several visits there. The mushrooms are nonetheless foremost in my mind, so I will comment on them also before closing this account.

The journey from Mexico City to Huautla de Jiménez takes one past beautiful volcanoes to the only village in Mexico where one feels free to drink the water—Tehuacán. Non-polluted springs are plentiful and the bottled water of Mexico comes from this town. The swimming pool at the hotel was filled with naturally carbonated water, so diving in presented quite a challenge. Floating, however, was easy due to the water's natural buoyancy.

From Tehuacán we traveled to a smaller town and then began our long mule trips. As the bird flies, the distances covered on muleback were not very long. But the trails went up and down ravines and around the mountains in such a way that it took quite a while to travel from one place to another. The Indians measured these distances by the number of baskets which could be woven en route. If asked how far it was to someplace in particular, they might reply, "Two baskets."

We encountered many mule trains on these trails, and it was here that I first heard the whistling language used by the *arrieros,* the mule guides or handlers, to communicate in the mountains. They would whistle when approaching a curve in the narrow, winding path to let people ahead know they were coming. I soon learned that the Indian women did not whistle, though they did understand what the whistling meant. Once I whistled at my mule to get it going and apparently communicated something rather shocking by mistake. Our arriero turned around so abruptly, in utter amazement, that he actually fell off his mule! Later, when I repeated the whistle for some women in Huautla, they giggled but did not explain what I had said.

The spoken language of Huautla also fascinated me. Not only was it tonal, but I learned that one had to respond in the pitch or key established by the first person to initiate a conversation. Otherwise, an entirely different message was communicated.

I shall always remember with fondness the people of Huautla, especially Herlinda, in whose house we stayed and who translated for us. Their warmth and curiosity remain with me. Among many other memories, I vividly recall the time we brushed our teeth with carbonated water (or perhaps it was beer), which made us look like we were foaming at the mouth. It was a strange sight for the villagers to see—at the time, they did not brush their teeth—and several of the town's children gathered to watch with interest.

The native clothes, or *huipiles,* worn by the women were colorful and beautiful. Their cloth was woven on a backstrap loom, a device new to me.

I can still hear the clapping of the women as they made fresh tortillas. The food was good, though I thought it a bit heavy with garlic.

In general, the peoples we visited led a hard life in the mountains. The image of their *milpas,* or corn fields, growing on the mountainsides at incredible angles is still very clear in my mind. So is the memory of sliding down the mountainsides in mud from the torrential rains. And I recall how the Indians carried heavy loads upon their backs when they traveled the dangerous mountain paths. They had gotten so accustomed to this burden that when they dropped off a load of goods they filled their packs with rocks to maintain their balance on the return trip.

The first mushroom rite I attended was that conducted by Aurelio Carreras in 1953. I remember my father's excitement when it became clear that we were going to be able to participate in this *velada,* though we were not allowed to take the mushrooms. We helped prepare for it by joining the Indians in search of the mushrooms, which were gathered quietly with great reverence in an atmosphere of secrecy.

Not until 1955 did I take the mushrooms myself, a few days after my father and Allan Richardson became the first non-Indians allowed to do so in a *velada.* I was not prepared in advance for the experience, except for certain writings my parents had shared with me: Sahagún's historical account and more recent papers by Schultes and Jean Bassett Johnson, the late husband of our friend Irmgard Weitlaner Johnson, who had witnessed the Mexican mushroom rite in the 1930s.

My mother and I first took the sacred mushrooms on the afternoon of Tuesday, 5 July 1955, in Herlinda's house in Huautla. My father has written of this event, in *Mushrooms, Russia and History* (p. 303): "This was the first occasion on which white people ate the mushrooms for purely experimental reasons, without the aura of a native ceremony."

The first thing I remember about the experience is that the mushrooms tasted dreadful. This is due to psilocybin, the active ingredient, which has a very distinc-

38

tive taste. Later, in the field, we could tell if the mushrooms we found were hallucinogenic by breaking off a piece and either smelling or tasting it. The presence of psilocybin was unmistakable. It tasted so awful that when I first ate them I chewed only one, then swallowed whole the rest of the four pairs I was given. They still took effect very quickly.

What transpired has always been personal to me. I have never before disclosed these experiences to anyone except my parents.

The visions I saw were primarily multi-colored geometric patterns, extremely vivid. They seemed to be originating from the back of my head, even though I could see them in front of me through open eyes. At the time, I did not know physiology or anatomy, but later, after becoming a nurse, I learned that my impression was correct. The optic nerve ends in the occipital lobe at the back of the brain, so that is where visual images are processed.

The geometric patterns were sharply defined, but if I tried to focus on anything else in the hut my vision (and the visions) became blurred. This was because an effect of the mushrooms is dilation of the pupils. All sense of distance is distorted, whereas only the visions are seen with absolute crystal clarity.

The reader can therefore imagine my reaction when my father persisted in asking me such questions as: "Who was king of England" during this time or that time? Initially I turned my attention away from the visions and answered him. But finally I told him he was bothering me. (My mother later wrote, in "I Ate the Sacred Mushroom," her article published in the May 19, 1957 issue of *This Week* magazine: "From a distance I heard my daughter Masha say impatiently, 'Oh, Father, I'm having too good a time to bother talking to you!' ") For although I could answer his questions I preferred the wonderful visions.

Clinically speaking, when the mushrooms take effect one at first feels nausea, and cold. Basal temperature drops and the pulse slows. There is also a diuretic effect, which can be problematic when one is unable to leave the room in which a *velada* takes place.

My body felt shortened. As I was curious about the effect of the mushrooms on pain, I bit my finger and found I could feel the pressure but no pain.

The visions lasted about five hours, passing all too quickly. Thereafter we slept for a short while. When I awoke, I felt more rested and refreshed than ever before.

I have sometimes been asked if I was not afraid to take the mushrooms that first time. The answer is no. I knew that my father had taken them and he looked fine, as did the Indians, who had used them for centuries. Neither apprehension nor anxiety ever crossed my mind.

I took them again on several occasions, both as mushrooms and as pills prepared by Albert Hofmann. The experience seemed charged with more significance when a session was conducted by María Sabina, who was truly a great woman of this earth. Words are inadequate to describe her presence.

On one occasion, in my visions, I seemed able to hop all over the world. I would come down and alight to visit friends far away. There was a quality of truth about it. However, I told no one, for I did not know what to make of it. My parents and I were aware that the mushrooms appeared to have a certain extrasensory potential. But we did not want to publicize this fact. My father explained our reasoning in his last book, *Persephone's Quest: Entheogens and the Origins of Religion* (1986), in which he writes, on page 33 of the trade edition: "I had always had a horror of those who preached a kind of pseudo-religion of telepathy, who for me were unreliable people, and if our discoveries in Mexico . . . were to be drawn to their attention we were in danger of being adopted by such undesirables." Among these discoveries, or rather experiences, was the occasion when my mother saw a city at a distance in her visions and described it to us vividly. Later, while returning on a different path through the mountains than we had taken previously, we suddenly saw Mexico City far below us, whereupon my mother said: "That's the vision." It was just as she had described it.

It is difficult for me to say what lasting effect the mushrooms have had on my life, for I cannot imagine my life without them and the search for the role they played in other peoples' cultures. To this day, I find myself looking for mushrooms in paintings and tapestries. And though I do not go out of my way to hunt for wild mushrooms, I have a tendency to watch for them. When I go for a stroll with my friends I am always the one looking down!

My father never told me why he felt it would be good for me to try the sacred mushrooms. But he genuinely wished to share his life with me and all his experiences.

In the end, he shared his death with me as well. I now relate, for those who cared about him, the most difficult part of my essay: an account of his last days and how he died.

It was something of a fluke that he happened to be with me at the end. We always tried to be together on New Year's Eve, the anniversary of my mother's death in 1958. But Christmas was a holiday we did not usually spend together.

We had intended to stick to this schedule in 1986 as well, a year when Christmas happened to fall on a Thursday. I had already made certain other commitments related to the nursing school where I am an administrator. (The dean was away due to illness in her family, and I was filling in for her.) So it came as a surprise when Yvonne, my father's housekeeper, phoned on the Thursday or Friday before Christmas to confirm that she was planning to spend the holidays in Florida. That made it mandatory for me to bring my father from his home in Danbury, Connecticut, to mine in Binghamton, New York—a drive of three and a half hours. I could not leave him alone in Danbury, and he would not allow me to hire a nurse to stay with him. (I had done so in the past and he always fired nurses as fast as I could hire them.)

I therefore made the drive down to Danbury on Sunday, 21 December, picked him up and returned to my home the same day.

On Monday I put in a day at the school, and that evening my father took me out for dinner. I had wanted to stay home, but he insisted on taking me out. We had an absolutely lovely time together.

Later Monday night, however, my father became ill with what seemed to be signs of impending heart trouble. So Tuesday morning I insisted that he see a doctor and arranged it for him, but later that day my father asked me to cancel the appointment. He said he was feeling much better and would rather not go for a check-up until the day after Christmas. I protested; he was adamant. So finally I did as he requested and rescheduled his visit with the doctor.

Gordon at Masha's home in December 1986. This is the last photograph taken of him. *Masha Wasson Britten.*

At dinner Tuesday night he had one of his terrible coughing spells that troubled him during the last years of his life. I felt badly for him and wished that he did not have to suffer so. Then we talked about his plans for the months ahead, including a trip to Belize he expected to make in the first few weeks of 1987. Reluctantly but honestly, I told him I felt he should not make the trip; his health was not equal to it. He could barely get out of a chair by himself, and I would not be able to go with him to help. His coughing continued, so I advised him not to eat anymore until it subsided. Then I helped him upstairs to his room.

Two hours later my children returned from the airport. My son had gone to pick up my daughter, who had flown in from California for the holidays. Shortly after their return, at about 11:00 Tuesday evening, 23 December, I went back to my father's room to check on him.

He told me he was cold, and I noticed that his skin was very gray—signs of possible heart failure. I therefore hurried to another room to call for the ambulance. Meanwhile, my son, who has medical training, stayed with him.

I was finishing the phone call when my children called me back into his room. My father was having a grand mal seizure. There was nothing I could do for him but hold him. He died in my arms.

When the ambulance arrived the medics wanted to try to revive him using cardio-pulmonary resuscitation, which my father had explicitly stated should not be attempted. I had a difficult time preventing them from doing it. But at last, once again, my father's will prevailed.

He died of a heart attack or massive stroke or both. It was painful to be with him at the end—but also wonderful, for otherwise he would have died alone.

In accordance with his wishes, my father was cremated and interred with his brother's remains in the columbarium of the Washington Cathedral, on 2 January 1987. The lovely, simple ceremony, held in the Bethlehem Chapel, included a Nunc Dimittis sung a cappella by the boys' choir; this was also as he had requested.

Such are my memories, recorded here in honor of my father and my mother. I am grateful to them both for having raised me to appreciate life as a real adventure, and for giving me the courage to explore it, as they did, in a never-ending quest for greater knowledge.

Notes

1. *Editor's note:* In his book *The Wondrous Mushroom: Mycolatry in Mesoamerica* (New York: McGraw-Hill Book Co., 1980, p. 29), Gordon mistakenly wrote that Masha was 13 years old in 1955.

A Vote for Gordon Wasson

MICHAEL D. COE

I do not exactly remember when I first met Gordon Wasson, but it must have been in the early 1970s. He was already a legendary figure to me, for I had heard much of him from the equally legendary and decidedly colorful Steve Borhegyi, director of the Milwaukee Public Museum before his untimely death. Steve, who claimed to be a Hungarian count and dressed like a Mississippi river boat gambler, was a remarkably fine and imaginative archaeologist who had supplied much of the Mesoamerican data for Gordon and Valentina Wasson's *Mushrooms, Russia and History*, particularly on the enigmatic "mushroom stones" of the Guatemalan highlands. His collaboration with the Wassons proved even to the most skeptical that there had been a sort of ritual among the highland Maya during the late Formative period involving hallucinogenic mushrooms.

Gordon must have wanted to pick my brain for some detail of Mesoamerican culture, so we arranged to lunch at some now-defunct eatery in New Haven. It turned out that he had spent part of his young manhood in this city, during which time he had courted a young lady and been a political reporter for one of the local newspapers. Among his journalistic activities, he had several times interviewed the famous Hiram Bingham, who, while he was a Yale Professor, had discovered Machu Picchu (as a matter of fact, at the time of our lunch I had curatorial charge of the Machu Picchu collections at the Peabody Museum). Bingham had gone on to become a highly successful Connecticut politician, serving first as Governor of the state and then as a member of the United States Senate. In the latter role he achieved the dubious distinction of being the second senator ever to be censured by his colleagues (his successor in infamy was Joseph McCarthy). I asked Gordon, whom I immediately held in awe as a scholarly, silver-haired gentleman of the old school, what kind of person Bingham was, since I had never met anyone who knew him personally. Gordon emphatically replied, "He was a *horse's ass!*" This marked the beginning of my strong affection for Gordon.

Gordon often came to our house on Saint Ronan Street, and Sophie and I visited him in Danbury. He was a connoisseur of good food and wine, and we often shared information on the culinary pleasures of a country dear to all three of us: Italy. I doubt that I have ever known anyone more erudite, with a more incredible stock of information on the most out-of-the-way subjects, all of it, as far as I know, completely accurate.

His Danbury study was lined with rare scholarly books from floor to high ceiling. Gordon need not have moved very far from his own library in his quest for knowledge. He was a stickler on detail. When Gordon Whittaker and I sent him our edition of *The Treatises on Superstitions* by the 17th Century Mexican priest Hernando Ruiz de Alarcón, he thanked us and then pointed out, in the kindliest fashion, that we had not grasped the difference between monks and friars—a major lapse on our part which no one else had noticed. Gordon would indeed have been a formidable editor!

Gordon's influence on my own field, Mesoamerica, was profound. Although anyone reading such important primary sources as Fray Bernardino de Sahagún (a friar, not a monk!) would have known of the great ritual importance of peyote, mushrooms and ololiuqui in Aztec culture, it was really not until the Wassons' research in both field and library, and their collaboration with scientists like Roger Heim and Richard Evans Schultes, that Mesoamericanists realized the pervasiveness of entheogens in the high cultures of Mexico and Central America, both ancient and modern. I was impressed by all of this. For me, his most elegant analysis centered on the beautiful, strange Aztec statue of the seated Xochipilli or "Flower Prince," god of summertime and of pleasurable activities, his body covered with entheogenic plants. He was rightfully proud of that paper.

While Gordon was enormously enthusiastic about his subject (he frankly enjoyed the experience of taking mushrooms under the guidance of his Mazatec mentor, María Sabina), he was no hippie guru. It would have been easy for him to have turned into another Timothy Leary during the decade of the '60s, but this was entirely contrary to his nature. He was first and foremost a scholar, and he was not prepared to warp scholarship in the interests of the current *zeitgeist*. Many of my Yale students reacted positively, usually for the wrong reasons, to Gordon's work. Two of them—simon-pure hippies in lifestyle and outlook—had been introduced to psychotropic *Stropharia* mushrooms while visiting the Classic Maya site of Palenque, in the heart of Chol Maya country. A kind of dropout community had arisen at Palenque among expatriate lotus-eaters. Notwithstanding the fact that these fungi were growing on "cow pies" in pastureland, the idea got about (and into print) that the ancient lowland Maya and their Chol descendants ate the "magic mushrooms." Gordon steadfastly maintained that this was pure poppycock, pointing out that there was not the slightest evidence from archaeology, ethnohistory or ethnobotany that mushrooms had ever been used ritually in the Maya lowlands. As usual, he was right—despite much wishful thinking by others.

It is rare to find anyone, above all a scholar, who grows old gracefully. Gordon was one who did. With advancing age, his mental acuity and delightful sense of humor seemed to grow stronger. Perhaps it was due to his custom of sleeping outdoors at night, even in the coldest weather (his bed was on the back porch of his Danbury house). He was a wonderful raconteur, and his stories got better as he grew older; I often regret I did not tape some of them.

It was a pleasure to be present when honor came to him in his eighties. Yale's Peabody Museum of Natural History bestows its Verrill Medal at very rare intervals. When Gordon became a medalist, it was in recognition by the curators of his accomplishments both as a scholar and scientist. He was a true disciplinarian, who had brought various lines of inquiry to bear on a cultural riddle: the entheogen-induced numinous states so important in the ritual life of early cultures around the world. The curators were quite pleased with his selection, and I think Gordon was pleased as punch.

There are some fortunate ones for whom fame and recognition only increase as they reach true old age, instead of being banished to generational oblivion. My friend the artist Josef Albers was one of these lucky octogenarians. Gordon was another. I suppose one reason for this is that he continued to be productive until the end. In his final year, he was still fussing with publishers, and concluded a successful negotiation with Yale University Press for his farewell volume.

In Fall 1987, only a few months before his death, Gordon was invited to speak at the Master's Tea in Yale's Silliman College. The whole point of these "teas" is to give the undergraduates in the residential colleges a chance to listen to, and question, outstanding figures from the "outside world" in a setting more relaxed than a lecture hall. On the guest roster that year were notables like Susan Sontag, Julia Child, Helen Gurley Brown, various artists, and sports luminaries.

It must have been a revelation to these young people to find out that a very aged person had a great deal to say, especially on his life in scholarship, that was both illuminating and exciting. At the end of the year, we asked our son Peter, who had been co-chairman of the series, how the Silliman students had informally ranked their guests. We were not surprised to learn that Gordon had won their vote, hands down.

A Bibliophile's View of Gordon Wasson's Books and Bookplates

ROBERT DEMAREST

R. Gordon Wasson's books hold a special fascination for the antiquarian, for the lover of books, for anyone who understands the effort required to produce a beautiful yet readable typographic work. This was not coincidental, for Gordon loved beautiful books. He appreciated fine typography and the bindings of the artisan. In many ways, he felt that the way a book looked, its *presentation,* was as important as its content, the ideas it contained.

All of Gordon's books demonstrated a respect for the almost-lost Old World tradition of fine printing and binding. His first, *The Hall Carbine Affair* (1941), written shortly after the start of his career with the J. P. Morgan Co., was a thing of beauty. But he managed to surpass it with the first of his mushroom monographs, *Mushrooms, Russia and History* (1957), co-authored with his wife Valentina Pavlovna. Like most of his limited editions, it was designed by the Italian master printer Hans Mardersteig and printed on handmade paper by the Stamperia Valdonega of Verona, Italy, one of the world's foremost presses. This book's collotypes—its multi-colored illustrations—were exceptional in quality, tone and registration. Since its price when published was a hefty $125, almost all 512 copies of the book were purchased by either institutional libraries or collectors of fine books. Even Gordon's inexpensive hardcover edition of *The Road to Eleusis: Unveiling the Secret of the Mysteries* (1978), which retailed for the bargain price of $12.95, was beautifully printed by the Stamperia Valdonega.

I early realized that I had to meet the man who created these works of scholarship and art. My "quest" began with *Soma: Divine Mushroom of Immortality* (1968), the first of his books I encountered, in the early 1970s. It was a copy of the trade hardcover, not the deluxe, edition. But its content so impressed me that whenever, in my travels, I visited libraries, mostly those of universities, I looked for material dealing with psychoactive mushrooms. I found that they typically kept their Wasson books not on the open shelves but in special collections, restricted "stacks;" would-be readers thus had difficulty getting to them. This led me to believe that there was something rather special about them, which in turn heightened my interest in books and eventually led to my interest in book selling.

So it was that I started a mail order book business out of my home in Naples, Florida, around 1977. The books I handled dealt primarily with hallucinogens—a combination of antiquarian and in-print books. At first it was only a hobby; I had a Bachelor's degree in math and was making my living selling real estate. But as I continued to collect and sell fine books, I found that I wanted to learn more about the craftsmanship involved in their making. The best option open to pursue such studies was in library science, which I undertook in 1978, eventually earning a Master's degree and becoming a library administrator. So in a sense, Gordon Wasson helped shape my career.

I finally met him in 1981. I had written to him expressing my interest in his work. In subsequent correspondence I advised that I would be pleased to assist him in locating any out-of-print books he might require for his research. He told me he needed copies of his own books, including the trade edition of *Soma*, which in due time I was able to supply. He then invited me as a guest in his home for a couple of days between Christmas and New Year 1981—the first of several visits.

At the time, I observed that his health was not good. His doctor had advised him to stop sleeping on the open back porch of his studio. (Gordon had chosen to sleep there for many years in the belief it was healthy for him.)

I was of course curious to see his book collection, which was housed in the studio building behind his home, where he did all his writing and research. The studio's walls were lined with bookshelves from floor to ceiling. But the first of his collection I saw was in the main house: a complete set, 17 volumes of text and 11 volumes of engravings, of Diederot's and D'Alembert's *Encyclopédie,* the first comprehensive French encyclopedia. Gordon's set was of course a first edition, published between 1751 and 1772. It intrigued me because, at the time, it was worth more than my house—at least $45,000.

Later he showed me his main collection. His bookshelves were a gold mine, reflecting both the man's intellectual capacity and love of books. I was even more impressed by his ability, unusually rare these days, to read—and write—in a variety of foreign languages than I was by the quality and scope of his specialized collection of books on mushrooms. He spoke French, certainly, but was conversant in several other languages as well.

We corresponded periodically between occasional visits for the next few years, usually when he wished to obtain a book for research purposes, or needed additional copies of his works that could only be obtained through antiquarian book dealers. I always arrived with at least two cartons of books when I visited him—books he had asked for or books from my collection that I thought he might enjoy. He greatly admired the elaborate custom leather bindings I had commissioned for some of his Harvard *Botanical Museum Leaflets* and for *Les Champignons Hallucinogènes du Mexique/Nouvelles Investigations,* two books on which he had collaborated with Roger Heim, published with illustrated stiff wraps in 1958 and 1967, that I had combined in one binding.

My most memorable visit to his home impressed upon me the strong attachment he had for his books. It was in August 1982, the day before his personal collection was to be packed for transportation to the site of the new Tina and Gordon Wasson Ethnomycological Collection at Harvard University's Botanical Museum. That he chose to donate them to Harvard was certainly not surprising.

His respect for that institution was apparent to all who knew him and to many who only knew *of* him. Many of the scholars on whom he depended for translations or data were Harvard-educated, and of course he was a close friend of the famous, long-time director of the Botanical Museum, Richard Evans Schultes, who had been very helpful to Gordon in his research.

However, it was clear when I saw Gordon that August afternoon that he had very mixed emotions. His books were still on the shelves; no crates or boxes were lying about. But a work crew was apparently expected the next day to start boxing them. He was clearly depressed and tearful about it. I thought it exceptional for him to demonstrate so openly the depth of his emotions. He especially lamented the difficulty of writing his "next book or two" without this valuable reference collection nearby; he felt that he would be severely hampered in his work.

"How will I write without them?," he asked—a painful, rhetorical question. For his collection included many rare, old volumes long since out of print.

I could imagine and appreciate, if not exactly share, his sense of loss. There are certain men who have a real talent for taking a book maybe 300 or 400 years old and extracting enough information from it—*new* information that others had either dismissed as unreliable or overlooked—to extend current knowledge and expand our conceptual framework. Gordon clearly had the knack. His basic premise was that much information in old books, however unlikely it might seem today, was accurate. That is how he felt about the stories written down by the friars in Mexico hundreds of years ago, which told about local Indians eating hallucinogenic mushrooms. Most scientists up to the 1930s believed that the friars were wrong; that they had mistaken peyote for mushrooms. The scholars refused to admit there were any such things as hallucinogenic mushrooms. But Gordon was willing to say: "I believe that these fellows recorded this accurately; therefore I have to go out and find the mushrooms." And of course he did just that. Much of his scholarship, and much of the work he commissioned that resulted in new and more accurate translations from the original source documents, was a sign of his respect for the ability of those who came before him to record information accurately. The same view served him well again when he set out to prove his theory that the Soma described in the ancient Rig Veda was also an hallucinogenic mushroom.

Gordon Wasson the bibliophile, the lover and respecter of fine old books, also recognized in them a key to preserving his own research against the passions and ravages of a rapidly evolving culture. It was largely for this reason that he lavished so much care on the production of his limited editions, and no doubt also explains his desire to see them retain a significant monetary value. At his request, I would bring, when I visited, photostats of auction price records, listings from *American Book Prices Current, Bookman's Price Index,* antiquarian book dealers' catalogs and other indications of the current market price for his limited editions. In 1982, I think it was, he saw that a copy of *Mushrooms, Russia and History* had sold for $1,750. On a subsequent visit I brought him a copy of a British book seller's catalog listing a presentation copy of the same book, signed by him, for $2,500. He was obviously pleased that his books were escalating in value. As valuable works, Gordon realized, they would be sought after, saved—even venerated; they would not be discarded as dated material.

There are few ways to accomplish this today in the modern publishing environment, other than by carefully and lovingly creating works that obviously represent the efforts of artists and artisans. Gordon's limited edition books, for example, occasionally reproduced illustrations using the expensive *pochoir* process. This process, using hand-cut stencils, yields superb illustrations with a beautiful texture. He obviously felt that the benefits gained by such touches, including frequent use of handmade paper and aesthetic, unusual bindings, outweighed the expense.

In light of Gordon's devotion to quality printing and binding, it is interesting to note that his first experience in the world of publishing involved what is surely the cheapest, most ephemeral of publishing media: the newspaper. Before becoming a banker, ethnomycologist or book author, Gordon worked for several years as a business reporter for a newspaper in New England. Of further interest, Gordon's article, "Seeking the Magic Mushroom," which was published in the May 13, 1957 issue of *Life* magazine, was the only extensive and substantial piece dealing with his research that was ever to appear above his name in a large circulation periodical. In large measure, this article was responsible for his (and the Mexican shaman María Sabina's) notoriety within the 1960s counterculture, a phenomenon for which he felt some responsibility but of which he clearly did not approve. In contrast to what he considered the far more satisfying results of his privately printed first editions, this experience appears to have had a profound affect on Gordon's publishing inclinations. Thereafter, he devoted himself almost exclusively to hardcover monographs priced in the hundreds of dollars when published.

Gordon knew of my keen interest in how his books were made, which led to an unexpected result. Less than a month before his death in December 1986, there arrived at my door a rather large box from him. It contained two bound galley proof copies of his last book, *Persephone's Quest: Entheogens and the Origins of Religion* (1986), and a short note explaining that he thought I would like them. I certainly did! They were two of five or six working copies he had circulated among the book's three other authors—Stella Kramrisch, Jonathan Ott and Carl A. P. Ruck—for their final corrections before publication. I felt honored as I lifted them out of the box and paged through them, inspecting the handwritten notes the authors had made here and there. I was also surprised, because authors do not normally like others to see the corrections they make in producing their works; they would rather keep their manuscripts a private affair and let the final product be definitive. But thinking back, I recall that in every discussion I asked Gordon about his works in progress. So this may have been his way of saying, "Thanks for being interested." Or perhaps, sad to tell, he may have sensed that he did not have long to live, and simply wanted to make sure that the galley proofs went to a person he knew would appreciate their value and preserve them.

Gordon Wasson's Bookplates

Tristram Shandy's warning against the excesses of those who advance hypotheses merits a page in *Persephone's Quest*:

> It is in the nature of a hypothesis when once a man has conceived it, that it assimilates everything to itself, as proper nourishment, and from the first moment of your begetting it, it generally grows stronger by everything you see, hear or understand.

Such excesses can plague those who advance new hypotheses that later are shown to be correct, as well as those whose hypotheses are later proved wrong. Gordon's scholarly pre-occupation with his hypothesis of the religious implications of hallucinogenic mushrooms and their forgotten, but continuing, effects on human attitudes toward fungi was so woven through the fabric of his life that tracing the evolution of the theory is tantamount to telling his life story. That is precisely what he does in *Persephone's Quest,* a book that convincingly marshals the evidence gathered by him and his colleagues to vindicate his hypothesis. Nevertheless, or perhaps for this reason, it is easy to see where he has stepped on the toes of professional scholars—or, more precisely, certain academics.

Gordon Wasson did not like to be wrong and, given sufficient time, could support his hypotheses with an overwhelming body of supporting evidence. So it was with Professor John Brough of Cambridge University in an extended disagreement, conducted in print, over the identity of Soma. So it was, too, with the history of the "rediscovery" of the Mazatec use of the hallucinogenic *Psilocybe* and the naming of one particular species.

I learned of the latter dispute in an unusual manner: from Gordon's bookplates. The first time I saw them was during my first visit to his home, in December 1981, while looking through his library. As a bibliophile, I naturally have an interest in bookplates and soon noticed that those in Gordon's personal book collection were of three different designs—all featuring hallucinogenic mushrooms, and all designed by Margaret Seeler, a German artist well-known as an enameler.

The first bookplate—that is, the earliest—was a rather large and handsome plate measuring 12.4 × 8.0 cm, and containing three panels. From left to right, these panels featured: a Mexican *Psilocybe* species, identified on the bookplate as *Psilocybe wassonii;* a terra cotta mushroom effigy that stood, for many years, as the centerpiece of Gordon's studio library; and *Amanita muscaria,* or fly-agaric, the mushroom that Gordon identified with Soma. I had seen all these images in one place or another previously, but the first panel's mushroom confused me a little. I recognized it from the illustrations on the cover and title page of *Les Champignons Hallucinogènes du Mexique,* where it is identified as *Psilocybe wassonii* Heim. But I had also seen it elsewhere, or something quite similar, identified as *Psilocybe muliercula,* a name bestowed on it by Rolf Singer and Alexander H. Smith.

Curiously, Gordon's second and third bookplates (again, chronologically speaking) were smaller, and the illustration of *Psilocybe wassonii* had been eliminated from both. I asked Gordon about the variation. He responded with an afternoon of tales concerning his dealings with Singer and Smith.

My subsequent investigation of this decade-and-a-half dispute led me to more than a half dozen publications revealing the details of an obviously bitter feud among learned men. Gordon expected of scholars a certain commitment to ethical standards and honorable practice that he felt Singer and Smith had transgressed in the process of seizing unfairly the right to name the mushroom at the

Gordon's original bookplate included a drawing of *Psilocybe wassonii* Heim. *Right,* bookplate adopted by Gordon after Rolf Singer nameed *P. wassonii* differently. A third plate used by Gordon was identical to this one but smaller.

center of this controversy. In his opinion, they had callously disregarded his long years of patient, groundbreaking research precededing his earlier discovery of this mushroom species.

A portion of the tale was set down by Jonathan Ott in a paper, "Mr. Jonathan Ott's Rejoinder to Dr. Alexander H. Smith," which responded to an earlier attack on Gordon's claim (and Ott's defense of it) published in the journal *Mycologia.* When the editors of this journal refused to publish Ott's rejoinder without some deletions, Gordon came to the rescue. For even when he lost an argument, Gordon wanted to have the last word, and he was therefore more than willing to arrange to have Ott's piece published in 1978 as a booklet by the Harvard Botanical Museum—No. 6 in Gordon's Ethnomycological Studies series. He then magnanimously published Singer's *A Correction*—effectively a rejoinder to Ott's rejoinder—in a subsequent number of the same series three years later. In *R. Gordon Wasson's Rejoinder to Dr. Rolf Singer,* published one week after that, Gordon had the last word.

The controversy centered around the disclosure and documentation of the Mexican mushroom cults and the description of a particular mushroom found by Gordon in use by the Mazatecs. On one side of the controversy stood Singer and Smith, both eminent mycologists and furthermore mycological "insiders" and academics with a thorough grounding in the botany, structure, classification and nomenclature of the field. They were allied, through long association, with the editors of *Mycologia,* the premier mycological periodical. On the other side stood Gordon, his champion Ott and the Mexican mycologist Gastón Guzmán.

Smith had published an article in a 1977 issue of *Mycologia* (69:1196–1200) critical of Ott's recent book *Hallucinogenic Plants of North America* (1976) and two articles published in 1976 by Guzmán, Ott et al., *Mycologia* 68:1261–67 and

68:1267–72. Smith's piece appeared to be a mean-spirited and petty questioning of the accuracy of all three works and, further, chided the authors for lacking "the ability of some people to say what they mean." Ott suggested that Smith's article was actually written in retaliation for a single footnote in Ott's book ascribing less than honorable motives to Singer's research trip to Oaxaca to study the psychotropic fungi of Mexico. Singer had followed Gordon's earlier itinerary step-by-step as described by Roger Heim, who had accompanied Gordon on some of his expeditions, in a number of scientific journals. He had purchased some of the mushrooms from the same vendors Wasson's party had encountered and spoke with many of the same people they had contacted. Amusingly, some of the itinerary of Gordon's earlier expedition and that of Singer's later trip have been found in previously-classified CIA documents. Both parties, it turns out, had been under observation by the CIA in an attempt to learn as much as possible about these newly discovered hallucinogens—a subject the CIA found interesting at the time, for possible use in developing "mind-control" drugs.

In naming a new species, priority is based upon the first published scientific description appearing in Latin. Heim had described the species accurately in French in his preliminary articles appearing in *Revue de Mycologie.* He had also written extensively of his and Gordon's expeditions and discoveries in *Comptes rendus* of the Academie des Sciences. Heim's Latin description appeared in *Mycologia* on April 29, 1958, but Singer's and Smith's Latin description of the same mushroom had been rushed into print 25 days earlier. Thus they had priority, "displacing" Heim's name, *Psilocybe wassonii,* with theirs, *Psilocybe muliercula.*

Given the circumstances of Singer's Oaxaca "field work" and comparing his and Gordon's respective contributions to an understanding of the ethnological use and habitat of these fungi, it was Ott's contention that an exception should be made to the rule stipulating priority in naming a mushroom. Gordon, of course, was in an awkward position. He enjoyed the honor of having a "sacred" mushroom named for him. He also felt strongly that Singer and Smith had behaved underhandedly and surreptitiously. But still, he felt ill at ease about "continuing the appearance of polemical publications." In the end, no exception was made to the priority rule and so—at least officially—*Psilocybe wassonii* became *Psilocybe muliercula.* In 1979, Guzmán and Steven Pollack named another Mexican species, *Psilocybe wassoniorum,* "in honor of Mr. and Mrs. Wasson for their valuable ethnomycological work on the hallucinogenic mushrooms."

When the name of the first mushroom changed, so did Gordon's bookplate. He did not remove the old ones already in place; nor did he correct them by pasting the new ones over or next to them. He simply stopped using the old plates and switched to the new for all subsequent acquisitions. By the time of my visit in 1981 the conversion was completed. He gave me loose samples of all three bookplates, and additionally duplicates of some of the books in his personal collection that included his bookplates. The tale of the mushroom-naming controversy now springs to life every time I open these volumes.

— PRE PSYC. Movement ... WAS SOCIETY DIFFERENT IN WAY OF INNER MIND?

'Somatic' Memories of R. Gordon Wasson

WENDY DONIGER

Surely it is scholarship, not politics, that makes the strangest "bedfellows" of all. A case in point is R. Gordon Wasson, the most *gentlemanly* person I have ever had the privilege to work with academically and to know as a friend.

He was an old-fashioned American aristocrat who almost seemed a holdover from the 19th Century, a great banker who looked and acted the part: elegant and suave; gracious in the special manner of one accustomed to command; fastidious about manners and haberdashery; prudish about four-letter words; pedantic about neologisms and grammar; and almost spartan in his attitude toward food and, even more so, drink. That this Gordon Wasson should have become a hero to the hippies and an unwilling guru to the "drug generation" was truly a great irony.

No less ironic, perhaps, is the fact that this man, whose formal education went no further than a Bachelor's degree and whose life had been spent not in the ivory towers of academia but in the dark canyons of Wall Street, should have made such great contributions to some of the major scholarly issues of our age.

A further, minor irony relates to Gordon's tendency to scorn colloquial neologisms. For he himself introduced several new words to our language, such as *mycophile, mycophobe* and *entheogen.* The latter, which Gordon proposed as a sub-stitute for *hallucinogen,* was never really accepted by his public, though he felt that its sacral allusion (it means literally "producing a god within") made this term much more appropriate in light of the important role played by psychoactive plants in many religious rites.

The mixture of banker and researcher Gordon embodied became physically incarnate in his books: works rich in information and ideas, but also in form. These magnificent, rare examples of great bookmaking soon became collector's items selling for several hundreds of dollars at auction—to wealthy book collectors, one assumes, rather than to assistant professors.

It is a shame that the combination of qualities Gordon embodied should nowadays be so rare as to surprise us. He was a man of whom it truly could be said, "He was a scholar and gentleman."

Indeed he was a scholar, but he had his own brand of scholarship—a scholar-ship through friendship. People, more than books, were his research sources. When he wanted to know something, he read all about it and then, if his questions remained unanswered (as they usually did, being questions no one had asked before), he simply wrote to the acknowledged world expert on the subject. And, oddly enough, the experts wrote back. They were intrigued by his projects; they liked him; and they became his informants, his allies and often his friends. Some of these alliances were logical, others surprising. Aldous Huxley, yes; Robert Graves, yes: one could see why he would like them and why they liked him. They too were amateur scholars—highly literate, learned amateurs who had built their international reputations on massive and impressive literary oeuvres. But John Cage? Not so obvious at first, though it makes sense when one thinks about it—especially considering Cage's mycophilia. And what about Timothy Leary and Richard Alpert (Ram Dass)? No, surely not! Yet Leary, by his own admission, once regarded Gordon as his spiritual guide.[1]

Exactly how did Gordon Wasson go from high finance to *High Times* magazine,[2] from high society to Leary's book *High Priest*? The answer, of course, was the drug connection. Gordon's trail-blazing work on hallucinogens, the true opiate of the people, caught the interest of artists and scholars alike.

But the process worked both ways. Gordon always was a "people person." So not only did it happen that the lure of magic mushrooms led him to other mycophiles, but his love of mushrooms came to him initially through the love of people—or, more precisely, the love of a person: his wife Valentina.

In my own relationship with him, I came to wonder which was the chicken and which the egg. I was glad I had met Gordon Wasson, partly because it was he who made possible my first publication[3]—one that brought me immediately into contact, fresh from graduate school, with an unexpectedly wide and favorable readership. With typical tact and concern, Gordon worried that this potentially controversial book *Soma* might endanger my fledgling career if my name were too closely identified with its maverick hypothesis that Soma, the sacred substance described in the ancient Rig Veda, was actually the fly-agaric mushroom. We therefore agreed that the book would be published in his name only, attributing to me the section on the history of the plant, which I had written entirely by myself. As the years went by and the book was taken far more seriously than either of us had ever expected, he generously asked if I would mind if he began to refer to me as the co-author; and of course I assented.

Actually, in an odd sort of way, it was I, not he, who stumbled serendipitously upon the Soma mushroom, as Gordon acknowledged himself in the book's introductory pages. In addition to sending him all the Vedic translations he had asked for, I one day sent him, purely for his amusement and partly to tease him for his over-nicety, a story from the *Mahabharata* that I happened to be working on separately as a personal project. It told about a god, in disguise, who urinated before a Brahmin, and referred to the urine—apparently sarcastically, but actually in truth—as Soma. Like Queen Victoria, Gordon was not amused;[4] instead, to my surprise, he was wildly excited—for he knew, as I did not, that Siberian shamans retained in their urine the drug properties of fly-agaric mushrooms when they ate them.

But whenever I thought that the reason I valued this man was because he had led me to the book, I always realized that actually the reverse was far more true: the main reason I was glad to have done the book with Gordon Wasson was

In *Soma: Divine Mushroom of Immortality,* Gordon and co-author Wendy Doniger O'Flaherty argued that the sacramental substance Soma described in the Rig Veda was actually the fly-agaric mushroom, *Amanita muscaria. Courtesy Richard Evans Schultes.*

because it launched my long epistolary friendship with him. ("Our correspondence was supreme," he once remarked.[5])

I suppose that he was, in his way, a genius. Certainly he had the inspiration; but he also had the other 99% of what genius demands. I mean by this the "perspiration," the "capacity for taking trouble" that Samuel Butler regarded as essential to genius. Gordon never gave up trying to prove that Soma was the fly-agaric mushroom, *Amanita muscaria.* While we were working on the Soma book—indeed, for many years *after* the book had been published, when I was no longer really interested in it at all—letters from Gordon arrived almost every other day: sometimes just a paragraph, sometimes just a sentence, each one querying some small point, correcting my English (sometimes even correcting my Sanskrit), querying my interpretations, offering new citations, asking about parts of the Rig Veda he had not considered before; asking, asking, asking. He replied to his critical reviews; he marshaled the support of his affirmative reviewers; he enlisted the help of other scholars; he adduced new evidence; he invented new arguments—he never, ever gave up. Finally, almost 20 years after publication of the *Soma* book, he published, almost posthumously, a long essay summarizing and, in some cases, reproducing the post-publication debates.[6] Indeed, even more than his "capacity for taking trouble," Gordon fit Butler's expanded definition of genius:[7]"A supreme capacity for getting its possessors into trouble of all kinds and keeping them therein so long as the genius remains."

One of the reasons that Gordon kept at me was that he never entirely succeeded in persuading me that the second half of his hypothesis—the argument that Soma was not only a hallucinogen but one in particular, the fly-agaric mushroom—was correct. One critic sharply noted this when he remarked:[8] "If his [Wasson's] ideas about the mushroom were so convincing, why didn't she [Wendy Doniger O'Flaherty] mention the *Amanita muscaria* in her commentaries when her translation of the Rig Veda was published years later?" I was pretty cagey about it at first: without admitting it outright to Gordon, I did not believe (and still do not) that we had proven Soma definitely *was* a mushroom, any more than I felt anyone could prove it was not. On the other hand, I did believe we had established a far more important hypothesis: Soma was a hallucinogen. If true, this casts an entirely new light on the whole of Indian religious history, particularly on the esoteric traditions and on yoga techniques. Gordon did not fully realize my reservations until he was exposed to them on several occasions. "I did not learn Wendy O'Flaherty's attitude until I read her note to 'The Last Meal of the Buddha'[9] in *The Journal of the American Oriental Society* and also in this book!," he eventually wrote.[10] He was not daunted in the least, but merely redoubled his efforts to persuade me.

My own work, however, was leading me toward another sort of hypothesis entirely: it did not really matter what Soma was, since it was lost so early in history; what actually played so important a part in Vedic civilization was the *idea* of Soma. Indeed, my more recent collaborations with Brian K. Smith on the subject of substitutions in the Vedic sacrifice[11] have inclined me in the direction of Smith's hypothesis that there may never have been an original Soma plant at all, and that *all* of Soma's "substitutes" (including, perhaps, the fly-agaric mushroom) were surrogates for a mythical plant that never existed save in the minds of the priests.

I wish that I could try out *that* one on Gordon. I am sure that he would take it seriously, oppose it energetically and build it into his next round of counter-arguments. He lived a very full life, but he still regretted dying when there were so many more things to learn, so many arguments to make, so many more books to write. He never did give up.

Notes

1. Leary testified to this in his book *High Priest* (New York and Cleveland: The World Publishing Co., 1968).
2. An interview with Gordon, conducted by Jonathan Ott and Steven H. Pollack, was published in the October 1976 issue of *High Times*.
3. *Soma: Divine Mushroom of Immortality,* by R. Gordon Wasson, with Wendy Doniger O'Flaherty (New York: Harcourt, Brace and World, 1968).
4. This though he proved capable, on many occasions, of overcoming his native delicacy in

the cause of scholarship, and eventually himself drank urine containing hallucinogenic residues. Perhaps his greatest moment of this sort came when he was the guest of honor at a banquet in Singapore. His host promised him a special treat, a dish that would bring long life and continuing virility. A live cobra was brought out; its stomach sliced open; and the still-quivering liver (or heart?) drawn out and offered to Gordon. Without hesitation, he ate it; his reluctance to insult his host—and also, I think, his unquenchable curiosity—far outweighed his reluctance to eat the cobra's liver.

5. To Robert Forte, as reported in "A Conversation with R. Gordon Wasson (1898–1986)," in *ReVision: The Journal of Consciousness and Change* Spring 1988; 10(4):13–30.

6. "Entheogens and the Origins of Religion," Part One (pp. 16–148) of *Persephone's Quest: Entheogens and the Origins of Religion,* by R. Gordon Wasson, Stella Kramrisch, Jonathan Ott and Carl A. P. Ruck (New Haven, Connecticut: Yale University Press, 1986).

7. Samuel Butler's "Genius," in his *Note-Books* (1912).

8. Gene Kiefer, "More About *ReVision* and the Sacred Mushroom." No. 5 in a series for the *SFF Newsletter,* September 1988. Published by the Kundalini Research Foundation.

9. Originally published in *The Journal of the American Oriental Society* October–December 1982; 102(4). Reprinted as pp. 117–139 in *Persephone's Quest,* including my epilogue on pp. 138–139.

10. *Persephone's Quest,* p. 29.

11. "Sacrifice and Substitution: Ritual Mystification and Mythical Demystification," by Brian K. Smith and Wendy Doniger. Forthcoming in *Numen.*

R. Gordon Wasson:
The Poet of Ethnomycology

WILLIAM A. EMBODEN

Eleusinians will always join together in contest and battle again and again.
—The *Homeric Hymn to Demeter* (267–8)[1]

The discipline of ethnomycology is but three score years old and now includes many among its numbers. This was not always so. It took a poet such as R. Gordon Wasson to give the discipline wings, to multiply her numbers, to increase her vocabulary and to uncover the mysteries so long dormant. His poetry was of the type described in *The Poet and His Muse,* a piece by the great French poet Jean Cocteau: for Cocteau, poetry was any excercise of the creative spirit. So Gordon, with the mushroom as his muse, drew from his creative well the science and mythology of ethnomycology. His Ethnomycological Studies series is now famous and will no doubt be cited many times in the essays comprising this anthology. We have all profited from his writings.

The line from the *Homeric Hymn to Demeter* that opens my essay is a personal case in point. It is from the translation appearing in *The Road to Eleusis: Unveiling the Secret of the Mysteries,* a book co-written by Gordon, Albert Hofmann and Carl A. P. Ruck. Together they argue persuasively that the hitherto unknown sacramental substance imbibed at Eleusis was concocted from a psychoactive fungus. Inspired by Gordon's ideas and questions, it is the work of a true scientist-mythologist and the scientific work I most enjoy—a 4,000-year-old puzzle at the point of being solved by Gordon Wasson and his colleagues, using ancient poetry, cultural patterns, comparative religion, hard science and cultural anthropology to produce a small volume that I call the poetry of science. That is why I took a line from the Homeric Hymn of Demeter appearing in the book as my motif.

There is also a reason I chose that particular line from the poem. My relationship to Gordon Wasson was sometimes adversarial, involving a good-natured battle of minds, an ongoing intellectual chess game. The thrust and parry we exchanged was that of friends enjoying the much ignored art of disputation. As with chess, it was played as a game for ourselves and a few spectators. We did this in letters, in meetings and in journals.

One is distinguished by having had Gordon Wasson as an opponent in any debate. His knowledge was formidable, his speech eloquent, and his vocabulary so stellar that he has added to our mycological lexicon.

In his earlier volume *Soma: Divine Mushroom of Immortality,* he sent most of us to libraries to begin a reading of the extensive Vedic hymns that for so long had puzzled Sanskrit scholars. In 1972 he made a presentation of that subject before a rapt audience of botanists, mycologists, anthropologists, religious scholars and the general public in a lecture series that was to culminate in the book *Flesh of the Gods,* published in 1972 under the editorship of Peter Furst, who brought us all together. At that convocation he united Greek, Indian and Middle American mycological lore, weaving a spell over all of us with his profound erudition. I was but one of several hundreds of individuals attempting to tap that reservoir, issue challenges and to walk away saying, "Of course I know Gordon Wasson." With him we traveled backward and forward in time, into "vistas beyond life." He could summon up Blake, St. John of Patmos, Aryan mystics, Aristides and a Mazatec shaman, integrating their visions and making of them a sensible pattern. For by ingesting the "magic mushrooms" he had experienced and understood the sacred agapé described by all these figures. "You feel that an indissoluble bond unites you with the others who have shared in the sacred agapé," he explained. "[It] is as though the world had just dawned . . ." This freshness of vision and newness of everything characterized the mind of Gordon Wasson more than any other quality. And it was this that helps substantiate my assertion that he was a true poet.

Unknown to many in that audience was the fact that he had made it possible for several mycologists to pursue their studies and travels. His personal generosity was not flaunted, rather he supported individuals and concepts he believed in. This done quietly with the decorum and the dignity that so typified the man.

I still have a keen recollection of that first encounter with a man who believed in the immediacy of vision and the personal encounter with one's self. In many respects he suggested the late Joseph Campbell, whose *Masks of the Gods* and such-like volumes had opened so many doors for us. But with Gordon there was the additional aspect that he was a self-taught scientist. Perhaps that is not quite accurate, for he surrounded himself with scientists of the highest calibre: Richard Evans Schultes, Roger Heim and Albert Hofmann. In linguistics he allied himself with Carl A. P. Ruck and Daniel H. H. Ingalls, Professor of Sanskrit at Harvard. This unification of great minds led to great works, and Gordon was a prodigious scholar.

Additionally, I had long admired him for his exquisitely realized book *Mushrooms, Russia and History,* one of the rarest items for any bibliophile or mycologist to own. It has been virtually impossible to find for quite some time; over 15 years ago, a copy sold for more than $2,000. In March of 1979, I wrote to him at his home in Danbury, Connecticut, asking if he would consider authorizing a reprint of that famous volume. His answer of 24 March 1979 was:

My Dear Emboden:

I have read so much of your writings that I feel as if I know you. Thanks for your letter. MR&H [*Mushrooms, Russia and History*] will not be reprinted, if I have anything to say about it.[2] In today's world it could not be reproduced for love or money. Further there are [so] many defects . . . omissions, commissions, better ways of saying what we had to say . . . that it would have to be rewritten. Moreover, everything I have published since then has really been an expansion, an improvement, on one passage or other of MR&H.

I have *not* seen your review of *The Road to Eleusis.* When did it appear? I was in Europe in December-January. In the pile of mail that confronted my return, if it was there, it may have been lost in the *mêlee.* Thank you for mentioning it to me. Have you a spare copy or can you xerox the part pertaining to *Eleusis?* A review has appeared in *Vuelta,* the journal owned and edited by Octavio Paz in Mexico, by a man that I have never heard of . . . Hanue García Terrés . . . a superb review. Have you seen it? It seems he is a Hellenist, a poet, man-of-letters, former Ambassador of his country to Greece. Tell me if you have not seen it.

I had already decided to attend your lecture in Harvard. Whether it will be possible I do not know. I look forward to listening to you and meeting with you.

You may have seen my recent Botanical Museum Leaflet (Harvard), Soma brought up to date, I suppose.

Cordially and in haste,

R. G. Wasson

Gordon's letter was certainly cordial and hardly in haste. His announcement that he would attend my lecture at Harvard (an invitation I had extended in my letter to him) was viewed by me with gratitude and alarm. Gratitude that he would dignify my presentation by making a special trip to Harvard, and alarm that I would be presenting a novel thesis involving mushroom iconography before the leading authority in the field.

The April 15th lecture was auspicious: portents were everywhere. It rained, snowed, hailed and lightning flashed between intermittent bursts of sunshine. Surely Gordon Wasson was about to appear!

The audience that gathered in the Botanical Museum lecture hall numbered 70 people. And among them, indeed, was R. Gordon Wasson, driven by his chauffeur from Connecticut. I lectured on aspects of the water lily iconography in early Maya codices, ceramics and stelae. With some elan and much trepidation I advanced the thesis that some of the images Wasson had interpreted as mushrooms were the peltate leaves of narcotic water lilies. At that time I stated, "I do not wish to critique this work [referring to the 1958 Heim and Wasson collaboration on hallucinogenic mushrooms of Mexico] but to extend it by implication to the sacred narcotic water lily, *Nymphaea ampla,* in some of the ritual presentations made by these authors. This adds an additional element of credibility to their hypothesis as well as my own." (It was presumptuous of me to say so. As if Wasson and Heim needed any additional credibility coming from me!) I presented Maya imagery from the aforementioned sources in defense of my hypothesis, then proceeded to critique the fresco of Tepantitla that recreates the garden of paradise. In every aquatic context I made the interpretation of a water

The Teopancaxco mural, dating from 300–600 A.D., includes pictographic glyphs along its borders that depict either mushrooms or the leaves of a narcotic water lily. *Courtesy Wasson Collection.*

lily where Heim and Wasson had found mushrooms.

The lecture was well received and several pertinent questions were asked. Then, from the back of the room, came the voice of Gordon Wasson. My instant reaction was to think, "Now I'm in for it." For about 10 minutes we exchanged ideas and finally agreed to disagree. Afterwards Gordon was most affable and generous in his appraisal. On 29 April of that year I received another letter from him:

> Dear Emboden:
> Thanks a million for your welcome letter of 22 April [I had sent an appreciation for his presence at the lecture with some references regarding mushrooms in opera libretti that he was unfamiliar with] and for the Shostakovitch references [*Katerina Ishmailova* is an opera in which Katerina poisons her father-in-law with mushrooms] and what you said about my talk [he had given an extraordinarily fine afternoon presentation at Harvard]. No, I have not assembled all of the literary references but they are . . . at least most of them in *Mushrooms Russia and History*. Your lecture was well worth going to Cambridge to hear. Where will you publish your findings? I was delighted, at last, to meet you.
> Sincerely yours,
> R. Gordon Wasson

I wrote to him indicating I had prepared a manuscript for *The Journal of Ethnopharmacology*, which finally appeared in Volume 3 of 1981 under the title "Transcultural Use of Narcotic Water Lilies in Ancient Egyptian and Maya Drug Ritual." After this appearance we began a series of fascinating (from our perspectives) critiques of one another's analyses, in the spirit and style of 18th Century pamphleteering. These published exchanges appeared irregularly in *The Ethnopharmacology Newsletter* (now sadly defunct), adding considerable spice to a publication that otherwise was a straightforward vehicle for communicating happenings in the area of ethnopharmacology. I pushed my point of view, and in his highly literate and exceedingly stylish writing, Gordon would counter with an opposing view. Many of our colleagues found this highly stimulating.

One day a few years later I said to Richard Evans Schultes, who for decades was Director of the Botanical Museum at Harvard University, "Don't you believe that Wasson sees mushrooms everywhere?" Schultes, one of the most droll and brilliant scholars I am privileged to know, answered with a wry smile. "The way in which I am able to tell you two apart," he said, "is that you are the one who finds water lilies everywhere and Gordon finds mushrooms in the same places." He had a tremendous admiration for Gordon Wasson and the feeling was reciprocated. Schultes' earliest work as a graduate student had been on the psychoactive mushrooms of Mexico, and he shared a great deal of information with Gordon. It is to Richard Evans Schultes that the splendid book *The Road to Eleusis* is dedicated.

Small wonder, then, that Gordon chose the Botanical Museum of Harvard University for his incomparable collection of books, art and artifacts relating to ethnomycology. In it are priceless treasures of every kind, including yet-unpublished manuscripts, thousands of books, Mayan ceramics, oriental jade mushrooms, Guatemalan mushroom stones and so forth. It is not merely the collection of a scholar, but of a scholar-poet. Each piece is the finest of its kind, each book carefully chosen so that the cohesive whole of the collection is unparalleled in any institution of the world. Through the efforts of Schultes, this unique collection remains intact for ethnomycologists of the world to utilize as a great tool—the finest monument to a great ethnomycologist, R. Gordon Wasson.

After the passing of a giant such as this we are compelled to ask ourselves: what of his legacy?

In the "vistas beyond the horizons of this life" (using one of his own expressions) we will associate the name R. Gordon Wasson with transcendent science. That is to say, the individual who comes to knowledge not merely by the accumulation and analysis of data, but by being a creator of new ways of knowing. This was always Gordon's greatest strength. He often thought his ideas might appear to be deliberately "cryptic" when they were not. Indeed, he led us to investigate the many levels of existence and of knowing. Gordon believed in seeing "without the intervention of mortal eyes." He demonstrated this in his ability to integrate the rivalry between seeing and hearing, attaining the transcendent state of the Greek temple healer and the Mexican shaman. The synthesis extended to his worldview: for Gordon there was no conflict between the immediate present and an eternity. Somewhere in the evolution of awareness man came upon the state of *ecstasis,* which for Gordon meant rising to the "highest pitch of which mankind is capable." An equivalent word and experience, he suggested, is *bemushroomed.*

R. Gordon Wasson approached life with undiluted awe, and he approached science in much the same fashion, viewing it as one of the many paths to wisdom. In the company of other scholars he was always a luminary, with his insights and his profound and diverse forms of expression. I always had the strange feeling that in addition to having attended the rites of the Mazatecas in their mushroom ceremony he had secretly attended the rites of Eleusis and had discourse with the high priestess.

In conclusion, we take comfort in observing that we have not lost Gordon the scholar, for his extraordinary works remain with us. We have lost instead Gordon the man, with whom we loved to spar; the gentleman to whom one could address a fairly brutish commentary only to receive an elevated, eloquent reply that was disarming in its perception, accuracy and added insights. In these special memories, Gordon, Eleusinian, we "will always join together in contest and battle again and again."

Notes

1. From Danny Staples' translation in *The Road to Eleusis: Unveiling the Secret of the Mysteries*, by R. G. Wasson, A. Hofmann, and C. A. P. Ruck (New York: Harcourt Brace Jovanovich, 1978, p. 66).
2. *Editor's note:* Gordon's resolve not to republish *Mushrooms, Russia and History* appears to have wavered over time. In "The Divine Mushroom of Immortality," his essay published in Peter Furst's *Flesh of the Gods* (New York: Praeger Publishers, 1972, pp. 185–200), he had earlier written: "If I live and retain my vitality, you may see published over the coming years a series of volumes, and, at the end of our road, there may be a new edition of our original work, reshaped, simplified, with new evidence added and the argument strengthened." Materials gathered for the proposed second edition are on file at the Tina and Gordon Wasson Ethnomycological Collection, Harvard Botanical Museum.

'Vistas Beyond the Horizon of This Life': Encounters with R. Gordon Wasson

PETER T. FURST

It is difficult for me now to imagine a time in my career as a professional anthropologist when Gordon Wasson was not some sort of presence in my research or writing, either as an invaluable factual or bibliographic source, a stimulating subject for a book or just a review, or someone off whom to bounce this or that germ of an idea. In fact, there was only a brief period after I turned to anthropology when, for some now inexplicable reason, I remained in total ignorance of Gordon and of his and Valentina's 1957 book *Mushrooms, Russia and History.* I say "inexplicable" because all my life I have decidedly been what Gordon called a *mycophile*—a lover of mushrooms, both wild and cultivated; since childhood they had been among my favorite dishes. Also, longer ago than I care to remember, like other German children of my generation I had recited the old folk riddle that asks the identity of the little fellow who stands alone in the forest, perched on one leg and wrapped in his bright red cape; and I had yelled out the answer, "Der Glückspilz, der Glückspilz!," never wondering why, of all the mushrooms, the fly-agaric, with its (undeserved) sinister reputation for deadly toxicity, should thus be accorded the attributes of happiness, bliss and good fortune. It was a question that also interested Gordon and to which he supplied a most convincing answer.

I like to think that eventually I would have found my way to the considerable corpus of Wasson literature even if I had not stumbled, by way of trying to understand the iconography and symbolism of pre-Colombian West Mexican funerary sculpture, into the living world of peyote and via peyote the whole, vast complex of Mesoamerican ritual "hallucinogens"—a value-laden term Gordon abhorred and tried hard, toward the end of his life, to have replaced with *entheogens,* meaning "the god within." (This latter term, coined a few years ago by Gordon, Jeremy Bigwood, Jonathan Ott and Carl A. P. Ruck, is certainly more in accord with how indigenous peoples see the plants possessed of such remarkable effects on the senses than *hallucinogens* or any of the other, often awkward, mostly unsatisfying and imprecise terms we have been using over the years.) Once I started to do research on peyote among the Huichols, the road inevitably led to Gordon—and from there to 20 years of continuing amazement at the fertility of his mind, the meticulousness of his interdisciplinary scholarship and the convincing logic of many of his arguments.

However, he did not convince me in *every* case. I never, for example, found all of the mushrooms he felt sure are present in the Tepantitla mural. And this is as good a place as any to start my reminiscences of Gordon.

It was in the mid-1960s, while taking photographs in the Teotihuacán exhibits of the National Museum of Anthropology in Mexico City, that my eye was caught by something peculiar about the famous 3rd Century mural painting which the late, great Mexican scholar Alfonso Caso had interpreted as a depiction of Tlalócan. The dominant configuration in Agustín Villagra Caleti's reconstructed museum replica of the original is a deity over whose image towers a giant, flowering plant that could be read, and has been, as a "tree of life" or *axis mundi*. The deity's face has some characteristics resembling, but not identical to, those associated with the rain and earth deity Tlaloc in Classic and Post-Classic Mexican art.

As described by the 16th Century Franciscan chronicler Fray Bernardino de Sahagún, in Aztec eschatology Tlalócan was the paradise where those who had died by drowning or lightning enjoyed a pleasant afterlife under the benign hand of Tlaloc. At first sight, the mural, which once adorned the walls of a sacred compound or palace, does seem to visualize just such a conception. The deity, who wears an enormous quetzal bird headdress, presides over the upper half of the mural, water flowing from the mouth and water and seeds from the hands. She[1] is attended by priests in similar ceremonial garb dispersing seeds. The lower register, presumably the watery underworld, is animated with happy little people—stylistically reminiscent of the small, unpainted ceramic "dancing" figurines typical of the early Classic at Teotihuacán—playing, singing and cavorting amid flowering plants, and diving and swimming in the waters of a lake.

Some scholars have reservations about the validity of using Spanish colonial data dealing with 16th Century Aztec beliefs to interpret the sacred art of other Mesoamerican peoples, especially those predating the Aztecs by centuries—in this case, by more than 1,000 years. On the other hand, why should the idea of a Tlalócan not have been more than 1,000 years old when the Aztecs founded their capital of Tenochtítlan on an island in Lake Texcoco in 1325? After all, five centuries after what was supposed to have been the spiritual as well as military and economic conquest of Mexico by Spain, Nahuatl-speaking inhabitants of the Sierra de Puebla still call their underworld *Talocan*!

But even if one is willing to assume a great deal of continuity in Mesoamerican religion and religious iconography, if not style, over time and across a wide area, as some of us are inclined to do, and accepts Caso's interpretation of the mural as being at least on the right track, there still seemed to me from the beginning something odd about the so-called tree. *Webster's Third International Dictionary* defines a tree as "a woody perennial plant having a single main stem that may be short but is usually considerably elongated, has generally few or no branches on its lower part, and is crowned with a head of foliage, or [as in palms] of foliage only." The Tepantitla "tree" has no main truck surmounted by a crown, but rather a pair of undulating leafy branches curving inward from behind the deity's headdress and sprouting numerous other leafy stems or branches with trumpet-

The Tepantitla mural. *Courtesy Wasson Collection.*

shaped flowers at their ends and streams of water adorned with disembodied eyes flowing from them. It looked to me more like a conventionalized representation of a flowering vine than a tree, inspired as much by myth as by nature, but still recognizable as a vine that is in some major way associated with water.

But a vine of what species? What vine sprouts flowers shaped like those in the mural, and what species is closely associated with water? And would a vine necessarily be inconsistent with the concept of a tree of life or a "world tree"—the tree as *axis mundi*?

Actually, questions of this sort, and the argument I eventually developed in favor of the morning glory *Turbina* (formerly *Ipomoea*) *corymbosa,* did not occur to me until later, when, after returning to Los Angeles from an extended stay in West Mexico, I went back to the color slides and black and white photographs I had taken of the mural over several of my periodic pilgrimages to the great museum. The writings of R. Gordon Wasson had something to do with that, together with the counsel of Richard Evans Schultes, the distinguished longtime director of Harvard's Botanical Museum and ranking authority on New World hallucinogens, and earlier still my association with the veteran, Austrian-born, Mexican ethnologist Roberto J. Weitlaner.

In 1966, shortly after I had begun my Huichol research, I contacted Weitlaner at his office at the Instituto Nacional de Antropología é Historia (INAH) in Mexico City and arranged a visit for the following day to talk about Huichols and peyote as a sacred focus of their religion, and to benefit, as a newcomer to Mesoamerican ethnology, from his long experience and counsel. I knew that "Papa" Weitlaner, as

69

his friends and colleagues fondly referred to him, had to be over 80 years old. And in fact, when I arrived at INAH, I found that the Instituto had just published a 669-page *homenáje* to honor him titled *Summa Antropológica en homenáje á Roberto J. Weitlaner,* initiated by a group of his friends, colleagues and former students three years earlier on the occasion of his 80th birthday. I had bought a paperback copy for him to autograph; it was a weighty tome indeed, in both the literal and scholarly sense, comprising some 60 essays in Mesoamerican social, cultural and physical anthropology, archaeology, religion, ethnohistory, ethnobotany and psychopharmacology.

"Aha," he said when he saw the book. "I have prepared a better one for you." And he handed me a copy of the much scarcer hardcover edition, which he had already inscribed.

I told him how honored I was—Weitlaner had 30 years of distinguished ethnographic and linguistic research among Mesoamerican Indians on me. For a man of 83 he was amazingly vigorous physically, and intellectually one of the most stimulating scholars I had ever met. I asked about his future plans, and his reply was admirable: he was planning another field trip for the following year, intending to travel by mule or burro into the remote mountain villages of southern Oaxaca to look one more time at indigenous calendrics and divinatory mushroom use among the Mixe Indians. He said he was hoping his friend Gordon Wasson would come down from New York again to go with him. Mindful that by then Weitlaner would be 84, I said I could only hope that when I reached his age I would still have even half his energy. (Sadly, Weitlaner died before he could make the trip.)

"I don't know much about the Mexican sacred mushrooms beyond what I've read in Sahagún," I told Weitlaner, which was true, since prior to 1966 I had no more than a peripheral interest in the cultural history, art, ethnography and ethnobotany of the psychoactive flora—fields that came to predominate much of my subsequent research and writing.

"Then I have to introduce you to Gordon Wasson's writings," Weitlaner said. "This is a man you must meet. He will be very interested in your reports about peyote among the Huichols."

He said the *homenáje* would be a good start in the Wasson corpus, because it included a significant article by him about some of the most important Mexican divinatory plants. Wasson had identified four plants used in prehispanic ecstatic divinatory ritual: *Piciétl* (*Nicotiana rustica* L.), tobacco; *peyotl* (*Lophophora williamsii* Lem. Coulter); the sacred mushroom the Aztecs called *teonanácatl,* meaning "divine (or god's) flesh" (*Psilocybe* spp.); and *ololiuqui,* the white-flowered morning glory *Rivea corymbosa.* To these Wasson added another species of morning glory, the blue-flowered *Ipomoea violacea,* whose seeds were, and still are, used in Oaxaca as *ololiuqui* was, and still is, employed elsewhere; and a member of the mint family, *Salvia divinorum,* which he felt sure was the same as the botanically as yet unidentified divinatory plant the Aztecs called *Pipiltzintzíntli.*

Weitlaner and Wasson, it turned out, had enjoyed a close and fruitful collaborative relationship from the time of the Wassons' first visit to Mexico in 1953 looking for survivals of divinatory mushroom intoxication described in the early Spanish colonial literature. Not coincidentally, Weitlaner himself had witnessed such usages among the Mazatecs of Huautla de Jiménez, Oaxaca, as early as 1936 and again in 1938, this time in the company of his daughter Irmgard Weitlaner Johnson, who became a distinguished scholar in the field of Mexican Indian textiles, and her husband, Jean Bassett Johnson. The latter (who was killed in North Africa in 1944) described the experience in a paper he read in 1938 at a meeting of the Sociedad Mexicana de Antropología, and in a longer paper published in 1939 by the Gothenburg Ethnographical Museum. The Wassons came across Johnson's account when they were doing the ethnomycological research that brought them to Mexico in 1953 and that in 1957 resulted in their collaborative *Mushrooms, Russia and History*.

The Weitlaners and Johnson had been allowed to witness the mushroom ceremony but did not partake of the sacred fungi. Gordon was determined to establish such close rapport with the Mazatec *curandera* and mushroom priestess María Sabina that she would eventually allow him to try the sacred fungi and experience its wondrous powers for himself. But it was Weitlaner who ultimately made this possible, as Gordon has readily and movingly acknowledged in print—appropriately, in the pages of the Weitlaner *homenáje*. It is worth quoting *in extenso* because it not only illustrates Gordon's gracious style and generosity of spirit, but also because the rich collection in which it appears has not been easily available:

> In the beginning we discovered Roberto J. Weitlaner. Without minimizing what we owe to others, I rejoice when this occasion presents itself that I may properly define my debt to him. He led us by the hand on our first excursion on muleback into the Indian country, to Huautla de Jiménez, on my second trip to Mazatlán de los Mixes, on my visits to San Agustín Loxicha in the Sierra Costera, and to the Mazahua country. For ten years I have had repeated recourse to him, to tap his immense knowledge of the Indians, their ways, their languages, their history. He has guided my steps in the libraries, unearthed apt quotations in the sources bearing on our theme, introduced me to others working in the field who could also pin down facts. His patience, good humor, and *joie de vivre,* in the sierra and in Mexico City, are unfailing. But above all else I have tried to learn from him his secret of dealing with the Indians. The Indians are simply living by the conventions of an orally transmitted culture such as our own forebears lived by only a little while ago. When you visit their villages you make allowances for this time lag. You do not treat them kindly as inferiors or children. You do not treat them *as though* they were equals. (The Indians are quick to see through you.) Ing. Weitlaner taught us to treat the Indians as equals. A secret simple yet elusive. As the poet said, truly, "this is the famous stone that turneth all to gold."

Wasson learned the secret well. He has characterized his personal experience of the sacred mushrooms as a "soul-shattering happening;" the mushroom affords one, he wrote, "more clearly than our mortal eye can see, vistas beyond the horizons of this life."

An article, "Seeking the Magic Mushroom," accompanied by dramatic photos, appeared in the May 13, 1957 issue of *Life,* and in Spanish in *Life en Español* on June 3 of that year. Valentina published her personal account of another mushroom experience in *This Week,* May 19, 1957. Of course, they reserved the full account of their experience and its background for the pages of their book. Over the following years Wasson was to speak publicly and write of it many times in some of his most memorable and moving prose. Many years later he also published a verbatim transcription, in the original and in translation, of an entire mushroom *velada* conducted by María Sabina, together with actual recordings of the ceremony. It was an elegant addition, published at great expense, to the literature on contemporary Mesoamerican religion, one not likely to be soon, or ever, duplicated in its comprehensiveness. I reviewed it in *Natural History* in 1975, the first time that journal accepted a review of a book whose high price (initially $275) made it unavailable to the general public. The editor made an exception because he agreed with me that the book had an historic importance offsetting its price, and because, in his opinion, it was time to call attention to other Wasson works never reviewed in *Natural History*.

In a very real sense, Gordon's 1955 experience with the sacred mushrooms changed his life. For, having spent most of his professional career on Wall Street, he went on to devote the next 30 years not just to library research on mushrooms, his favorite subject, but also to numerous field trips and painstaking ethnobotanical and ethnohistorical studies of the several species of sacred mushrooms and other divinatory plants still in use among some Mexican Indians. And of course there eventually followed the exhaustive multi-disciplinary investigation, beginning in 1963, that led him to propose a botanical identification for the mysterious Indo-European plant deity Soma, celebrated in the Vedic hymns first set down in the Sanskrit language around 1500 B.C. According to Gordon, this Soma was none other than *Amanita muscaria,* the spectacular red-capped, white-stemmed fly-agaric mushroom (the Glückspilz of German folklore) that was still used as an intoxicant in ecstatic Siberian shamanism in the early 20th Century, and whose place in European folklore and folklife more than hinted at a far more important role it may have played in ancient pre-Christian times. He discussed his hypothesis in several publications, raising the hackles of some conservative Vedic scholars but eliciting the strong support of others. Gordon's major published work on the subject, his exhaustive *Soma: Divine Mushroom of Immortality,* appeared in 1968. By this time Wasson, the ex-banker, had firmly established ethnomycology as a recognized new branch of botany and its subfield, ethnobotany (the study of the use and meaning of plants among different peoples), and himself as its ranking interdisciplinary scholar and historian.

But let us return now to Roberto Weitlaner, who, after all, started Gordon on his Mexican mushroom quest, and on whom Gordon, ever eager for scientific data, called for help in obtaining quantities of morning glory seeds sufficient for testing in order to solve the long-standing enigma of their chemistry, and thence an explanation for their potent psychological effects. The key figure here was Albert Hofmann, whose subsequent findings, to which Gordon made a crucial practical contribution, were one of the great surprises of psychopharmacological research.

I have written of this before in some historical detail (cf. Furst 1976), but it is worth brief repetition in the present context. Hofmann was a chemist at Sandoz Ltd., a major manufacturer of pharmaceuticals headquartered in Basel, Switzerland, who, with his associate Dr. A. W. Kroll, in 1938 for the first time synthesized *d*-lysergic acid diethylamide, a derivative of ergot. Ergot is a toxic fungal infestation of grasses, especially rye, that in the Middle Ages caused the

European outbreaks of "madness" and death known as St. Anthony's Fire. The newly synthesized compound was to become famous—or notorious, depending on one's point of view—as LSD-25, so named by its discoverers because it was the 25th in a series of lysergic acid compounds they were testing for possible medical application. Since animals on whom it was tried in 1938 showed no effects, it was put aside without being tested on humans. Five years later, on April 16, 1943, while working with other ergot derivatives, Hofmann suddenly experienced restlessness and dizziness, soon followed by what he described, in the Weitlaner *homenáje*, as "fantastic visions of extraordinary realness, and with an intense kaleidoscopic play of colors."

After determining that these extraordinary effects were due to accidental ingestion of a small amount of LSD-25, he deliberately took another dose three days later. The effects were even more astounding. Thus began the saga of LSD, the most powerful psychoactive compound known to that time, ushering in a whole new era of research into the nature of the unconscious and representing, in the view of some scholars, one of the two or three major scientific milestones of the 20th Century. Beginning with Wasson's work on mushrooms in the 1950s and accelerating with the field studies of the 1960s, it also inaugurated, on the one hand, increasing scholarly attention to the history and functions of so-called "hallucinogens" in the religious life of indigenous peoples and, on the other, and not coincidentally, near-messianic claims for LSD as a sort of universal panacea for what ailed society. To the dismay of such serious students of the phenomenon as Gordon, among others, it also triggered widespread experimentation with a whole range of "psychedelic" drugs by people who were often psychologically unprepared to deal with ecstatic experience.

The powers that be reacted in predictable ways. Throwing out the baby with the bathwater, apparently in ignorance or with deliberate disregard of the scientific evidence and the observations and experience of scholars, they passed stringent punitive laws. Most of them are still on the books, where these non-addictive psychoactive substances are uncritically lumped with dangerously addictive and physiologically damaging narcotics, thereby impeding scientific research of potential benefit to society while little deterring uncontrolled experimentation but raising its social costs. In the meantime, the twin ironies of alcohol's legality and the generous federal subsidies for another, highly addictive drug, nicotine—whose victims outnumber those killed by heroin and all the other illegal drugs combined by hundreds of thousands every year—was totally lost on the larger society, which viewed with undisguised horror the apparently new, but actually ancient, fascination with altered states of consciousness among its alienated young. What made the situation even more ironic is that at the time, and for many years after, when LSD was made illegal and its unauthorized manufacture, sale, use and even free distribution subject to long prison terms, the Pentagon and the CIA were subjecting hundreds of people (the Army's tests alone involved 1,500 individuals), many without prior knowledge or consent, to secret experiments with LSD, results of which—some highly damaging—were unavailable to the general scientific community. These experiments continued over a dozen years, until the late 1960s; but of this we had no inkling until the scandal was finally exposed in 1975.

Ethnobotanical and pharmacological research into the ritual "hallu-cinogens" nevertheless continued undeterred. Wasson had been working closely with the distinguished French mycologist Roger Heim, director of the Muséum National d'Histoire Naturelle in Paris, with whom he was, characteristically, in contact even before his Mazatec experience, and who came to Mexico to join him on further expeditions into the mountains of Oaxaca in search of new mushroom species being used by Indian peoples. Following these forays with Wasson, Heim, who in 1956 had identified one species of psychoactive mushrooms as *Psilocybe caerulescens*, was able to add another dozen or so species to the growing list of named mushrooms of the divinatory kind, mostly in the genus *Psilocybe*, others belonging to *Conocybe* and *Stropharia*. Thanks to his and Wasson's work and that of other Mexican and U.S. mycologists, by the end of the 1950s the taxonomy of the Mexican divinatory mushrooms was pretty much complete.

So the '50s were a crucial time for Gordon and the whole new field of ethno-mycology, though in justice some early pioneers in the story of the rediscovery of the sacred mushrooms need to be acknowledged here—as Gordon often did—as well. First there was Blas Pablo Reko, an astute, Austrian-born physician-turned-ethnobotanist, who for years in the 1930s and earlier openly asserted that the legendary hallucinogenic mushrooms really existed and were still in use. In this view he went up against the stubborn skepticism of a distinguished North Ameri-can botanist, William A. Safford, who flatly asserted the opposite: that there never was any such thing; that the early Spanish writers on the subject and their Aztec informants must have been mistaken; and that *teonanácatl* must have been peyote rather than a mushroom. Reko, incidentally, ran into the same problem with morning glory seeds, with the same Safford insisting that there was no such thing as psychoactive morning glories and that Aztecs who talked of them to the early Spanish chroniclers must have meant the seeds of *Datura*!

When Weitlaner first encountered the Oaxacan Indians who were using divinatory mushrooms in 1936, he sent a specimen to Reko, who forwarded it to the Harvard Botanical Museum. Unfortunately, it arrived so badly deteriorated that it could not be identified. But two years later the young Schultes, whose senior honors thesis in botany at Harvard in 1937 had been on peyote and its cultural uses, went to Oaxaca, where he and Reko scoured the countryside for evidence of intoxicating mushrooms as well as morning glories in Indian ritual. They found them, notwithstanding Safford's ethnocentric denials. Schultes carefully noted the morphology of the mushrooms and published their first scientific description in 1939, a feat he followed in 1941 with his seminal monograph, published by the museum of which he later became director, *A Contribution to Our Knowledge of* Rivea corymbosa, *the Narcotic Ololiuqui of the Aztecs.*

What remained unclear at the end of the 1950s was the chemistry of both the mushrooms and morning glory seeds. Hofmann was the chemist whose tests eventually solved these riddles, but again Gordon played a significant, helpful role. First, Gordon's friend Heim brought back to Paris from their Mexican forays sufficient quantities of *Psilocybe mexicana* spores to allow him to successfully grow a laboratory culture of sacred mushrooms. Attempts to isolate their active prin-ciples in Paris having proved unsuccessful, Heim decided to submit several specimens to Hofmann in Basel. Hofmann, who like Heim was in close and continuous contact with Gordon, and who also travelled to Mexico at the latter's invitation, was almost immediately successful in solving the puzzle of their chemistry, discovering the agents responsible for their extraordinary effects and soon thereafter reproducing them synthetically. The principle active agent, related to a chemical agent called serotonin naturally present in the brain, was named *psilocybin* by Hofmann. Also present as an unstable derivative was a compound he called *psilocin*. The chemistry is more complex than that, of course, but this description must suffice for the present discussion. Wasson was one of the first to learn of Hofmann's astonishing discovery that the active agents responsible for the extraordinary effects Gordon had experienced amount to only about 0.03% of the total weight of the plants. To achieve the effect of as many as 30 mushrooms (only a few comprise each person's dose in a ritual) would require only 0.01 gram of the crystallized powder dissolved in water. Pretty powerful stuff!

As for morning glory seeds, a number of experiments during the 1950s by reputable scientists, including ingestion by volunteers of as many as 125 seeds at a time, failed to produce any of the extraordinary out-of-body sensations that had been attributed to them. Was Safford right, after all? Of course he was not. Wasson himself had twice used the seeds of the blue morning glory, *Ipomoea violacea*, in his New York apartment. Their effects had been "extraordinary." But, he pointed out, he had prepared them exactly as did the Oaxacan Indians, who grind them up on a special *metate*, soak the flour in water, pass the water through a cloth strainer and, soon thereafter, drink the resulting liquor. If taken whole, or even if their hard shells—hard enough to allow the seeds to pass intact through the digestive system—are cracked, they have no effect. The failure to prepare the seeds in the traditional way, evidently, was the reason the earlier experiments had failed.

What remained, then, was to solve the riddle of the seeds' chemistry. In 1959 Wasson sent Hofmann two small bottles containing the seeds of two species of morning glory. One batch, he wrote in the accompanying letter, he had collected in Huautla de Jiménez; he thought these light-brown, roundish seeds were from *Rivea corymbosa*, the sacred *ololiuqui* ("little round thing") of the Aztecs, known in Oaxaca as *ololuc*. The other bottle contained seeds he had collected in San Bartolo Yautepec, a Zapotec town. These, black and angular, he thought had come from the blue morning glory *Ipomoea violacea*. The Zapotecs called them *badoh negro;* the Aztecs, *tlitliltzin*. Gordon's identifications proved correct.

Small though they were, these initial samples were to prove very exciting to Hofmann. His chemical analysis indicated, quite unexpectedly, the presence of indole compounds structurally related to, of all things, LSD and the ergot alkaloids. But before he could be sure, he needed seeds in much greater quantities Could Wasson send him more of these interesting seeds, and lots of them? Wasson turned to Weitlaner for help, and with his assistance and that of Weitlaner's daughter Irmgard, Wasson was able to ship 12 kilograms of *Rivea corymbosa* seeds and 14 kilograms of *Ipomoea violacea* seeds to Basel.

With these large quantities, which reached Hofmann in the early part of 1960, the chemist was able to isolate their main active principles and identify them securely as ergot alkaloids closely related to *d*-lysergic acid diethylamide—LSD! So, through his relationship with Wasson, a 20-year-long research series had closed for Hofmann in what he referred to as a "magical circle." It had begun with his discovery of LSD and concluded with this new and, from the perspective of psychopharmacology, quite unexpected—even sensational—discovery in a family of higher plants, the Convolvulaceae, of lysergic acid alkaloids that had previously been known only in the lower fungi of the genus *Claviceps*.

For me, plowing through all this new and exciting literature by Gordon and his colleagues was a journey of discovery. For one thing, it placed peyote use and sacred lore among the Huichols in a larger cultural and historical framework. For another, it kindled my interest in the general prehistory of hallucinogens; in their relationship to shamanism and the ecstatic trance as its hallmark; and in their iconography, especially their representation in pre-Colombian art, and what one might reasonably be able to reconstruct from such imagery about the belief systems and ritual practices of long-ago cultures.

Which brings me back to the Tepantitla mural and its "tree." To make a long story short, I called Schultes at Harvard, introduced myself, and asked if I might send him my photographs and my then still half-baked thoughts of the "tree" as a morning glory. Or might it be a Datura? Schultes was not then familiar with the mural, but he was enormously kind and patient, suggesting that instead of using the mails I bring the pictures in person—not to Harvard but to San Diego, where he would be participating in a conference.

On the appointed day I drove the 150 miles from Los Angeles to San Diego, and he graciously made time to look over the black and white and color slides. His preliminary verdict: the Tepantitla mural, as I had suspected, almost certainly depicts a morning glory. Since the mural came from central Mexico, the vine in question would have to be *Rivea corymbosa*. Had I read his 1941 monograph? No. Had I read his 1941 monograph? No. Had I read Ruíz de Alarcón? I had not. Ruíz de Alarcón was a Jesuit priest who in 1629 wrote an important manuscript in which he described and excoriated beliefs and practices of prehispanic origin among his flock in an Aztec-speaking village in Guerrero. Schultes recommended it highly because it contained much valuable material on the "infernal *ololiuqui*." If we were right, it meant that some of the things about which Ruíz de Alarcón complained, and which Sahagún and other Spanish priests had described in the previous century, could now be traced back at least to Teotihuacán.

"You have a lot of catching up to do on your reading to back this up, expand it and get it into print," Schultes said. "We have a big library at the Botanical Museum and you can find anything you want there. You also ought to send your photos and your ideas to Gordon Wasson. He's written a lot about *ololiuqui* and I think he's going to find this very exciting."

He did, indeed.

On my next trip to Mexico I found an 1892 edition of Ruíz de Alarcón's 17th Century manuscript, edited by Francisco del Paso y Troncoso and published in the *Anales* of the Mexican National Museum. The original manuscript was in the rare book collection of the Escuela Nacional de Antropología, housed in the Museum of Anthropology, but the librarian generously allowed me to make a xerox copy. It was this copy that years later formed the basis for the first published translation into English of the Ruíz de Alarcón manuscript by Michael D. Coe and Gordon Whittaker (1982). (Not surprisingly, the very first order, prepaid, came from Gordon Wasson.) The several chapters dealing with the deification and divinatory use of *ololiuqui* proved a gold mine, even though their author refused to identify the plants from which they came, for fear, as he put it, that "base people" in the non-Indian population might be seduced into trying them.

Eventually I accumulated enough information on the ethnohistory, mythology, iconography, botany and ecology of the Mexican morning glories, and enough support from the likes of Schultes and Wasson, to go into print with a reinterpretation of the Tepantitla "tree" as the deified *Rivea corymbosa,* whose association with a water goddess in the mural was entirely logical: a translation into myth and symbol of direct observation of the natural world and the ecology of morning glories (cf. Furst 1972a; 1976).

Gordon's fascination with the Tepantitla mural, however, turned not on the identification of the morning glory but rather the mushrooms he thought he could identify in several portions of the mural. It was they, he felt, that really held the key, so he focused on them, not *ololiuqui,* in *The Wondrous Mushroom* (Wasson 1980, pp. 161–167); the morning glory is mentioned in passing, displaced by Gordon's contention that the whole scene "is of the entheogenic vision of the Mesoamericans" (p. 165).

Well, perhaps. Nor do I deny that some of the images *might* be mushrooms, or, for that matter, that the Teotihuacános knew as much about sacred mushrooms as did their descendants at the time of the Conquest and as do some indigenous peoples to this day. There is good archeological evidence (e.g., the famous "mushroom stones") that some Mesoamerican Indians considered certain mushroom species sacred as long ago as 500 B.C. So why not also the Teotihuacános? Still, sometimes even a Gordon Wasson might see what he is sure must be there, not what is. But one does not need to follow Gordon wherever he might lead to appreciate the fertility of his mind.

The early discussions of such matters with Dick Schultes also led to a lecture series at UCLA on the ritual use of hallucinogenic plants. I asked Dick if he would participate, perhaps by giving the opening lecture to set the stage and by assisting with suggestions for other speakers. He agreed, provided we could schedule his lecture so as not to interfere with his periodic forays to the Amazon. When I asked if he thought we could get Gordon to participate, he answered, "You couldn't keep him away."

The series became a reality in the spring of 1970, under the sponsorship of the Social Science Department of UCLA Extension. Schultes led off, and, as the ranking authority on the subject, presented an overview of hallucinogens in the New World. He had at that time counted no fewer than 80 in past and/or present use, a list that now numbers around 100.

Gordon related his experiences with the Mexican mushrooms. It included some charming recollections about how he and Valentina first discovered, on their long-ago honeymoon, their cultural differences on matters fungal: he, the Anglo-Saxon, and she, the Russian. ("Repugnant fungal growths, manifestations of parasitism and decay" is how he regarded mushrooms then, Gordon told his utterly charmed audience, while Valentina's love for them was "a visceral urge, a love for mushrooms that passeth understanding." He then went on to relate how the discovery of that difference set them on their explorations in ethnomycology and inspired their classification of Indo-Europeans into two groups: "mycophiles"

77

and "mycophobes.") The Los Angeles botanist William Emboden discussed the cultural history of *Cannabis sativa* (marijuana). James W. Fernandez, then at Dartmouth, took us beyond the New World to sub-Saharan Africa and into the social and symbolic world of a psychoactive shrub called *Tabernanthe iboga*. Marlene Dobkin de Rios spoke on psychedelic healing with *ayahuasca (Banisteriopsis caapi)* among Mestizo slum dwellers in northern Peru. Weston La Barre, the distinguished ethnologist of religion at Duke University, offered a lecture on "Hallucinogens and the Shamanic Origins of Religion." My own talk was on the use of peyote among the Huichols. Last, but due to his growing fame not least, was Carlos Castaneda, the amanuensis and—as Richard de Mille would eventually establish beyond reasonable doubt in his two books, *Castaneda's Journey* (1976) and *The Don Juan Papers* (1980)—the inventor of the "Yaqui sorcerer Don Juan." Carlos's first book had been published the year before by the University of California Press.

With hallucinogens such a hot topic at the time, and considering the obvious need to place "mind-altering" substances into a larger socio-cultural and historical framework than the limited one of what had come to be called the "counterculture," the symposium obviously had the makings of a book. So I called the speakers and got agreement all around. Gordon went so far as to suggest that he provide not one but two chapters, the second on Soma/fly-agaric, with one condition: he would retain the copyright on both. The roster of topics and speakers was sufficiently impressive to convince Gladys Topkis, then an editor at Praeger Publishers, to sign a contract which substantially added to what UCLA Extension could afford to pay the lecturers beyond their travel expenses. I remember Gordon—who, as a former banker, was ever mindful of the value of money—being especially delighted at the prospect of receiving two fees instead of one for his two chapters. Praeger and I reserved the right to refuse any contribution that could not be salvaged by extensive rewriting and editing. Not surprisingly, no changes were required—not a single word—in either of Gordon's chapters. I remember Gladys calling me from New York after she had received them to say how much she liked them: "My God, what a charming man he must be if he's anything like his writing!" I assured her he was.

The book appeared in 1972 as *Flesh of the Gods: The Ritual Use of Hallucinogens*, and included 10 essays: seven by most of the original lecturers, led off by Schultes, and three by new recruits—the distinguished Colombian anthropologist Gerardo Reichel Dolmatoff; Johannes Wilbert, who, as director of the UCLA Latin American Center, had been my boss as well as my valued mentor and friend; and Douglas Sharon, who would later become the director of the San Diego Museum of Man. Reichel Dolmatoff wrote on the psychotropic jungle vine *Banisteriopsis caapi*—*ayahuasca* in Quecha and *yajé* among the Tukanoan Desana, whose society and ritual arts he had been studying. Wilbert's paper, on the use of tobacco as a shamanic intoxicant among the Warao of the Orinoco Delta, was the first in a series of publications on that topic that culminated this past year (1987) in his encyclopedic *Tobacco and Shamanism in South America*. Sharon wrote here for the first time about the famous Peruvian healer Eduardo Calderón and the use of the *San Pedro* cactus, *Trichocereus pachanoi*, in Peruvian folk medicine. These three replaced one of the original contributions that was dropped and another, by

Carlos Castaneda, that failed to materialize, thereby saving the editor a certain amount of embarrassment after the majority of professional anthropologists had come to the conclusion there was more imagination than ethnography in his writings. Before he was to have turned in his manuscript he signed a contract with Simon & Schuster for what was to become a whole series of bestselling Don Juan books. According to his new literary agent, the agreement precluded his client from publishing anything with anyone else. So that was that.

With respect to the Castaneda saga, I must admit that when I first read his draft for what was to become *The Teachings of Don Juan,* I, like many others, still thought that he was basically on the up and up, and that literary license and graduate student overenthusiasm and naivete could explain away some of the things about his tales I found worrisome: the bit about smoking mushroom dust, for example, with a resulting ecstatic experience, and inconsistencies in the chronology of encounters, dates and places, as well as the whole problem of the actual cultural context of "Don Juan" and his practice of "sorcery," which just didn't sound like any Mesoamerican Indian I'd ever met or heard or read about. The subject of the smoking of mushrooms that had crumbled into powder came up when Gordon was in Los Angeles for the UCLA lecture. He asked what I thought of Carlos and I told him that perhaps Carlos was inclined to embellish his stories to make them more entertaining, but that when I first listened to him discuss his field work with fellow students at UCLA I had no reason to think that the stories weren't essentially true.

"When I first read the book I thought I smelled a hoax," Gordon said. "Have you ever heard of anybody in Mexico smoking mushrooms in powder form?" I had not. Gordon told me he did not believe it was possible, pointing out that Castaneda has Don Juan smoking mushrooms he had bottled up in a gourd and that had turned to dust over a year's time, and after smoking the powder experiencing effects that are typical of *Psilocybe mexicana.* This was hard for him to believe, explained Gordon, because dried and pulverized mushrooms don't even ignite. And if that part wasn't true, how much else in the book was invention? He said he had written to Carlos in 1968 asking many questions about things that bothered him, and that he had not been well satisfied with the answers or won over by Carlos amending his story of having smoked mushroom dust to say he had "ingested" it. Of course, Gordon said, one must be fair, and perhaps Carlos was just not a very good ethnobotanist or had naively believed stories fed to him in jest by Don Juan. That there might never have been a Don Juan in the first place had not yet occurred to Gordon; but neither had it occurred to the rest of us who, from experience with Carlos as raconteur, should have known better.

De Mille (1980, pp. 319–333) analyzed the 1968 correspondence between Gordon and Carlos in detail in *The Don Juan Papers,* pointing out that at the time Wasson did not yet appreciate to what a degree his own writings, beginning with *Mushrooms, Russia and History,* had actually served to inspire Carlos and contributed to the creation of the Don Juan stories. Gordon's first reaction to *The Teachings of Don Juan*—that he "smelled a hoax"—was, of course, right. And though he wavered somewhat over time on the subject, he did come to realize that he should have paid more attention to his sense of smell.

Well, so far this has all been rather serious stuff. But I want to end these reminiscences on a lighter note: an incident involving Gordon, mushrooms—non-hallucinogenic and non-sacred—and myself; a story that is much funnier in retrospect than it was at the time. If it should come to his attention in whatever spirit world he may dwell, I hope it will not embarrass him for the small part he played in it but rather strike him, too, as being funny. It will be the first time he hears it, because I never told him about it during his lifetime.

During Gordon's Los Angeles stay in the spring of 1970, one of the people enrolled for the lectures, an enthusiastic lady who was a well-known former actress of Danish birth, brought me a gift, a boxful of sizable, white mushrooms somewhat resembling the common supermarket kind. She said she had collected them on a mushroom hunt the previous weekend, and that they were utterly delicious sauteed in butter, or scrambled with eggs or cooked in an omelette. I showed them to Gordon, who exclaimed, "Oh, you will enjoy these. They can't compare with chanterelles or the morel, but they are very fine, some of the best wild mushrooms I have eaten."

The following day, alone at home while my wife Dee (who died in an automobile accident in 1974) was touring Gordon around Los Angeles, I prepared an omelette with about half the mushrooms. Twenty minutes later I began to feel increasingly queasy. This was soon followed by violent stomach cramps, and I barely reached the bathroom before starting to vomit. I realized immediately that it had to be the mushrooms and that I had better get to the hospital fast. I didn't trust myself to drive and was just dialing for the ambulance when my wife returned. I told her what I thought was wrong. She promptly bundled me into her car, and we tore down the Hollywood Hills and east along Sunset Boulevard toward the hospital, passing red lights with her leaning on the horn and me retching and moaning and, frankly, very scared.

When we reached the hospital I was immediately given a tetanus shot—for what purpose, I have no idea, but I had the same experience once before in a Guadalajara hospital after having been stung in the hand by a scorpion of the highly toxic Durango variety. I was then bundled off to intensive care, the box of remaining mushrooms—which my wife had had the presence of mind to snatch from the kitchen counter for identification—clutched in my hand. Miracle of miracles, a resident on duty was an amateur mycologist who took one look inside the box and immediately identified the offending fungus as "good eating" for practically everybody—except about 10% of the population, who happen to be allergic to them! I had not been in any real danger, he assured us; these mushrooms are not deadly, but capable of causing severe gastrointestinal distress.

"Did you ask anybody if it was O.K. to eat these things?," the resident inquired when he came back a few hours later.

"Of course," I answered. "I asked THE authority—Gordon Wasson."

"Wasson!," he exclaimed in obvious awe. "You know him? I've read his work. You couldn't have asked a better man. But this time he made a mistake. You better tell him about that allergic 10%."

As I said, I never told Gordon about the incident, save to share with him the information that I'd heard that a few people, maybe 10% of the population, were allergic to those mushrooms he had so highly recommended.

"Oh," he replied, "I hadn't heard that. I must look into it. There must surely be something in the literature about that." Gordon never accepted any hearsay without turning to books or asking the experts himself. "But if that's true, perhaps you had better try just a very small amount to begin with to see if they agree with you."

Too late, dear Gordon, I thought; too late.

Notes

1. With some of my colleagues, I have long been convinced that this is a goddess, not a male god. Indeed, I think it likely that she represents a manifestation of the Great Goddess of Teotihuacán, as the giver of life, water and agricultural fertility.

References

Castaneda, C. 1969. *The Teachings of Don Juan: A Yaqui Way of Knowledge.* Berkeley and Los Angeles: University of California Press.

Coe, M. D., and G. Whittaker. 1982. *Aztec Sorcerers in Seventeenth Century Mexico: The Treatise on Superstitions by Hernando Ruíz de Alarcón,* Publication No. 7. Albany, New York: Institute for Mesoamerican Studies, State University of New York.

De Mille, R. 1976. *Castaneda's Journey.* Santa Barbara, California: Capra Press.

De Mille, R. 1980. *The Don Juan Papers.* Santa Barbara, California: Ross-Erikson.

Furst, P. T. 1972a. Morning glory and mother goddess at Tepantitla, Teotihuacán: iconography and analogy in pre-Colombian Art. In *Mesoamerican Archeology: New Approaches,* edited by N. Hammond. Austin, Texas: The University of Texas Press.

Furst, P. T. 1972b. *Flesh of the Gods: The Ritual Use of Hallucinogens.* New York: Praeger.

Furst, P. T. 1975. Drugs, chants and magic mushrooms. *Natural History* 1975; 84(10):74–79.

Furst, P. T. 1976. *Hallucinogens and Culture.* San Francisco: Chandler & Sharp.

Hofmann, A. 1976. The active principles of the seeds of *Rivea Corymbosa* L. Hall F. (Ololiuhqui Badoh), and *Ipomoea Tricolor* Cav. (Badoh Negro). In *Summa Antropológica en homenáje á Roberto J. Weitlaner.* Mexico, D. F.: Instituto Nacional de Antropología e Historia.

Schultes, R. E. 1941. *A Contribution to our Knowledge of* Rivea corymbosa, *The Narcotic Ololiuqui of the Aztecs.* Cambridge, Massachusetts: Botanical Museum of Harvard University.

Summa Antropológica en homenáje á Roberto J. Weitlaner. 1966. Mexico, D. F.: Instituto Nacional de Antropológica e Historia.

Wasson, R. G. 1967. Ololiuhqui and the other hallucinogens of Mexico. In *Summa Antropológica en homenáje á Roberto J. Weitlaner.* Mexico, D. F.: Instituto National de Antropología e Historia. Pp. 329–348.

Wasson, R. G. 1968. *Soma: Divine Mushroom of Immortality.* New York: Harcourt, Brace & World.

Wasson, R. G. 1980. *The Wondrous Mushroom: Mycolatry in Mesoamerica.* New York: McGraw-Hill.

Wasson, R. G., G. Cowan, F. Cowan, and W. Rhodes. 1974. *María Sabina and her Mazatec Mushroom Velada.* New York: Harcourt Brace Jovanovich.

Wasson, V. P., and R. G. Wasson. 1957. *Mushrooms, Russia and* History. New York: Pantheon Books.

Wilbert, J. 1987. *Tobacco and Shamanism in South America.* New Haven, Connecticut: Yale University Press.

Wasson and the Development of Mycology in Mexico

GASTÓN GUZMÁN

Introduction

It is difficult and rare for an amateur scientist to become a respected expert in his field of study. One who earned this distinction was R. Gordon Wasson, the author of many important papers and books on the sacred mushrooms of Mexico.

Wasson started his career as a journalist. He then entered the banking profession and was later named vice president at an important New York bank. Simultaneously, his insatiable curiosity compelled him to devote long hours of personal time to the study of fungi as an avocation. His wife Valentina, a pediatrician, shared his interest. Together they founded an important new branch of mycology, the field of ethnomycology, which specifically concerns the role of fungi in the cultural development of people.

The story of how they accomplished this has frequently been told. I would now like to supplement part of that story, concerning Wasson's work in Mexico, by recounting my personal involvement with him and his research, followed by a review of subsequent developments in Mexican mycology.

Encounters with Wasson in Mexico

In 1955, when I was a graduate student at the Polytechnic Institute in Mexico City, I was assigned to catalog the plant collection of the Botany Laboratory. I organized the specimens of all other families except for the fungi. These had been stored in alcohol or formaldehyde, with no information available on their condition when fresh. Furthermore, in many cases, the labels had fallen off the containers and students had replaced them randomly. So I decided to discard everything in the fungi collection and start a new one.

Thus I became a mycologist almost by chance, collecting mushrooms for the collection in the forests near Mexico City in the summer of 1955. I was surprised to find so many species of mushrooms there, yet no books on mushrooms in libraries

83

or bookstores in Mexico City. This led me to resolve that I would one day write a book on mushrooms.

In 1956, the professor in charge of the botanical laboratory resigned and I was assigned to replace him. Also that year, I was required to declare the topic of my professional thesis. As I was surrounded by mushrooms in the laboratory, and in light of all the field work I had done gathering them in recent months, I decided to make mushrooms the subject of my thesis. I therefore spent all of 1956 continuing to collect them while teaching botany at the Institute.

At the end of that year I was asked by some Swiss pharmaceutical laboratories to collect hallucinogenic mushrooms, of which I then knew nothing. (The request was conveyed by Dr. Gonzálo Halffter, who was working for these laboratories in Mexico City.) I accepted the assignment. To prepare myself, I read two important papers on the subject written by Roger Heim (1956a; 1956b). From these papers I learned, to my surprise, that Wasson had been doing field research on the sacred hallucinogenic mushrooms of Mexico since 1953, and in 1956 had done so with Heim's collaboration.

Early in 1957, Teófilo Herrera, a researcher at the University of Mexico, invited me to collaborate with Rolf Singer on an expedition planned for that summer to study hallucinogenic mushrooms in the field. I accepted this fortunate invitation, and in July 1957 accompanied Singer and his assistant Miguel Palacios on a trip to the Mazatec mountains in the State of Oaxaca (Figs. 1 and 2). I had visited that region in 1953, collecting specimens of "barbasco" (*Dioscorea* spp.) for Syntex Laboratories while working as a botanical explorer in Southeastern Mexico.

Prior to my departure with Singer in July, Wasson (1957) published his first popular article on sacred mushrooms in the May 13, 1957 issue of *Life* magazine. This proved very useful to me in my field work. In that article, "Seeking the Magic Mushroom," Wasson first publicly announced the existence of the magic mushroom cult in Mexico, which opened the door to subsequent investigations of these fungi. (Unfortunately, another effect of this publication was that the sacred mushrooms were profaned, reduced to simple commercial objects. It also inspired a rash of popular books and pamphlets on the subject, some of which included many errors or which simply repeated information from previous mycological writings.)

On the July 1957 expedition with Singer to Huautla de Jiménez, and in company with Herrera to Popocatéptl and the Nevado de Toluca mountains in the central part of Mexico, I learned much about mushrooms. Singer really was my teacher on mycology.

The last day of the expedition was especially interesting. We met Wasson in a small village on our way to a crude rural airstrip located on top of a hill—the only flat land in the area—near San Andrés (Figs. 3 and 4). He had been in the village doing mushroom research. I was surprised to find him speaking with me in perfect Spanish, with Singer in English and with the local Indians in the Mazatec language. After this unfortunately short, but agreeable, meeting, I began a very satisfying correspondence with Wasson and received copies of almost all his published works.

Fig. 1. Huautla de Jiménez from the air (1957). *Gastón Guzmán.*

Fig. 2. The main marketplace in Huautla (1957). *Gastón Guzmán.*

Fig. 3. The landing strip near San Andrés, with deep ravines on three sides and a sheer rock cliff on the fourth (1957). *Gastón Guzmán*.

Fig. 4. Rolf Singer's assistant Palacios (*left*), Singer and Gordon near Huautla (July 1957). *Gastón Guzmán*.

While collecting mushrooms with Singer in the Huautla de Jiménez region in 1957 I was introduced to *Psilocybe mexicana* Heim. It looked familiar; I thought I remembered having gathered this fungus some time previously in the Xalapa region in the State of Veracruz. Upon returning to my laboratory in Mexico City I found it among specimens I had collected in 1956 at Rancho Lucas Martin near Xalapa and which I had identified as *Panaeolus* sp., describing it as "a rare species." In connection with this, it is interesting to note that the first hallucinogenic fungus reported in Mexico was *Panaeolus sphinctrinus* (Fr.) Quél. [as *P. campanulatus* L. var. *sphinctrinus* (Fr.) Bres.] (Schultes 1939), and that the first hallucinogenic species described by Heim (1956a) was *Psilocybe mexicana*. Also noteworthy is that a specimen of *Psilocybe mexicana* collected by Wasson from the Mixe region in the State of Oaxaca was identified by Heim as *Panaeolus fimicola* (Fr.) Quél., as I later observed in the Museum National d'Histoire Naturelle in Paris.

But especially significant, it seemed to me, was that Huautla de Jiménez and Xalapa—two distinct geographical regions—have the same ecological conditions in terms of climate and vegetation. These facts inspired me to undertake a phytogeographic study of the hallucinogenic mushrooms of Mexico. Thus it was that in 1957–58 I traveled through the mountains of Southern and Central Mexico, often following in Wasson's steps as I studied the distribution of magic mushrooms and gathered much important information on the subject. Always, in these travels, I made contact with the Indians (Fig. 5) in search of information.

Fig. 5. Guzmán's contacts with the Indians of Mexico proved valuable in his field work. This healer lived in the Zapotec region of San Agustín Loxicha, State of Oaxaca (1958). *Gastón Guzmán.*

During and after this period, I stayed in touch with both Wasson and Singer. I sent Singer many collections of fungi for his help in identifying them. In 1959 I met Heim during one of his visits to Mexico and established an exchange of letters and publications with him also. Finally, I made contact with Schultes, who provided much important information over the years.

In 1958 I published my first work on hallucinogenic mushrooms (Guzmán 1958), a study of the ecology of *Psilocybe muliercula* Singer & Smith (= *P. wassonii* Heim). I then published a revision of the known hallucinogenic species and their distribution in Mexico (Guzmán 1959a), followed shortly by my professional dissertation (Guzmán 1959b). The latter was dedicated to Singer, Heim, Herrera and Wasson in recognition of their help in my research. In 1983, my monograph on the genus *Psilocybe,* started in 1957, was finally published (Guzmán 1983a). It was supported by a grant from the Guggenheim Memorial Foundation of New York, which I had gotten thanks to Schultes' invitation in 1971.

In 1974, Wasson visited my laboratory at the Polytechnic Institute and Herrera's at the University of Mexico. The three of us then dined at a restaurant (Fig. 6), where I told Wasson about a church in Chignahuapan (Fig. 7), in the State of Puebla, where a fungus is venerated. Wasson, greatly interested, said that he knew of no similar church in all the world, and that he was anxious to see it. We then laid plans for an expedition to Chignahuapan, which Wasson, Herrera and I visited in the summer of 1975 in the company of Irmgard Weitlaner Johnson, R. Hernandez, J. Cruz and J. Ott. Our research there was conducted with the help of local people, especially Cándido Arroyo and Macario Arroyo (Fig. 8), the church caretakers.

Fig. 6. Gordon first heard of the church of Chignahuapan while dining with Guzmán (*center*) and Herrera (*right*) at a Mexico City restaurant in 1974. *Gastón Guzmán.*

Fig. 7. The church of Chignahuapan (1975). *Gastón Guzmán*.

Fig. 8. Gordon with Cándido Arroyo (*left*) and Marcio Arroyo (*right*) outside the doors of the church of Chignahuapan (1975). *Gastón Guzmán*.

The church of Chignahuapan was built in honor of the venerated fungus. I had initially visited this church in 1961, after learning of it from Alfredo Barrera, one of my professors at the Polytechnic Institute. Barrera had accompanied me on that occasion, but no study resulted. Our 1975 expedition identified the fungus as *Ganoderma lobatum* (Schw.) Atk., a non-hallucinogenic species (Guzmán et al., 1975).

The *Ganoderma* enshrined in the church in Chignahuapan has, on its face, an arresting sketch of the crucified Christ, with the sun and moon on either side of him (Fig. 9). At the time of my visit in 1961, it was kept in a simple glass case and could be touched. By 1975, however, it had been moved to a closed metal case that had a magnifying glass built into its front panel to facilitate viewing. The metal case itself is incorporated into a large metal cross (Fig. 10). In 1961, the fungus had been situated prominently in the main part of the church; but by 1975, the cross with the fungus had been relegated to a corner of the building, as if to establish that now it was a thing of only secondary importance. Yet the Indians, when visiting the church, first stop to venerate the cross in its corner and only then go to the principal altar, where an image of a Virgin is located.

Fig. 9. Close-up of a photograph on a postcard purchased in 1961 by Guzmán at the church of Chignahuapan shows the revered fungus bearing a sketch of the crucified Christ with the sun and moon on either side. *Gastón Guzmán.*

Fig. 10. The sacred fungus of the church of Chignahuapan is kept in a sealed glass container in the crux of this metal cross (1975). *Guzmán*.

It seems likely that the Chignahuapan Indians once ate hallucinogenic mushrooms that grew in nearby ravines, using them for religious purposes. This is consistent with the practice of Indians living in and near Necaxa, which is also in the State of Puebla. Among the latter I had earlier found evidence of a *Psilocybe* cult: improvised Christian altars in the walls of ravines; and local reference to the sacred mushrooms *P. caerulescens* Murr. and *P. mexicana* Heim as "teotlaquilnanácatl" (Guzmán 1960), as I will discuss later. Wasson, Herrera and I therefore hypothesized (see Guzmán et al., 1975) that the religious of the church in downtown Chignahuapan had once invited local Indians to visit their church, but had found that the Indians preferred to worship Christ in the ravines by ingesting sacred mushrooms. The religious of the church in Chignahuapan then attracted the Indians to the church by means of the drawing of Christ on the *Ganoderma*, which led them to believe that they must go to the church in order to properly worship Christ.

My fourth encounter with Wasson was in 1977, during a conference on psychoactive fungi held in Olympia, Washington. Wasson, Schultes, Albert Hofmann and I were all invited lecturers.

Our fifth and final meeting was in 1981, when I joined Wasson in Mexico City at a "round table" seminar organized in his honor by the Colegio Nacional. During this two-day event, Wasson delivered two lectures on his experiences with the sacred mushrooms. I spoke on known species of *Psilocybe* and their distribution. Other speakers at the meeting were Casas Campillo, Octavio Paz, Garca Terrés, R. de la Fuente, B. de la Fuente, León Portilla, F. Benitez and I. Bernal.

My last communication with Wasson was a letter I received from him in 1983, in which he congratulated me on my *Psilocybe* monograph and encouraged me to continue my research.

Wasson was, for me, a model to emulate, especially in his active and enthusiastic approach to research. He also exemplified scholarly teamwork, for his studies on mushrooms were always collaborative efforts. Thus *Mushrooms, Russia and History* (Wasson & Wasson 1957), *The Wondrous Mushroom* (Wasson 1980) and the books he wrote with Heim on the hallucinogenic mushrooms of Mexico (Heim & Wasson 1958; Heim et al., 1967) reflect the input of numerous specialists in fields as diverse as anthropology, linguistics, archaeology, mycology and ethnobotany.

Heim in 1957 described a species of *Psilocybe* which he named, in honor of Wasson, *Psilocybe wassonii,* though its correct name was established to be *P. muliercula* Smith & Singer when certain nomenclatural and taxonomic problems were later resolved (Singer & Smith 1958; Guzmán 1983a). Subsequently, my colleague Steven Pollack and I described another hallucinogenic species, found in the Naolinco region of the State of Veracruz, as *Psilocybe wassoniorum,* in honor of Wasson and his wife Valentina (Guzmán & Pollack 1979).

The Mexican Mycological Society nominated Wasson in 1981 as an Honorary Member. It was my privilege to present this nomination in recognition of Wasson's great contributions to the field of Mexican mycology.

My Experience Eating the Sacred Mushrooms

Before the publication of Wasson's *Life* article in May 1957, the effects of consuming sacred mushrooms were almost unknown outside of certain Mexican cults that used them for religious purposes. I first ate the sacred mushrooms in July 1958, on the last night of an expedition to the Mazatec region that was preceded by several earlier expeditions to the Oaxaca mountains. My motivation was simple scientific curiosity about the native hallucinogenic mushrooms, which I had collected in the field several times in the previous year. I was then working on the taxonomy, ecology and distribution of *Psilocybe,* and felt it proper to experience the mushrooms firsthand, as had Wasson, to corroborate reports of their hallucinogenic effects.

I therefore asked Don Isauro Nava, an intelligent, bilingual Mazatec Indian who was head of the family in whose house I was staying, if it were possible for me to try the sacred mushrooms. My interest, I told him, was to take them because I believed in them and also to possibly help my digestive problems. He thought it over for a while before answering in the affirmative.

Some background will explain why Don Isauro was willing to grant my request. I had met him the previous year when Singer, Palacios and I, during one of our trips, had been his guests. In subsequent months I made several visits alone, and once with my friend Xavier Madrigal, to Don Isauro's house in a place called Rancho El Cura, about an hour's journey by foot from Huautla de Jiménez. The house had earthen floors, walls of rough-hewn wood, a roof constructed of palm fronds and wood, and two doors but no windows. There was only one large room, with a kitchen and dining area at one end. At least three families shared this humble dwelling—all Don Isauro's relatives—but so warmly did they welcome me into the house that I decided to establish my work center there. I slept on a sleeping mat located in a corner of the room directly opposite a mushroom dryer. (The dryer, a wire-frame structure, had several shelves lined with newspaper. Fresh mushrooms were placed on the shelves and dried from below by the heat of a simple oil lamp.) During one of my stays, Don Isauro had even invited me to be his daughter's godfather, an honor I accepted. Her baptism and a convivial party took place on my next visit, cementing my friendship with Don Isauro.

In this cordial atmosphere I finally tried the sacred mushrooms. A native ceremony for that purpose was convened in the house on the night of July 12, 1958. Eerde (1981) later published a paper based on my notes, in which she mistakenly reported that this ceremony happened in 1957 and that I took 24 mushrooms. The actual dose was six pairs of fresh specimens of "San Isidro"— *Psilocybe cubensis* (Earle) Singer—which I had collected that morning. Presiding over the event was Don Isauro's 80-year-old mother. She handled the mushrooms reverently, passing them first over incense in front of an altar. The prayers she sang were a mixture of Spanish and Mazatec. Earlier, she had asked two of her eldest sons to also take the mushrooms that evening in order to accompany me during the ceremony. (Don Isauro explained that it is not wise to try the mushrooms by oneself; rather, one should take them in the company of an experienced friend.)

As instructed by Don Isauro, the only Indian present who spoke Spanish, I ate the mushrooms carefully and respectfully. Also as instructed, I had not eaten dinner that evening in order to be able to ingest them on an empty stomach. I only drank water while eating the mushrooms to aid in digestion and dilute their bitter taste. Still, I felt ill while taking the second half of my dosage and had to modulate my breathing before I could continue.

After the mushrooms were eaten, I barely chatted with Don Isauro, wondering when the hallucinogenic effects would start. I looked at my watch and saw it was late—about 9:00 p.m. (the family usually went to bed at about 8:00 p.m.)—so I decided to retire to my sleeping mat. When I told Don Isauro of my decision, he and his mother also decided to sleep. The two brothers remained seated at the kitchen table, where they stayed all night.

As soon as I laid down on my mat, all the lights in the house—several candles and one oil lamp—were extinguished. The only light left burning, as on other nights, was the oil lamp in the mushroom dryer. The mushrooms took effect half an hour after eating them. I was watching the illuminated mushroom dryer from my mat when it suddenly seemed to resemble a castle with human features. This was not an hallucination per se, but rather an illusion which I recognized as such. The castle smiled at me and said, "Come, come, don't be afraid." I felt surprised and somewhat frightened in spite of the castle's comforting words. Then, as I reached for my eyeglasses, the castle laughed loudly. I decided to turn my back on it and try to get some sleep, but this proved impossible, for now I saw many attractive, bright colors wherever I turned, regardless of whether my eyes were open or closed. These colors gradually transformed into very tall Negro men who danced around me and my sleeping mat, singing. I then felt very comfortable, as if I were on a magnificent mattress, and also felt like laughing as the castle continued to call me. I almost asked it to keep silent, for I was very busy looking at the spectacle of the dancing men.

When I tried to understand why I was seeing such strange sights, the dancers disappeared. A dog then appeared, following a cat. Both circled my mat very quickly as the castle continued to call me and laugh. Again I tried to analyze these "hallucinations," but found it too difficult. I could not think clearly about anything.

Next, I noticed my legs making motions as if walking, though I was still lying on by back. I scolded them for this, but they ignored me and continued, like children at play.

Suddenly I moved to a cavern under my mat, where I observed many curious, sparkling little animals surrounded by very strange music. Their unreality was obvious. I began to feel bad and somewhat crazy. When I tried to get back on my sleeping mat I found it to be almost out of reach near the ceiling of the cavern. There were stairs nearby, but these were broken. Eventually, however, I succeeded in getting back onto the mat. There I found two of my cousins speaking affectionately to me. One of them, a woman, stretched her hand toward me in greeting. Suspecting that she and her hand were not real, I balked at responding, wondering: "Who would fail to acknowledge a woman with her hand extended?" It then occurred to me that this was not proper behavior for a gentleman; I therefore stretched my hand out to reciprocate, still lying on my back. At that point I suddenly realized how ridiculous my actions and thoughts were, for all the strange things I was seeing were surely not real. And yet they seemed so credible!

Later, I remembered something Wasson had written about *Amanita muscaria*: its psychoactive constituents pass through the body and are expelled—still psychoactive—in urine. Perhaps the same was true of *Psilocybe* mushrooms. This inspired me to go outside and urinate, for doing so, I reasoned, might alleviate the intensity of my experience. Rather than walking, however—for walking was impossible due to the independent motions of my legs—I began to cross the room on all fours. As I reached the door, someone said, "The engineer has escaped!" (The Indians call anyone in work clothes who comes to that region from Mexico City an engineer.) I was at once surrounded—this time by real people.

Don Isauro then asked me, "What is happening, engineer?" I assured him I was all right and explained that I needed to urinate. He wanted to know why I was creeping about on all fours, and then pointed out it was raining outside. I told him I was glad about the rain, for I needed a shower.

Upon returning to my mat, I saw that the mushroom dryer still resembled a castle. This led me to conclude, at first, that Wasson was wrong about the mushrooms' psychoactive effects passing out of the body in urine. I subsequently learned that what he had written was correct for *Amanita* but does not apply to *Psilocybe*.

These experiences lasted almost through the night. I fell asleep at about 4:00 a.m. An hour later, I was awakened by the sound of people in the house making tortillas and coffee as they prepared for work. Immediately, I looked in the direction of the animated castle and saw that it had reverted to a mushroom dryer. I felt tired and hungry, but otherwise fine.

Later that morning I started preparing my things to return to Mexico City. Don Isauro asked me not to leave. He wanted me to stay another week, in accordance with the Indians' religious ideas concerning the sacred mushrooms. Only after I explained to him several times that important work awaited me back home did he reluctantly agree to let me go.

The following night, as I tried to fall asleep in a rustic hotel in the town of Teotitlán del Camino, where I stopped on my long way back to Mexico City, I noticed a very persistent buzzing noise in my ears. This made me nervous and kept me awake, for I associated the sound with the hallucinations I had seen the night before: I wondered if they were about to return. They did not, but the buzzing continued for about a week—revealing, perhaps, why Don Isauro had wanted me to stay with him for that length of time.

Though this *velada* took place more than 30 years ago, I still remember it as though it happened yesterday. The experience taught me that what Wasson and the Indians had reported about the effects of hallucinogenic sacred mushrooms is entirely true—and mysterious. I also learned that we must be very respectful of the Indians' credibility.

Mexican Mycology After Wasson

It is possible to divide the history of Mexican mycology into two parts: before and after Wasson. The first part began with the 16th Century writings of Bernardino de Sahagún and continued through the first half of the 20th Century, when papers by Schultes (1939) and others described the religious use of ostensibly psychoactive mushrooms by native Mexican Indians. Their findings, however, were all but ignored by other scientists, in part because World War II diverted attention for several years.

Among the 25 oldest writings on this subject, those of Sahagún (1555–1590) are most interesting for their exactitude. His several volumes on the ethnography of Mexico, written during the Spanish occupation, comprise a monumental historical document. He wrote bilingually, with parallel texts in Nahuatl and Spanish. The Spanish text has since been translated into other languages and published in several Spanish editions, one of which is listed in my references at the end of this essay.

There are at least three brief passages in the work of Sahagún regarding sacred mushrooms, besides other passages in which he refers to edible mushrooms. Twice he described a sacred mushroom called by the Indians *teonanácatl* or *nanácatl*. His description of this mushroom and its effects was so exact that it is identifiable as a species of *Psilocybe* closely related to *P. aztecorum* Heim emend. Guzmán, found by Wasson in the town of Nexapa, State of Mexico, on the slopes of the volcano Popocatépetl in 1955; and later by me in the forests of

Desierto de los Leones near Mexico City (Guzmán 1968) and high in the mountains of Mexico (Guzmán 1978). It seems very probable that the sacred mushroom mentioned by Sahagún which was used by the Indians in Tenochtitlán (now Mexico City) was *P. aztecorum.*

After Wasson called attention to the word *teonanácatl* it was used by some writers indiscriminately to describe any of the Mexican hallucinogenic mushrooms. But Sahagún applied it only to those mushrooms used by the Nahuatl tribe. It no longer has currency; at any rate, neither Reko, Weitlaner, Schultes nor Wasson found the term *teonanácatl* being used by modern Indians. However, I discovered that a similar term, *teotlaquilnanácatl,* was the popular name applied to sacred mushrooms such as *Psilocybe mexicana* Heim and *P. caerulescens* Murr. by the Indians who use them in the region of Necaxa, State of Puebla—a Nahuatl region (Guzmán 1960). This naturally raises the question: was the Nahuatl word *teotlanquilnanácatl* misconstrued by Sahagún to be *teonanácatl?* Unfortunately, a definitive answer has so far proven elusive.

It was near the end of the first part of the history of Mexican mycology that Wasson initially took notice of pre-Columbian mushroom stones. These small statues normally measure about 30 cm high and are shaped anthropomorphically or zoomorphically, with a round hand that strongly resembles the cap of a mushroom. According to Mayer (1977), about 400 have been found in Mesoamerica; they mainly derive from Maya culture. One was the key to Wasson's discovery of the true sacred mushrooms, which followed a letter he received from his friend Giovanni Mardersteig in the early 1950s. In it, Mardersteig told of a mushroom-shaped stone (Fig. 11) on display at a Zurich museum. This particular specimen, found by Sapper in El Salvador in 1898, was the first to have caught the attention of learned society. Wasson's conviction that this and similar figures were emblems of an ancient mushroom cult in Maya culture, or at least religious symbols related to mushrooms, helped inspire his first expeditions to Mexico.

I have had three interesting experiences with mushroom stones. The first, in 1977, occurred when I was organizing the second national mushroom fair in Mexico, through the Mexican Mycological Society. While looking for a site to have the fair, I and other members of the organizing committee approached the Director of the National Museum of Anthropology in Mexico City. When we started explaining our plan to the Director he reacted with surprise and disapproval. "Why mushrooms in this museum?," he asked. When I mentioned, among other reasons, the importance of the mushroom stones in Maya culture, he immediately stopped the conversation. According to him, these stones "absolutely do not have any relationship with mushrooms." He also could not fathom why we thought his museum would be an appropriate place for a mushroom fair. It was finally held at another location in Mexico City, the Museum of Natural History, thanks to the kindness of its Director, Dr. Halffter.

My second experience with mushroom stones involved an embarrassing moment with Dr. Kobayasi, a distinguished mycologist from the National Museum of Natural History in Japan. He came to Mexico in 1977 in order to improve his understanding of the mushroom stones and also, by visiting Popocatéptl, *Psilocybe aztecorum.* During his stay, Kobayasi made several visits to the great Anthropology Museum in Mexico City before asking me, very con-

Fig. 11. A Japanese facsimile of the famous mushroom stone from El Salvador that now resides in the Rietberg Museum, Zurich.
D. Martinéz-Carrera.

fused: "Where are the mushroom stones exhibited? Am I in the wrong museum?" I regretfully informed him that not only was he in the wrong museum, but in the wrong country: he would have to go to Switzerland in order to investigate the mushroom stones. Six years later I met him again at the Third International Mycological Congress in Tokyo. There, I was surprised to see the Zurich museum's famous mushroom stone displayed at Kobayasi's ethnomycology exhibit. When I asked him how he managed to obtain the valuable artifact, he told me with great satisfaction: "This is not from Switzerland. This is from Japan." Then he turned the figure upside down to show me the base, where was written the legend: *Made in Japan.* Upon seeing this, I told Kobayasi that I was very interested in getting a similar replica of the original mushroom stone. He then showed me many such replicas which he kept beneath the table of his exhibit booth. I bought one, and this statue in my personal collection is perhaps the only mushroom stone in Mexico today. Every year I show it at the mushroom fairs held around the country, where it attracts much attention.

My third experience with mushroom stones, involving a small one found in the State of Michoacan, is described later in this essay.

The second part of the history of Mexican mycology began when Wasson read all existing material on Mexican mushrooms, going back to Sahagún, and made his first expedition to Mexico in 1953. This marked the beginning of a systematic scientific interest in Mexican fungi that introduced modern mycology to Mexico. Within a decade, Wasson (1962) was able to write: "The past six years have seen unprecedented activity in the study of the hallucinogenic mushrooms of Mexico. So diverse and extensive has this activity been, and so numerous are the publications about these mushrooms and their derivatives, that we believe a bibliography on the subject is timely." These comments accompanied Wasson's bibliography of more than 300 references on hallucinogenic mushrooms, of which about 50 dated from the 16th to 19th centuries, no more than six were scientific papers from the first half of the 20th Century, and 50 were mycological publications by Heim, Singer and others from 1956 to 1962. Also published since 1956 were 10 books and papers by Wasson and his collaborators, and 120 related publications on ethnomycology, anthropology, chemistry, pharmacology, physiology and psychology.

Wasson has elsewhere described in great detail the early historical background of ethnomycology in Mexico (e.g., Wasson & Wasson 1957; Heim & Wasson 1958; Wasson 1974; Wasson 1980). I will therefore not repeat this information in my essay, except to observe that, according to Wasson, nine Indian tribes in Mexico use sacred mushrooms or did in the past. Five of these—the Mazatec, Mixtec, Mixe, Chatin and Zapotec Indians—are located in the State of Oaxaca. Four others dwell elsewhere: the Nahuatl, from the center of Mexico to the Pacific in Colima and Chiapas and to the Gulf of Mexico in Veracruz; the Mazahua and the Otomi, in the center of Mexico; and the Tarascan, or Purepecha, in the State of Michoacan. Of these, the Mazahuan, Otomi and Tarascan Indians apparently no longer use the sacred mushrooms, having abandoned them at some point in the forgotten past.

It is also worth noting that not all the sacred mushrooms used in Mexico are *Psilocybe* (Guzmán 1959a; 1959b; 1983b). In addition to *Stropharia cubensis* Earle, also known as *Psilocybe cubensis* (Earle) Sing. following Singer's (1949) reclassification, Heim and Wasson (1958) identified *Conocybe siligenoides* Heim as another hallucinogenic sacred mushroom of Mexico.

Furthermore, not all sacred mushrooms in Mexico are hallucinogenic. Non-hallucinogenic varieties are also used for divinatory or religious purposes. According to Heim and Wasson (1958), these include: *Dictyophora indusiata* (Vent. ex Pers.) Desv. (as *D. phalloides* Desv.), *Clavariadelphus truncatus* (Quél.) Donk (as *Clavaria truncata* Quél.), *Gomphus floccosus* (Schw.) Singer [as *Neurophyllum floccossum* (Schw.) Heim], *Cordyceps capitata* (Holmsk. ex Fr.) Link and *C. ophioglossoides* (Ehrenb. ex Fr.) Link (both on *Elaphomyces variegatus* Vitt., also sacred mushrooms), *Lycoperdon mixtecorum* Heim and *L. marginatum* Vitt. (Heim et al., 1967).

I confirmed, in my subsequent research, that *Lycoperdon* are neither hallucinogenic nor narcotic (Guzmán in Ott et al., 1975). In fact, they are edible, as are *Clavariadelphus truncatus* and *Gomphus floccosus*. But the identification of *Cordyceps* as non-hallucinogenic is problematic. Wasson and Heim reported that it has no hallucinogenic properties because Hofmann had found it contains only trace amounts of an indolic compound. But *Cordyceps* is closely related to *Claviceps purpurea* (Fr.) Tul., the source of the "ergot" from which Hofmann isolated LSD, the most powerful hallucinogen yet discovered. It seems likely that the samples of *Cordyceps* which Wasson and Heim sent to Hofmann for his analysis were old, perhaps too old for reliable testing, having been purchased from the market in Tenango del Valle, State of Mexico, where the Indians sell them dried.

Another non-hallucinogenic sacred mushroom so identified by Heim and Wasson is the edible *Elaphomyces* species. It was reportedly used in the Alta Mixteca, in the State of Oaxaca, for the "treatment of grave wounds" and to "rejuvenate the organism" (Guzmán in Trappe et al., 1979). In the Nevado de Toluca region, Wasson observed the use of *Elaphomyces variegatus* with *Psilocybe muliercula, Cordyceps capitata* and *C. ophioglossoides* in nocturnal ceremonies: the Indians said that *Elaphomyces,* which they sometimes ate, presided over the ceremony when they ingested *Psilocybe* or *Cordyceps.* Unfortunately, these ceremonies seem to have been discontinued and are lost to us, making it difficult for us now to study the role of *Elaphomyces,* which, as I observed in the Tenango del Valle region in 1958, the Indians called "the great world."

Completing Heim's and Wasson's list of non-hallucinogenic sacred mushrooms is *Dictyophora indusiata,* reported by them to be used as a divinatory mushroom in the Chinantla region. Its strange form and odor almost certainly explain why it captured the attention of the Indians there. This fungus is closely related to *Clathrus crispus* Turpin, which in Yucatan is called "colador del brujo" ("colander of the witch") and used as a popular medicine for the eyes (Guzmán 1983b).

The invitation Wasson extended to Heim in 1956 to study the taxonomy of the sacred mushrooms was the key to getting scientists interested in these rare fungi. The cultures of *Psilocybe mexicana* which Heim grew in Paris were studied by Hofmann and his colleagues Brack and Kobel at Sandoz Laboratories in Switzerland. They managed first to isolate the psychoactive constituents psilocybin and psilocin from *P. mexicana* in 1958, then synthesized these chemicals later that year (Hofmann et al., 1958). This was a very important step in the research on hallucinogenic mushrooms, one that paved the way for subsequent studies by others on medical and psychological applications of these fungi.

Heim did not limit his studies of Mexican fungi to hallucinogenic mushrooms, but also researched others such as Ascomycetes, Aphyllophorales and Gasteromycetes. He found Mexican mycology an interesting, open field of study because virtually no one else was working on it then except for Herrera and me, who were just beginning.

Only two years after Heim's first expedition to Mexico, Heim and Wasson (1958), in collaboration with Hofmann, Brack, Kobel, Cailleux, Cerletti, Delay, Pichot, Lamperiére and Nicolas-Charles, published their important book *Les Champignons Hallucinogènes du Mexique.* It reviewed all the research conducted on hallucinogenic mushrooms up to that time, including archaeological, anthropological and ethnomycological studies. In it, Heim described more than 10 species of *Psilocybe.* He followed this in 1967 with a book presenting new information on these mushrooms; titled *Nouvelles Investigations sur les Champignons Hallucinogènes,* it was written in collaboration with Wasson, Cailleux and Thévenard (Heim et al., 1967).

Singer's arrival was also very important to the development of mycology in Mexico. He was the first to state in print that the genus *Psilocybe* includes hallucinogenic species (Singer 1949), following his examination of collections gathered by Schultes in 1938. He identified *Psilocybe cubensis* (Earle) Singer, a species that Earle had originally found in Cuba and described, in 1906, as *Stropharia cubensis* without including information on its hallucinogenic properties. Unfortunately, Singer's 1949 report on these matters, which was published as a small note in an 800-page book on the taxonomy of Agaricales, suffered the same fate as Schultes' earlier paper on teonanácatl (Schultes 1939): it was initially ignored.

Singer (1958a) eventually published a complete review of information on the ethnomycology of hallucinogenic mushrooms, and later that same year, in collaboration with Smith (Singer & Smith 1958), an important taxonomic monograph on these fungi that discusses one species from Europe and several from the United States, South America and Southeastern Asia in addition to Mexico. Singer (1957; 1958b) also published additional papers on other Mexican agarics in his series *Fungi Mexicana,* in which he described several new species.

Among other "post-Wasson" works on the ethnomycology of the sacred mushrooms, Hoogshagen (1959), Miller (1966), Ravciz (1960) and Lipp (1971) presented interesting findings from their studies in the State of Oaxaca. Caso (1963) published his interpretation of certain images found in the *Vindobonensis Codex*—pairs of mushrooms in the hands of Indians—as representations of sacred mushrooms, along with a delightful representation of "The Hill of the Mushrooms," also symbolized by a pair of mushrooms.

In several publications on ethnomycology, Lowy (1968; 1972; 1974) discussed the famous mushroom stones of ancient Mesoamerica and the importance of *Amanita muscaria* in the Maya culture. He related *A. muscaria* with the legend of the thunderbolt in Guatemala and Mexico, where this mushroom is called thunder by the Indians, but found that it was not used there for religious purposes. On the contrary, it is considered diabolical, a very toxic mushroom, as I observed many times among the Indians of Mexico.

Possibly, however, *Amanita* had once been regarded by the Mayas as a sacred mushroom, as Lowy concluded after studying the Maya codices that survived the destruction of many other Maya writings by the Spanish Conquistadors. For instance, in the *Madrid Codex* Lowy found what he interpreted to be a stylized representation of *Amanita muscaria.* It may be that this fungus, which is common throughout Guatemalan and Mexican pine forests, was once used by several tribes in Mesoamerica for religious purposes, as Wasson proposed had been true of northern Indian tribes of the Kamchatka Peninsula and elsewhere (Schultes & Hofmann 1979; Wasson 1980). It is interesting to note that Europeans long ago attributed to *Amanita* the ability to kill flies. This explains the mushroom's common name, "fly-agaric," by which even Linneaus referred to it. Mexican farmers also know of *Amanita*'s fly-killing ability. They use it for this purpose in their kitchens by putting small pieces of the fungus in a dish with sugar water—a practice I observed in Central Mexico that also is reported from Zacatecas (Acosta & Guzmán 1984).

Also strengthening the case that Mexican Indians may have once used *Amanita* is the small stone carving (Fig. 12) found at an archaeological site in the State of Michoacan among the Purepecha culture, which I studied and reported on (in Mapes et al., 1981). On one side, its form suggests an *Amanita* button; on the

Fig. 12. Front and back views of a mushroom stone found in an archeological dig in the Purepecha region in Patzcuaro, Michoacan. Note resemblance to *Amanita muscaria* on one side and a death's head on the other. *Gastón Guzmán*.

other, a death's head. This interpretation is consistent with the possible existence of an *Amanita* cult in ancient Mexico, perhaps linking the Purepecha Indians with the Indians of northern North America and Siberia, as well as with the Maya culture of Guatemala in accordance with Lowy's aforementioned observations. It seems to me probable that *Amanita muscaria* was widely used among different Indian tribes from Alaska to Central America who recognized its properties as an inebriant.

Elsewhere in the State of Michoacan I found *Psilocybe caerulescens, P. mexicana* and *P. cubensis* in different locations. This lends credence to Fray Maturino Gilberti's translation, in his Tarascan (Purepechan) lexicon of 1559, of the Indian expression *cauiqua terequa*, which according to him meant "the mushroom that inebriates" (Wasson & Wasson 1957; Gilberti 1975/1559; Mapes et al., 1981). It would seem that long ago the Purepechans were familiar with the psychoactive properties of *Psilocybe*, if not *Amanita*—or possibly both.

Other ethnomycological works on the sacred mushrooms were written by Guzmán and López González (1970) and Escalante and López González (1971), who studied the use of *Psilocybe muliercula* Singer & Smith (= *P. wassonii* Heim) among the Matlazincas in the region of Nevado de Toluca. Later, in Guzmán (1982), I described as hallucinogenic another species from this region, *P. sanctorum* Guzmán, and subsequently added the hallucinogenic species *P. angustipleurocystidiata* (Fig. 13). De Avila et al. (1980) studied the use of edible mushrooms in Hueyapan, in the State of Morelos, where I subsequently described *P. angustipleurocystidiata* Guzmán from a collection made by V. Mora. This mushroom is used by the Indians to prevent stomach-aches or tooth-aches and for "seeing visions" (Guzmán 1983a).

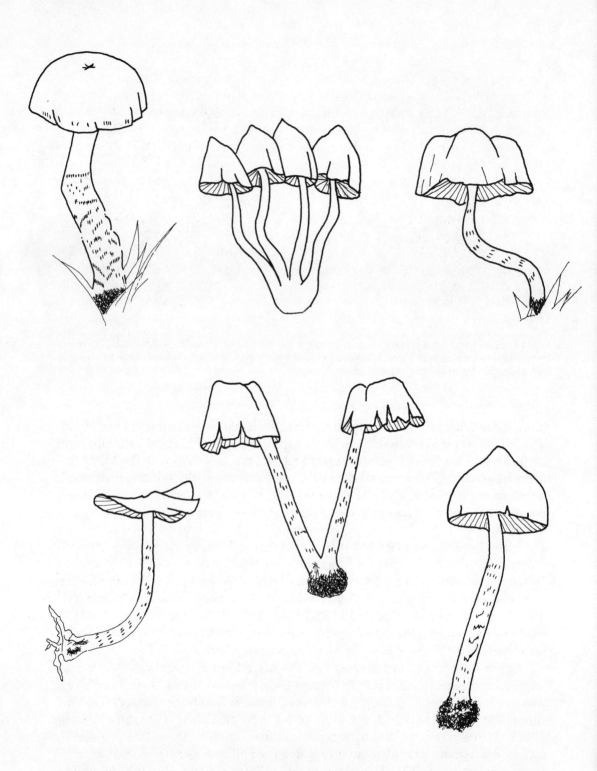

Fig. 13. Drawing of *Psilocybe angustipleurocystidiata* Guzmán, an hallucinogenic mushroom discovered by Guzmán in the State of Morelos and later found in the State of Mexico. *Gastón Guzmán.*

Mapes et al. (1981) studied the ethnomycology of the Purepecha of the Patzcuaro region, where they found that the mushrooms play a very important role among these people. The Purepecha can distinguish morphological and ecological features of several mushrooms, mainly the edible species. They consider the fungi "soil flowers," calling them "echeri uetsikuaro enganaka," which translates to "born from the soil." These concepts concur with the literal meaning of the Latin word *fungus* (plural *fungi*), "flowering from the soil," a term adopted long ago by mycologists. The Indians' terminology thus demonstrates the astuteness of their culture. While Mapes' group did not discover sacred mushrooms in use among the Purepecha, they did find the interesting *"Amanita"* mushroom stone described above.

The first chemical study in Mexico on the hallucinogenic mushrooms was by Chávez de la Mora (1961), who, guided by Herrera, isolated psilocybin from *Psilocybe cubensis*. Later, during the '70s, Casas Campillo's research team at the Polytechnic Institute studied the esteroids of several species of *Psilocybe* obtained from strains I had gathered in 1957 and 1958 (see Martínez Mancilla 1971; Esparza García 1979; Lazcano Hernández 1979; Gándara Acosta 1976). Their work opened the door for the possible use of hallucinogenic mushrooms in the esteroids industry.

Carvajal (1959) had earlier predicted another application for these mushrooms: neuropsychiatric research. That same year, a psychotherapeutic study by Nieto (1959) at the University of Mexico used mushrooms obtained from a culture of *Psilocybe cubensis*. This particular culture was obtained from Zenteno and Herrera (1958), based on a strain they had obtained from Singer in Huautla de Jiménez in 1957. Herrera (1967) subsequently ate the sacred mushrooms in a ceremony conducted by María Sabina in Huautla. He was accompanied by Ulloa and Isikawa, who later discussed the event (Isikawa 1973).

Studies on the culture and development of hallucinogenic *Psilocybe* mushrooms were carried out by several of Herrera's students at the University of Mexico. These included: Dubovoy on the morphogenesis of *P. caerulescens* (Dubovoy & Herrera 1967; 1968a; 1968b); Ulloa on aspects of the growth of *P. cubensis* and *P. mexicana* (Ulloa & Herrera 1967); Cuevas on the morphology and variability of *P. zapotecorum* and *P. muliercula* (Cuevas & Herrera 1971); and Olmos Fuentes on selecting media for *P. cubensis* and *P. mexicana* (Olmos Fuentes & Herrera 1973). Related research by Badham (1980) in the U.S. studied the effect of light on Mexican strains of *P. cubensis*.

Mine was the first ecological work on the sacred mushrooms (Guzmán 1958). This resulted from my studies on the habitat of *Psilocybe muliercula* Singer & Smith (= *P. wassonii* Heim) while researching its taxonomy and synonymy. Both Singer and Heim had based their descriptions of this fungus on dried specimens bought in the marketplace of Tenango del Valle in the Nevado de Toluca region of the State of Mexico, and had reported that it commonly grows in *Pinus* forests surrounding the town. However, after several expeditions to this region, I finally found it 10 kilometers from Tenango del Valle, in an *Abies* forest on the slopes of the Nevado de Toluca.

Mexican Mycology Today

The modern era of Mexican mycology was established by the studies undertaken in the 1950s through the '70s. Before the contributions of Wasson, Heim, Singer and others, mycology in Mexico was restricted to microscopic fungi, including: fungi with phytopathological significance, studied by Gándara and others in the 1930s; fungi of medical interest, researched by González Ochoa in the '40s; and yeasts, which were studied by Manuel Ruíz Oronoz and Sánchez Marroquín in the '40s and '50s.

Herrera and I began our mycological studies in the '50s, starting slowly because there was no one to guide us nor any good specialized research libraries, laboratories or herbaria. Together, we established the Mexican Mycological Society in 1968, started publication of its bulletin in 1968 and *Revista Mexicana de Micología* in 1985. These activities bridged the period of earlier, foreign works on Mexican fungi and the current situation in which Mexican researchers have assumed a more dominant role in this field. Thus the emblem of the Mexican Mycological Society, a mushroom stone, was chosen as a symbol of our strong Indian roots and heritage.

Mycology in Mexico has also been developing apace in a number of the country's institutions. The Instituto de Biología at the University of Mexico and the Escuela Nacional de Ciencias Biológicas at the Polytechnic Institute were the main centers for general research in this field in the 1950s and '60s, while the Hospital de Enfermedades Tropicales in Mexico City was the most important center for research on pathogenic fungi. Significant phytopathological studies on fungi were conducted at the Escuela de Agricultura at Chapingo, now known as the Colegio de Postgraduados de Chapingo. And both the Instituto de Biología and the Escuela Nacional de Ciencias Biológicas have distinguished themselves in the area of yeast research.

Of all these centers, the Colegio de Postgraduados has been especially influential. It produced several enthusiastic specialists on plant pathology, including Galindo, Isla de Behuer, Romero, Fuchicovsky and others, who later established the Mexican Society of Phytopathology, which in turn inspired Herrera's idea of forming the Mexican Mycological Society. (He described its history in the first issue of the *Bulletin of the Mexican Mycological Society* in 1968.)

The development of Mexican mycology proceeded step by step throughout the country. It was advanced by the establishment of new, important research centers such as the Mycological Laboratory at the University of Nuevo León in Monterrey by Castillo, which was founded in the 1960s and subsequently transferred to the Instituto Tecnológico at Ciudad Victoria. Also, new herbaria were built at the University of Veracruz in Xalapa; the University of Morelos in Cuernavaca; the University of Baja California in Ensenada; the Facultad de Ciencias of the University of Mexico; the Instituto de Ciencias in Tuxtla Gutiérrez (Chiapas); the Metropolitan University in Ixtapalapa; and the Department of Biology at Tlaxcala University. Additionally, the Institute for Experimental Biology was established, as a branch of the Centro de Estudios Avanzados of the Polytechnic Institute, at the University of Guanajuato. And important research laboratories were set up for the study of pathogenic fungi at the University of Mexico's Faculty of Medicine; the Centro Dermatológico Pascua in Mexico City; and the Division of Experimental Pathology of the Instituto Mexicano del Seguro Social in Guadalajara.

Mycological laboratories were also established at the University of Guadalajara's Instituto de Botánica and the Instituto Nacional de Investigaciones sobre Recursos Bióticos in Xalapa. Both institutions later contributed much to the development of edible mushroom cultures on agricultural waste, a line of research initially started in Xalapa in 1983.

At the Second International Congress of Mycology, held in Tampa, Florida in 1977, Zenteno, Ulloa, Herrera and I presented an outline of the development of mycology in Mexico and a list of 92 research projects on mycology at 32 institutions, including some in the United States and Canada (Zenteno et al., 1977).

The Future of Mexican Mycology

This "panoramic" view seems very encouraging, but actually Mexican mycology would benefit from more full-time mycologists to study the many diverse aspects of Mexican fungi. Knowledge of these mushrooms is still not complete and in fact, in some cases, is poor—most notably relating to those growing in the tropical regions. In a lecture I gave at the Mexican Mycological Society meeting in 1987, I estimated that 30,000 species of fungi grow in Mexico, of which only 4,000 have so far been studied. This indicates the magnitude of work that lies ahead of us.

Fortunately, the Mexican government has established a number of agencies to help foster scientific projects. These agencies, which represent an important base of support for Mexican mycology, include: Consejo Nacional de Ciencia y Tecnología (CONACYT); Secretaría de Educación Pública (SEP); Secretaría de Programación y Presupuesto (SPP); Secretaría de Agricultura y Recursos Hidráulicos (SARH); Secretaría de Desarrollo Urbano y Ecología (SEDUE); and Sistema Nacional de Investigadores (SNI). They have opened the door to good future development of work begun in the past. Truly, as Wasson (1980) observed in his excellent book *The Wondrous Mushroom:* "New perspectives in ethnomycology are beckoning to us."

Distribution of Hallucinogenic Psilocybe in Mexico

Since publication of my monograph on *Psilocybe* (Guzmán 1983a), little new information on these mushrooms has surfaced. Descriptions of *P. barrerae* Cifuentes & Guzmán (1981) in the States of Guerrero and Hidalgo and *P. sanctorum* Guzmán (1982) in the State of Mexico were published while my monograph was in production. Both species belong to the Section Zapotecorum Guzmán, along with *P. zapotecorum* Heim emend. Guzmán, *P. muliercula* Singer & Smith and *P. angustipleurocystidiata*.

While revising herbaria collections at the Instituto Nacional de Investigaciones sobre Recursos Bióticos in Xalapa and the University of Guadalajara's Instituto de Botánica, and studying some collections given to me by colleagues, I found several interesting new records of *Psilocybe* in Mexico. These were recently discussed by me and my colleagues (Guzmán et al., 1988). Twelve species from 19 locations are mentioned, of which 19 reports were entirely new.

Fig. 14 is a map showing all the locations in Mexico where hallucinogenic *Psilocybe* mushrooms have been found. About 10 were first reported by Wasson in the 1950s and '60s, and 80 by me and my colleagues subsequently (Guzmán et al., 1988). Note that these locations are more numerous in the subtropical, meso-phytic or deciduous forests, and less so in tropical and coniferous forests—an interesting fact which I first mentioned 30 years ago (Guzmán 1959a; 1959b).

Psilocybe angustipleurocystidiata Guzmán has been found in the State of Morelos (Guzmán 1983a; Mora & Guzmán 1983) and recently in the State of Mexico (Guzmán et al., 1988). Notably, these two separate areas of subtropical forests belong to the same geographic zone of the Pacific Ocean slopes of the Eje Neovolcanico Mountains. A drawing of this interesting species' fruiting bodies is presented here for the first time (Fig. 13).

Acknowledgments

I gratefully acknowledge CONACYT, the Instituto de Ecología and Sistema Nacional de Investigadores for supporting my research on fungi, and, for help received in the past, the National Polytechnic Institute, the Guggenheim Memorial Foundation, the Mission Arqueologique et Ethnomycologique of France in Mexico, Flora Neotropica and Tulane University.

Leticia Montoya Bello read the manuscript and improved the English text. Laura Guzmán Dávalos of the Instituto de Botánica in Guadalajara loaned impor-tant collections on *Psilocybe*. Guzmán Dávalos also helped revise and improve the English text. María Eugenia López Ramírez typed the text. To all these friends and colleagues I express my gratitude, as well as to my laboratory assistants Leticia Montoya Bello and Víctor M. Bandala Muñoz for their valued help in the laboratory. Finally, I recognize Arturo Trejo Leal from the Natural History Museum of Mexico City, who made the mushroom drawings and map that accom-pany this essay.

References

Acosta, S., and G. Guzmán. 1984. Los hongos conocidos en el Estado de Zacatecas (Mexico). *Bol Soc Mex Mic* 19:125–128.

Badham, E. R. 1980. The effect of light upon basidiocarp initiation in *Psilocybe cubensis*. *Mycologia* 72:136–142.

Carvajal, G. 1959. La quimioterapia racional y su futuro. *Acta Politécnica Mexicana* 1:29–37.

Caso, A. 1963. El paraiso terrenal en Teotihuacán. *Estudios de Cultura Nahuatl UNAM* 4:28.

Cifuentas, J., and G. Guzmán. 1981. Descripción y distribución de hongos tropicales (Agaricales) no conocidos previamente en Mexico. *Bol Soc Mex Mic* 16:35–61.

Chávez de la Mora, A. 1961. *Investigación Química de un Hongo Alucinante (Stropharia cuben-sis)*. Thesis. Mexico City: Centro Universitario Mexico.

Cuevas, F. J., and T. Herrera. 1971. Variaciones morfológicas de los micelios de *Psilocybe muliercula* y *P. zapotecorum* en diversos medios de cultivo. *Bol Soc Mex Mic* 5:37–46.

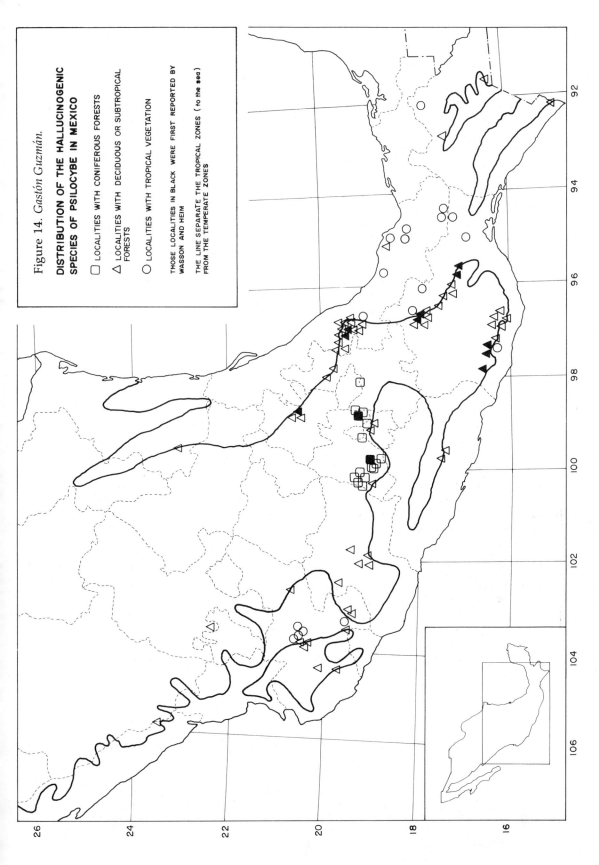

Figure 14. *Gastón Guzmán.*

De Avila, A., A. L. Welden, and G. Guzmán. 1980. Notes on the ethnomycology of Hueyapan, Morelos. *J Ethnopharmacol* 2:311–321.

Dubovoy, C., and T. Herrera. 1967. Estudio morfológico de los micelios de *Psilocybe caerulescens* Murrill en diversos medios líquidos de cultivo. *An Inst Biól UNAM* 38:111–150.

Dubovoy, C., and T. Herrera. 1968a. Morfogénesis de fíbulas, I. Desdicariotización de micelios de *Psilocybe caerulescens* Murrill en diversos medios líquidos de cultivo. *An Inst Biól UNAM* 39:45–76.

Dubovoy, C., and T. Herrera. 1968b. Influenica de factores físico-químicos en la morfogénesis de estructuras asexuales en medios líquidos de cultivo de *Psilocybe caerulescens* Murr. *An Inst Biól* 39:77–110.

Eerde, E. 1981. Fatal, mind boggling and delicious. *R & D Mexico CONACYT* 1(6):6–9,32.

Escalante, R., and A. López González. 1971. *Hongos Sagrados de los Matlazincas*. Mexico City: Mus. Nac. Antropol., Sec. Linguistica.

Esparza García, E. 1979. *Transformación de Esteroides por Psilocybe caerulescens var. mazatecorum*. Thesis. Mexico City: Escuela Nacional de Ciencias Biológicas, IPN.

Gándara Acosta, V. M. 1976. *Transformación de Derivados Androgénicos y Esterogénicos por la Fase Micelial de Psilocybe caerulescens var. mazatecorum*. Thesis. Mexico City: Escuela Nacional de Ciencias Biológicas, IPN.

Gilberti, M. 1975/1559. *Diccionario de la Lengua Tarasca o de Michoacán*. A reprint of the 1559 edition. Morelia: Editions Facsimilar, Bassal.

Guzmán, G. 1958. El hábitat de *Psilocybe muliercula* Singer & Smith (= *Ps. wassonii* Heim), agaricáceo alucinógeno mexicano. *Rev Soc Mex Hist Nat* 19:215:229.

Guzmán, G. 1959a. Sinopsis de los concocimientos sobre los hongos alucinógenos mexicanos. *Bol Soc Bot Mex* 24:14–34.

Guzmán, G. 1959b. *Estudio Taxonómico y Ecológico de los Hongos Neurotrópicos Mexicanos*. Thesis. Mexico City: Escuela Nacional de Ciencias Biológicas, IPN.

Guzmán, G. 1960. Nueva localidad de importancia etnomicológica de los hongos neurotrópicos mexicanos. *Ciencia, Mex* 20:85–88.

Guzmán, G. 1968. Nueva especiede *Psilocybe* de la Sección Caerulescentes de los bosques de coníferas de Mexico. *An Esc Nal Cienc Biols* 17:9–16.

Guzmán, G. 1978. Variation, distribution, ethnomycological data and relationships of *Psilocybe aztecorum,* a Mexican hallucinogenic mushroom. *Mycologia* 70:385–396.

Guzmán, G. 1982. Nuevos datos sobre el género *Psilocybe* y descripción de una nueva especie en Mexico. *Bol Soc Mex Mic* 17:89–94.

Guzmán, G. 1983a. *The Genus Psilocybe: A Systematic Revision of the Known Species Including the History, Distribution and Chemistry of the Hallucinogenic Species*. Vaduz: Beih. Nova Hedwigia 74, Cramer.

Guzmán, G. 1983b. Los hongos de la Península de Yucatán, II. Nuevas exploraciones y adiciones micológicas. *Biotica* 8:71–100.

Guzmán, G., and A. López González. 1970. Nuevo hábitat y datos etnomicológicos de *Psilocybe muliercula. Bol Soc Mex Mic* 4:44–47.

Guzmán, G., R. G. Wasson, and T. Herrera. 1975. Una iglesia dedicada al culto de un hongo, 'Nuestro Señor del Honguito' en Chignahuapan, Puebla. *Bol Soc Mex Mic* 9:137–197.

Guzmán, G., and S. H. Pollack. 1979. Tres nuevas especies y dos nuevos registros de los hongos alucinógenos en Mexico y datos sobre su cultivo en el laboratorio. *Bol Soc Mex Mic* 13:261–270.

Guzmán, G., L. Montoya Bello, and V. M. Bandala Muñoz. 1988. Nuevos registros de los hongos alucinógenos del género *Psilocybe* en Mexico y análisis de la distribución de las especies conocidas. *Rev Mex Mic* 4:255–265.

Heim, R. 1956a. Les champignons divinatoires utilises dans les rites des indies mazatéques, recuellis an cours de leur primer voyage au Mexique, en 1953, par M. V. P. Wasson et M. R. G. Wasson. *Comp Rend Séan Acad Sci* 242:965–968.

Heim, R. 1956b. Les champignons divinatoires recueillis par M. V. P. Wasson et M. R. G.

Wasson au cours de leurs missions de 1954 et 1955 dans les pays mije, mazatéque, zapotéque et nahua du Mexique meridional et central. *Compt Rend Sé*an Acad Sci 242:1389–1395.

Heim, R., and R. G. Wasson. 1958. *Les Champignons Hallucinogènes du Mexique.* Paris: Editions du Muséum National d'Histoire Naturelle.

Heim, R., R. Cailleaux, R. G. Wasson, and P. Thevenard. 1967. *Nouvelles Investigations sur les Champignons Hallucinogènes.* Paris: Editions du Muséum National d'Histoire Naturelle.

Herrera, T. 1967. Consideraciones sobre el efecto de los hongos alucinógenos mexicanos. *Neurología-Neurocirugía-Psiquiatría* 8:101–124.

Hofmann, A., R. Heim, A. Brack, and H. Kobel. 1958. Psilocybin, ein psychotroper wirkstoff aus dem mexikanischen ravchpilz *Psilocybe mexicana* Heim. *Experientia* 14:107–109.

Hoogshagen, S. 1959. Notes on the sacred (narcotic) mushrooms from Coatlan, Oaxaca, Mexico. *Oklahoma Anthropol Soc Bull* 7:71–74.

Isikawa, M. 1973. Experiences of the ceremonia de los hongos alucinógenos in Mexico and its considerations. Presented at the Ninth International Congress of Anthropological and Ethnological Sciences, Chicago.

Lazcano Hernández, I. 1979. *Metabolismo de la Testosterona por Psilocybe caerulescens var. mazetecorum.* Thesis. Mexico City: Escuela Nacional de Ciencias Biológicas, IPN.

Lipp, F. J. 1971. Ethnobotany of the Chinantec Indians, Oaxaca, Mexico. *Economic Botany* 25:234–244.

Lowy, B. 1968. Un hongo de piedra preclásica de Mexico Viejo, Guatemala. *Bol Inf Soc Mex Mic* 2:9–15.

Lowy, B. 1972. Mushroom symbolism in Maya Codices. *Mycologia* 64:816–821.

Lowy, B. 1974. *Amanita muscaria* and the thunderbolt legend in Guatemala and Mexico. *Mycologia* 66:188–191.

Mapes, C., G. Guzmán, and J. Caballero. 1981. Etnomicología Purépecha. *El Conocimiento y Uso de los Hongos en la Cuenca del Lago de Patzcuaro, Michoacán.* Mexico City: Cuadernos Etnobiología 2, Dir. Culturas Populares, SEP, Soc. Mex. Mic., UNAM.

Martínez Mancilla, M. del C. 1971. *Actividad Enzimática de Extractos Libres de Células de Psilocybe caerulescens var. mazatecorum, Sobre Algunos Esteroides.* Thesis. Mexico City: Escuela Nacional de Ciencias Biológicas, IPN.

Mayer, K.-H. 1977. *The Mushroom Stones of Mesoamerica.* Ramona, California: Acoma Books.

Miller, W. S. 1966. El tonalanatl mixe y los hongos sagrados. In *Summa Antropológica en Homenaje a Roberto J. Weitlaner.* Mexico: Instituto Nacional de Antropología e Historia, Secretaría de Educacíon Pública.

Mora, V., and G. Guzmán. 1983. Agaricales poco conocidos en el Estado de Morelos. *Bol Soc Mex Mic* 18:115–139.

Nieto, D. 1959. Psicosis experimentales. Efectos psiquicos del Hongo *Stropharia cubensis* de Oaxaca. *Neurología-Neurocirugía-Psiquiatría* 1:6–16.

Olmos Fuentes, C., and T. Herrera. 1973. Comportamiento de micelios de los géneros *Agrocybe, Panus y Psilocybe* (Agaricales) en diferentes medios de cultivo. *Bol Soc Mex Mic* 7:151–159.

Ott, J., G. Guzmán, J. Roman, and J. L. Díaz. 1975. Nuevos datos sobre los supuestos Licoperdáceos psicotrópicos y dos casos de intoxicación provocados por hongos del género *Scleroderma* en Mexico. *Bol Soc Mex Mic* 9:67–76.

Ravicz, R. 1960. La Mixteca en el estudio comparativo del hongo alucinante. *An Inst Nac Antrop e Hist* 13:73–92.

Sahagún, Fray B. 1955. *Historia General de las Cosas de Nueva España.* A modern edition, in three volumes, of the original work from 1555–1560. Mexico City: Editions Alfa.

Schultes, R. E. 1939. Plantae Mexicanae, II. Identification of Teonanácatl, a narcotic Basidiomycete of the Aztecs. Harvard *Botanical Museum Leaflets* 7:37–55.

Schultes, R. E., and A. Hofmann. 1979. *Plants of the Gods: Origins of Hallucinogenic Use.* New York: McGraw-Hill Book Co. Subsequently published, in 1982, as *Plantas de los*

Dioses: Orígenes del Uso de los Alucinógenos, the Spanish edition translated by A. Blanco, G. Guzmán, and S. Acosta. Mexico City: Fondo de Cultura Económica.

Singer, R. 1949. The Agaricales (mushrooms) in modern taxonomy. *Lilloa* 22:5–832.

Singer, R. 1957. Fungi Mexicana, series prima, Agaricales. *Sydowia* 11:354–374.

Singer, R. 1958a. Mycological investigations on Teonanácatl, the Mexican hallucinogenic mushroom, I. The history of Teonanácatl, field work and culture work. *Mycologia* 50:239–261.

Singer, R. 1958b. Fungi Mexicani, series secunda, Agaricales. *Sydowia* 12:221–243.

Singer, R., and A. H. Smith. 1958. Mycological investigations on Teonanácatl, the Mexican hallucinogenic mushroom, II. A taxonomic monograph of *Psilocybe* Sect. Caerulescentes. *Mycologia* 50:262–303.

Trappe, J. M., G. Guzmán, and C. Vázquez Salinas. 1979. Observaciones sobre la identificación, distribución y usos de los hongos del género *Elaphomyces* en Mexico. *Bol Soc Mex Mic* 13:145–150.

Ulloa, M., and T. Herrera. 1967. Factores que influyen en el crecimiento de los micelios de *Psilocybe mexicana* y *P. cubensis. An Inst Biol UNAM* 38:165–192.

Wasson, R. G. 1957. Seeking the magic mushroom. *Life* May 13, 1957. Spanish edition, "En busca del hongo mágico," published in *Life en Español* June 10, 1957.

Wasson, R. G. 1962. The hallucinogenic mushrooms of Mexico and psilocybin: a bibliography. Harvard *Botanical Musueum Leaflets* 20:25–73.

Wasson, R. G. 1974. The rôle of "flowers" in Nahuatl culture: a suggested interpretation. *J Psychedelic Drugs* 6:351–360.

Wasson, R. G. 1980. *The Wondrous Mushroom: Mycolatry in Mesoamerica.* New York: McGraw-Hill Book Co.

Wasson, V. P., and R. G. Wasson. 1957. *Mushrooms, Russia and History.* New York: Pantheon Books.

Wasson, R. G., G. Cowan, F. Cowan, and W. Rhodes. 1974. *María Sabina and her Mazatec Mushroom Velada.* New York: Harcourt Brace Jovanovich.

Zenteno, M., and T. Herrera. 1958. Hongos alucinantes de Mexico. Datos bibliográficos, obtención de carpóforos de *Psilocybe cubensis. An Inst Biol UNAM* 29:49–72.

Zenteno, M., M. Ulloa, T. Herrera, and G. Guzmán. 1977. Estado actual de la micología en Mexico. *Bol Soc Mex Mic* 11:137–144.

The Mushroom Conspiracy

JOAN HALIFAX

In the 1920s, a new discipline in the combined fields of anthropology and botany was created: ethnomycology. This important area of inquiry and scholarship was brought into being by R. Gordon Wasson, a banker and businessman who had the curious ability of bringing various disciplines together in the little known or neglected area of mushroom research.

Wasson's entry into this area during his honeymoon with his Russian bride Valentina Pavlovna in August 1927 is a well-known story. It is one he enjoyed telling friends and students, for Wasson was also a bard of the mushroom. On an afternoon walk in the Catskills, he often recalled,[1] the newly-wed couple had chanced upon "a forest floor carpeted with mushrooms." Valentina reacted by happily scooping them up in her dress to take back to the honeymoon lodge, where she cooked and ate them. Gordon, meanwhile, recoiled in fear of the "repellant" things, believing he was soon to be a widower. The evident contrast between their reactions provided food for thought—in fact, a feast. "This episode," Gordon explained, "made so deep an impression on us that from then on, as circumstances permitted, we gathered all the information that we could about the attitude of various peoples toward mushrooms—what kinds they know, their names for them, the etymology of those names, the folklore and legends in which mushrooms figure, references to them in proverbs and literatures and mythology." Eventually, at first with Valentina, then alone after she died of cancer in 1958, Gordon followed the thread of discovery he had picked up on his honeymoon, relentlessly pursuing a solution to the botanical and cultural mystery all his life while inspiring others as well.

I was one of the people whom Gordon inspired. I met him in the early 1970s when Stanislav Grof and I went to visit him in his home in Danbury, Connecticut. Grof and I, then married, were involved in LSD research with people dying of cancer. We knew of Gordon's work and felt his understanding of the use of hallucinogens by indigenous peoples could help us in our work with the dying. As a medical anthropologist, I was especially interested in how these plants were used in healing practices, and especially in the practices of shamans.

I had no idea at the time of our meeting that a long and personal friendship between Gordon and me would develop. This relationship was to become a kind

111

of conspiracy shared by a fascinating network of individuals—young and old, scholar and student, shaman and mystic. Some of us would have the opportunity over the years to meet in the New England calm of Gordon's dining room. Some of us would find ourselves conspiring at meetings, and some of us found our way to each other at Gordon's suggestion or insistance. All of us were on the track of the mystery, aware that the mushroom had hidden itself from the Western researcher, historian, anthropologist and archeologist, and also that it was a key to understanding the traditions of various old shamanic cultures. Moreover, many of us had not only been moved by the ethnographic beauty of those cultures which employed the mushroom as sacrament, but ourselves had also tasted of this mystery and felt a commitment to bringing it to light.

Gordon's and Valentina's early work on mushrooms over a span of 30 years culminated in their first book on the subject, *Mushrooms, Russia and History* (1957). In addition to exploring the significant role played by mushrooms in Russia (of which there had been almost no information in English published previously) and other Old World countries, it includes a magnificent section describing Gordon's subjective experience of being one of the first two "white outsiders" (along with his photographer Allan Richardson) to take the sacred mushrooms in a Mexican Mazatec Indian ceremony.

This work was an important contribution—not only in the field of mushroom research, but also in the world of publishing. Luxuriously crafted using handmade paper, quality binding and other fine touches, it set the tone for all of Gordon's later works. For Gordon loved beauty. He often told me that he had a passion to produce books of both aesthetic and scholarly value; as his research into mushrooms and their use was an obsession for him, so was the creation of exquisitely designed books. Gordon wanted every aspect of his work to be well-done, to be truly worthwhile. And so he created a series of books that are classics of the art of book production.

It was only about five years before publication of *Mushrooms, Russia and History* that the use of mushrooms as a sacrament began to fire Gordon's imagination and enthusiasm for experiential research. (One recalls his observation[2] that the psychoactive mushrooms "must have been a mighty springboard for primitive man's imagination.") The documentation of the use of *Amanita muscaria* by Siberian peoples in the early 1900s, as recorded by Russian ethnographers employed by the American Museum of Natural History, had already proved to be one among several important ethnographic resources for the Wassons. But then, beginning with letters received "almost in the same mail" in September 1952— one from the poet Robert Graves in Majorca and the other from Hans Mardersteig, a printer in Verona whose family published all of Gordon's books—the Wassons were alerted to the fact that sacred mushrooms were still being used by indigenous peoples in Mexico.[3] They went there to investigate a few months later, returning every year for several years. In the course of these visits and after, Gordon's field work with the Mazatec *curandera* María Sabina of Huautla de Jiménez would drastically change the lives of not only himself and this healer, but also the lives of the villagers in this tiny hamlet and of many Westerners who found their way south to Oaxaca and Huautla, where they participated in mushroom ceremonies.

Gordon himself first took the wondrous mushroom with María Sabina in June 1955. His initial experience with this woman, later followed by numerous other encounters with her, seemed to have been a powerful one. He continued to care about her fate. So it was that I, in the mid-1970s, went to Huautla at Gordon's insistence to visit her. His own health then was fragile, and he felt that it would prove to be too difficult for him to make the strenuous pilgrimage to her village. He had heard that she was ill and in trouble. At the time, I was working for the mythologist Joseph Campbell on his book *The Way of the Animal Powers.* The timing thus seemed right for the journey. I invited the composer John Watts, whose music seemed to model the mind in certain high states, to accompany me on this pilgrimage; Gordon and I both felt that it would be an asset to have John participate.

To assist us on our journey, Gordon wrote an introductory letter on Harvard stationary, and off we went. As any traveler knows, these letters mean little to nothing when you are "on the road." Most of the people you encounter cannot read any language, certainly not English, and they have not heard of Harvard. Yet this little piece of paper, when presented to the police on our way to Huautla, was our ticket to an archaic and puzzling world of politics and, ultimately, vision. The police were there to keep the "hippies" and curious out of the village. From the late '60s to the mid-'70s, Huautla had been besieged by seekers. The government had responded, for one reason or another, by deciding to close off the village to outsiders. So with some trepidation, at midnight, John and I presented the letter from Gordon—and "magically" passed through the gate.

Gordon's suspicions were confirmed by our visit. The old woman was living in abject poverty on the far side of her village. She had recently been assaulted by one of her daughters and now had neither food nor money. We found her in a desparate state. When she learned that we had been sent by Gordon—"Señor Wasson"—she brightened considerably and agreed to come to the house of our mutual friend Alvaro Estrada. There, safe from the rapacious villagers, she seemed to relax and told us the story of her extraordinary encounter with Gordon. It was clear that, although her life had been immensely complicated by the fame resulting from the publication of Gordon's accounts of his experiences with her, she felt that it was destiny that she and Gordon had met. She spoke of the years before her encounters with Gordon as having been characterized by great suffering. That was her life; and so it continued. She did not feel that Gordon was to blame for what was happening now. It was simply her fate. It was *their* fate.

Upon my return to New York, Gordon and I got together on many occasions to work on his book *The Wondrous Mushroom: Mycolatry in Mesoamerica* (1980). Often we met in the red leather environment of the Harvard Club, where our discussions ranged from the splitting of infinitives to the sacred ballcourt game. Behind his mild manner was a kind of freedom of mind and spirit that allowed Gordon to look into the hidden history of culture in a wild and nonconforming way.

In his books *Soma: Divine Mushroom of Immortality* (1968), which included a section by the Indologist Wendy Doniger O'Flaherty, and *The Road to Eleusis: Unveiling the Secret of the Mysteries* (1978), co-authored by Albert Hofmann and the Classic Greek scholar Carl A. P. Ruck, Gordon took the classical world by storm in suggesting an entheogenic basis for the Vedic and Greek mysteries. He also brought to light the contemporary use of *Amanita muscaria* in North America. And for several years before his death he was doing research on the cause of death of the Buddha. It was Gordon's contention that Buddha died not from pork poisoning but from mushroom poisoning, a theory he approached, as in all his work, historically and linguistically as well as botanically.

In truth, Gordon's work on the death of the Buddha was for him a way to complete his long journey through the world of the humble mushroom. He knew that he did not have long to live. For most of his life, the mushroom had been the foundation for a rich, rewarding life. Now it was time for the mushroom to teach him more about the ground of dying and death. What historical and religious figure could have been a more appropriate subject of study than Gautama Buddha for this last chapter of Gordon's life?

Many memories of Gordon make me smile with affection. For example, his good humor and deep tolerance stood by him on his first visit to Esalen, where he suddenly found himself in the midst of a gaggle of nude, young, female dancers on the dining deck. I also remember that his famous hospitality was unfailing on every count—even to the point that he did not live in his own house, but rather stayed in his nearby studio amidst the books and artifacts of a lifetime of research and collecting, leaving his house always free for his guests. One night I discovered that Gordon was actually sleeping on the studio's exposed back porch. He later told me that he always slept there—it was "better" for him, Gordon claimed. And I remember him as a mischievous dissembler in his grey suits and proper manner, delighted to do, say and think things less conservative types dared not.

Gordon Wasson opened a big door of perception for scholar and explorer alike. He created, or at any rate helped promulgate, the aforementioned mushroom conspiracy that ultimately so affected academia as well as the Haight. He and his close friends Weston La Barre, Richard Evans Schultes and Albert Hofmann together contributed much to our knowledge of ethnobotanical psychoactive substances. His accessible writings helped popularize their discoveries, forming the ground for the social revolution of the '60s. Gordon knew the power of his work. He was amused by, as well as respectful of, his part in the play. The mushroom always had a certain quality of secrecy and of conspiracy about it. And many have been the conspirators who entered the sanctum sanctorum of this secret through the door that Wasson opened.

Notes

1. This particular telling is from: Wasson, R.G. 1959. The hallucinogenic mushrooms of Mexico: an adventure in ethnomycological exploration. *Transactions of The New York Academy of Sciences,* Series II, February 1959; 21(4):325–339.
2. Ibid.
3. Wasson, V. P., and R. G. Wasson. 1958. The hallucinogenic mushrooms. *The Garden Journal* January–February 1958:1–6.

Ride Through the Sierra Mazateca in Search of the Magic Plant 'Ska María Pastora'

ALBERT HOFMANN[1]

R. Gordon Wasson, with whom I had maintained friendly relations since the investigations of the Mexican magic mushrooms, invited my wife and me to take part in an expedition to Mexico in the fall of 1962. The purpose of the journey was to search for another Mexican magic plant. Wasson had learned on his travels in the mountains of southern Mexico that the expressed juice of the leaves of a plant, which was called *hojas de la Pastora* or *hojas de María Pastora,* in Mazatec *ska Pastora* or *ska María Pastora* (leaves of the shepherdess or leaves of Mary the shepherdess), were used among the Mazatecs in medico-religious practices, in a way similar to the *teonanácatl* mushrooms and the *ololiuqui* seeds.

The question now was to ascertain from what sort of plant the "leaves of Mary the shepherdess" derived, and then to identify the plant botanically. We also hoped, if at all possible, to gather sufficient plant material to conduct a chemical investigation on the hallucinogenic principles it contained. On 26 September 1962, my wife and I accordingly flew to Mexico City, where we met Gordon Wasson. He had made all the necessary preparations for the expedition, so that in two days we had already set out on the next leg of the journey to the south. Mrs. Irmgard Weitlaner Johnson (widow of Jean B. Johnson, a pioneer of the ethnographic study of the Mexican magic mushrooms, killed in the Allied landing in North Africa) had joined us. Her father, Robert J. Weitlaner, had emigrated to Mexico from Austria and had likewise contributed to the rediscovery of the mushroom cult. Mrs. Johnson worked at the National Museum of Anthropology in Mexico City, as an expert on Indian textiles.

After a two-day journey in a spacious Land Rover, which took us over the plateau, along the snow-capped Popocatépetl, passing Puebla, down into the Valley of Orizaba with its magnificent tropical vegetation, then by ferry across the Popoloápan (Butterfly River), on through the former Aztec garrison Tuxtepec, we arrived at the starting point of our expedition, the Mazatec village of Jalapa de Díaz, lying on a hillside.

There we were in the midst of the environment and among the people that we would come to know in the succeeding two and a half weeks.

There was an uproar upon our arrival in the marketplace, the center of this village widely dispersed in the jungle. Old and young men, who had been squatting and standing around in the half-opened bars and shops, pressed suspiciously yet curiously about our Land Rover; they were mostly barefoot but all wore a *sombrero*. Women and girls were nowhere to be seen. One of the men gave us to understand that we should follow him. He led us to the local president, a fat *mestizo* who had his office in a one-story house with a corrugated tin roof. Gordon showed him our credentials from the civil authorities and from the military governor of Oaxaca, which explained that we had come here to carry out scientific investigations. The president, who probably could not read at all, was visibly impressed by the large-sized documents equipped with official seals. He had lodgings assigned to us in a spacious shed, in which we could place our air mattresses and sleeping bags.

I looked around the region somewhat. The ruins of a large church from colonial times, which must have once been very beautiful, rose almost ghost-like in the direction of an ascending slope at the side of the village square. Now I could also see women looking out of their huts, venturing to examine the strangers. In their long, white dresses, adorned with red borders, and with their long braids of blue-black hair, they offered a picturesque sight.

We were fed by an old Mazatec woman, who directed a young cook and two helpers. She lived in one of the typical Mazatec huts. These are simply rectangular structures with thatched gabled roofs and walls of wooden poles joined together, windowless, the chinks between the wooden poles offering sufficient opportunity to look out. In the middle of the hut, on the stamped clay floor, was an elevated, open fireplace, built up out of dried clay or made of stones. The smoke escaped through large openings in the walls under the two ends of the roof. Bast mats that lay in a corner or along the walls served as beds. The huts were shared with the domestic animals, as well as black swine, turkeys and chickens. There was roasted chicken to eat, black beans and also, in place of bread, *tortillas,* a kind of cornmeal pancake that is baked on the hot stone slab of the hearth. Beer and tequila, an *Agave* liquor, were served.

Next morning our troop formed for the ride through the Sierra Mazateca. Mules and guides were engaged from the horsekeeper of the village. Guadelupe, the Mazatec familiar with the route, took charge of guiding the lead animal. Gordon, Irmgard, my wife and I were stationed on our mules in the middle. Teodosio and Pedro, called Chico, two young fellows who trotted along barefoot beside the two mules laden with our baggage, brought up the rear.

It took some time to get accustomed to the hard wooden saddles. Then, however, this mode of locomotion proved to be the most ideal type of travel that I know of. The mules followed the leader, single file, at a steady pace. They required no direction at all by the rider. With surprising dexterity, they sought out the best spots along the almost impassable, partly rocky, partly marshy paths, which led through thickets and streams or onto precipitous slopes. Relieved of all travel cares, we could devote all our attention to the beauty of the landscape and the tropical vegetation. There were tropical forests with gigantic trees overgrown with twining plants, then again clearings with banana groves or coffee plantations, between light stands of trees, flowers at the edge of the path, over which wondrous

Albert Hofmann (*left*) with Gordon (c. late 1970s). *Masha Wasson Britten.*

butterflies bustled about.... We made our way upstream along the broad riverbed of Rio Santo Domingo, with brooding heat and steamy air, now steeply ascending, then again falling. During a short, violent tropical downpour, the long broad ponchos of oilcloth, with which Gordon had equipped us, proved quite useful. Our Indian guides had protected themselves from the cloudburst with gigantic, heart-shaped leaves that they nimbly chopped off at the edge of the path. Teodosio and Chico gave the impression of great, green hay ricks as they ran, covered with these leaves, beside their mules.

Shortly before nightfall we arrived at the first settlement, La Providencia ranch. The patron, Don Joaquin García, the head of a large family, welcomed us hospitably and full of dignity. It was impossible to determine how many children, in addition to the grown-ups and the domestic animals, were present in the large living room, feebly illuminated by the hearth fire alone.

Gordon and I placed our sleeping bags outdoors under the projecting roof. I awoke in the morning to find a pig grunting over my face.

After another day's journey on the backs of our worthy mules, we arrived at Ayautla, a Mazatec settlement spread across a hillside. En route, among the shrubbery, I had delighted in the blue calyxes of the magic morning glory *Ipomoea violacea,* the mother plant of the *ololiuqui* seeds. It grew wild there, whereas among us it is only found in the garden as an ornamental plant.

We remained in Ayautla for several days. We had lodging in the house of Doña Donata Sosa de García. Doña Donata was in charge of a large family, which included her ailing husband. In addition, she presided over the coffee cultivation of the region. The collection center for the freshly picked coffee beans was in an adjacent building. It was a lovely picture, the young Indian woman and girls returning home from the harvest toward evening, in their bright garments adorned with colored borders, the coffee sacks carried on their backs by head-bands. Doña Donata also managed a type of grocery store, in which her husband, Don Eduardo, stood behind the counter.

In the evening by candlelight, Doña Donata, who besides Mazatec also spoke Spanish, told us about life in the village; one tragedy or another had already struck nearly every one of the seemingly peaceful huts that lay surrounded by this paradisiacal scenery. A man who had murdered his wife, and who now sits in prison for life, had lived-in the house next door, which now stood empty. The husband of a daughter of Doña Donata, after an affair with another woman, was murdered out of jealousy. The president of Ayautla, a young bull of a *mestizo,* to whom we made our formal visit in the afternoon, never made the short walk from his hut to his "office" in the village hall (with the corrugated iron roof) unless accompanied by two heavily armed men. Because he exacted illegal taxes, he was afraid of being shot to death. Since no higher authority sees to justice in this remote region, people have recourse to self-defense of this type.

Thanks to Doña Donata's good connections, we received the first sample of the sought-after plant, some leaves of *hojas de la Pastora,* from an old woman. Since the flowers and roots were missing, however, this plant material was not suitable for botanical identification. Our efforts to obtain more precise information about the habitat of the plant and its use were also fruitless.

The continuation of our journey from Ayautla was delayed, as we had to wait until our boys could again bring back the mules that they had taken to pasture on the other side of Rio Santo Domingo, over the river swollen by intense downpours.

After a two-day ride, on which we had passed the night in the high mountain village of San Miguel-Huautla, we arrived at Rio Santiago. Here we were joined by Doña Herlinda Martinez Cid, a teacher from Huautla de Jiménez. She had ridden over on the invitation of Gordon Wasson, who had known her since his mushroom expeditions, and was to serve as our Mazatec and Spanish-speaking interpreter. Moreover, she could help us, through her numerous relatives scattered in the region, to pave the way to contacts with *curanderos* and *curanderas* who used the *hojas de la Pastora* in their practice. Because of our delayed arrival in Rio Santiago, Doña Herlinda, who was acquainted with the dangers of the region, had been apprehensive about us, fearing we might have plunged down a rocky path or been attacked by robbers.

Our next stop was in San José Tenango, a settlement lying deep in a valley, in the midst of tropical vegetation with orange and lemon trees and banana plantations. Here again was the typical village picture: in the center, a marketplace with a half-ruined church from the colonial period, with two or three stands, a general store, and shelters for horses and mules. We found lodging in a corrugated iron barracks, with the special luxury of a cement floor on which we could spread out our sleeping bags.

In the thick jungle on the mountainside we discovered a spring, whose magnificent fresh water in a natural rocky basin invited us to bathe. That was an unforgettable pleasure after days without opportunities to wash properly. In this grotto I saw a hummingbird for the first time in nature, a blue-green, metallic, iridescent gem, which whirred over great liana blossoms.

The desired contact with persons skilled in medicine came about thanks to the kindred connections of Doña Herlinda, beginning with the *curandero* Don Sabino. But he refused, for some reason, to receive us in a consultation and to question the leaves. From an old *curandera*, a venerable woman in a strikingly magnificent Mazatec garment, with the lovely name Natividad Rosa, we received a whole bunch of flowering specimens of the sought-after plant, but even she could not be prevailed upon to perform a ceremony with the leaves for us. Her excuse was that she was too old for the hardship of the magical trip; she could never cover the long distance to certain places: a spring where the wise women gather their powers, a lake on which the sparrows sing, and where objects get their names. Nor would Natividad Rosa tell us where she had gathered the leaves. They grew in a very, very distant forest valley. Wherever she dug up a plant, she put a coffee bean in the earth as thanks to the gods.

We now possessed ample plants with flowers and roots, which were suitable for botanical identification. It was apparently a representative of the genus *Salvia*, a relative of the well-known meadow sage.[2] The plants had blue flowers crowned with a white dome, which are arranged on a panicle 20 to 30 cm long, whose stem leaked blue.

Several days later, Natividad Rosa brought us a whole basket of leaves, for which she was paid 50 pesos. The business seemed to have been discussed, for two other women brought us further quantities of the leaves. As it was known that the expressed juice of the leaves is drunk in the ceremony, and this must therefore contain the active principle, the fresh leaves were crushed on a stone plate, squeezed out in a cloth, the juice diluted with alcohol as a preservative, and decanted into flasks in order to be studied later in Basel.[3] I was assisted in this work by an Indian girl, who was accustomed to dealing with the stone plate, the *metate*, on which the Indians since ancient times have ground their corn by hand.

On the day before the journey was to continue, having given up all hope of being able to attend a ceremony, we suddenly made another contact with a *curandera*, one who was ready "to serve us." A confidante of Herlinda's, who had produced this contact, led us after nightfall along a secret path to the hut of the *curandera*, lying solitary on the mountainside above the settlement. No one from the village was to see us or discover that we were received there. It was obviously a betrayal of sacred customs, worthy of punishment, to allow strangers, whites, to take part in this. That indeed had also been the real reason why the other healers whom we asked had refused to admit us to a leaf ceremony. Strange birdcalls from the darkness accompanied us on the ascent, and the barking of dogs was heard on all sides. The dogs had detected the strangers. The *curandera* Consuela García, a woman of some 40 years, barefoot like all Indian women in this region, timidly admitted us to her hut and immediately closed up the doorway with a heavy bar. She bid us lie down on the bast mats on the stamped mud floor. As Consuela spoke only Mazatec, Herlinda translated her instructions into Spanish for us. The *curandera* lit a candle on a table covered with some images of saints, along with a variety of rubbish. Then she began to bustle about busily, but in silence. All at once we heard peculiar noises and a rummaging in the room—did the hut harbor some hidden person whose shape and proportions could not be made out in the candlelight? Visibly disturbed, Consuela searched the room with the burning candle. It appeared to be merely rats, however, who were working their mischief. In a bowl the *curandera* now kindled *copal,* an incense-like resin, which soon filled the whole hut with its aroma. Then the magic potion was ceremoniously prepared. Consuela inquired which of us wished to drink of it with her. Gordon announced himself. Since I was suffering from a severe stomach upset at the time, I could not join in. My wife substituted for me. The *curandera* laid out six pairs of leaves for herself. She apportioned the same number to Gordon. Anita received three pairs. Like the mushrooms, the leaves are always dosed in pairs, a practice that, of course, has a magical significance. The leaves were crushed with the *metate,* then squeezed out through a fine sieve into a cup, and the *metate* and the contents of the sieve were rinsed with water. Finally, the filled cups were incensed over the *copal* vessel with much ceremony. Consuela asked Anita and Gordon, before she handed them their cups, whether they believed in the truth and the holiness of the ceremony. After they answered in the affirmative and the very bitter-tasting potion was solemnly imbibed, the candles were extinguished and, lying in darkness on the bast mats, we awaited the effects.

After some 20 minutes Anita whispered to me that she saw striking, brightly bordered images. Gordon also perceived the effect of the drug. The voice of the *curandera* sounded from the darkness, half speaking, half singing. Herlinda translated: did we believe in Christ's blood and the holiness of the rites? After our "creemos" ("we believe"), the ceremonial performance continued. The *curandera* lit the candles, moved them from the "altar table" onto the floor, sang and spoke prayers or magic formulas, placed the candles again under the images of the saints—then again silence and darkness. Thereupon the true consultation began. Consuela asked for our request. Gordon inquired after the health of his daughter, who immediately before his departure from New York had to be admitted prematurely to the hospital in expectation of a baby. He received the comforting information that mother and child were well. Then again came singing and prayer and manipulations with the candles on the "altar table" and on the floor, over the smoking basin.

When the ceremony was at an end, the *curandera* asked us to rest yet a while longer in prayer on our bast mats. Suddenly a thunderstorm burst out. Through the cracks of the beam walls, lightning flashed into the darkness of the hut, accompanied by violent thunderbolts, while a tropical downpour raged, beating on the roof. Consuela voiced apprehension that we would not be able to leave her house unseen in the darkness. But the thunderstorm let up before daybreak, and we went down the mountainside to our corrugated iron barracks, as noiselessly as possible by the light of flashlights, unnoticed by the villagers, but dogs again barked from all sides.

Participation in this ceremony was the climax of our expedition. It brought confirmation that the *hojas de la Pastora* were used by the Indians for the same purpose and in the same ceremonial milieu as *teonanácatl*, the sacred mushrooms. Now we also had authentic plant material, not only sufficient for botanical identification, but also for the planned chemical analysis. The inebriated state that Gordon Wasson and my wife had experienced with the *hojas* had been shallow and only of short duration, yet it had exhibited a distinctly hallucinogenic character.

On the morning after this eventful night we took leave of San José Tenango. The guide, Guadelupe, and the two fellows Teodosio and Pedro appeared before our barracks with the mules at the appointed time. Soon packed up and mounted, our little troop then moved uphill again, through the fertile landscape glittering in the sunlight from the night's thunderstorm. Returning by way of Santiago, toward evening we reached our last stop in Mazatec country, the capital Huautla de Jiménez.

From here on, the return trip to Mexico City was made by automobile. With a final supper in the Posada Rosaura, at the time the only inn in Huautla, we took leave of our Indian guides and of the worthy mules that had carried us so surefootedly and in such a pleasant way through the Sierra Mazateca. The Indians were paid off, and Teodosio, who also accepted payment for his chief in Jalapa de Díaz (where the animals were to be returned afterward), gave a receipt with his thumbprint colored by a ballpoint pen. We took up quarters in Doña Herlinda's house.

A day later we made our formal visit to the *curandera* María Sabina, a woman made famous by the Wassons' publications. It had been in her hut that Gordon Wasson became the first white man to taste of the sacred mushrooms, in the course of a nocturnal ceremony in the summer of 1955. Gordon and María Sabina greeted each other cordially, as old friends. The *curandera* lived out of the way, on the mountainside above Huautla. The house in which the historic session with Gordon Wasson had taken place had been burned, presumably because she had divulged the secret of *teonanácatl* to strangers. In the new hut in which we found ourselves, an incredible disorder prevailed, as had probably also prevailed in the old hut, in which half-naked children, hens and pigs bustled about. The old *curandera* had an intelligent face, exceptionally changeable in expression. She was obviously impressed when it was explained that we had managed to confine the spirit of the mushrooms in pills, and she at once declared herself ready to "serve us" with these, that is, to grant us a consultation. It was agreed that this should take place the coming night in the house of Doña Herlinda.

In the course of the day I took a stroll through Huautla de Jiménez, which led along a main street on the mountainside. Then I accompanied Gordon on his visit to the Instituto Nacional Indigenista. This governmental organization had the duty of studying and helping to solve the problems of the indigenous population, that is, the Indians. Its leader told us of the difficulties that the "coffee policy" had caused in the area at that time. The president of Huautla, in collaboration with the Instituto Nacional Indigenista had tried to eliminate middlemen in order to shape the coffee prices favorably for the producing Indians. His body was found, mutilated, the previous June.

Our stroll also took us past the cathedral, from which Gregorian chants resounded. Old Father Aragon, whom Gordon knew well from his earlier stays, invited us into the vestry for a glass of tequila.

A Mushroom Ceremony

As we returned home to Herlinda's house toward evening, María Sabina had already arrived there with a large company, her two lovely daughters, Apolonia and Aurora (two prospective *curanderas*), and a niece, all of whom brought children along with them. Whenever her child began to cry, Apolonia would offer her breast to it. The old *curandero* Don Aurelio also appeared, a mighty man, one-eyed, in a black-and-white patterned *serape* (cloak). Cacao and sweet pastry were served on the veranda. I was reminded of the report from an ancient chronicle which described how *chocolatl* was drunk before the ingestion of *teonanácatl*.

After the fall of darkness, we all proceeded into the room in which the ceremony would take place. It was then locked up—that is, the door was obstructed with the only bed available. Only an emergency exit into the back garden remained unlatched for absolute necessity. It was nearly midnight when the ceremony began. Until that time the whole party lay, in darkness sleeping or awaiting the night's events, on the bast mats spread on the floor. María Sabina threw a piece of *copal* on the embers of a brazier from time to time, whereby the stuffy air in the crowded room became somewhat bearable. I had explained to the *curandera* through Herlinda, who was again with the party as interpreter, that one pill contained the spirit of two pairs of mushrooms. (The pills contained 5.0 mg synthetic psilocybin apiece.)

When all was ready, María Sabina apportioned the pills in pairs among the grown-ups present. After solemn smoking, she herself took two pairs (corresponding to 20 mg psilocybin). She gave the same dose to Don Aurelio and her daughter Apolonia, who would also serve as *curandera*. Aurora received one pair, as did Gordon, while my wife and Irmgard got only one pill each.

One of the children, a girl of about 10, under the guidance of María Sabina, had prepared for me the juice of five pairs of fresh leaves of *hojas de la Pastora*. I wanted to experience this drug that I had been unable to try in San José Tenango. The potion was said to be especially active when prepared by an innocent child. The cup with the expressed juice was likewise incensed and conjured by María Sabina and Don Aurelio, before it was delivered to me.

All of these preparations and the following ceremony progressed in much the same way as the consultation with the *curandera* Consuela García in San José Tenango.

After the drug was apportioned and the candle on the "altar" was extinguished, we awaited the effects in the darkness.

Before a half hour had elapsed, the *curandera* murmured something; her daughter and Don Aurelio also became restless. Herlinda translated and explained to us what was wrong. María Sabina had said that the pills lacked the spirit of the mushrooms. I discussed the situation with Gordon, who lay beside me. For us it was clear that absorption of the active principle from the pills, which must first dissolve in the stomach, occurs more slowly than from the mushrooms, in which some of the active principle already becomes absorbed through the mucous membranes during chewing. But how could we give them a scientific explanation under such conditions? Rather than try to explain, we decided to act. We distributed more pills. Both *curanderas* and the *curandero* each received another pair. They had now each taken a total dosage of 30 mg psilocybin.

After about another quarter of an hour, the spirit of the pills did begin to yield its effects, which lasted until the crack of dawn. The daughters, and Don Aurelio with his deep bass voice, fervently answered the prayers and singing of the *curandera*. Blissful, yearning moans of Apolonia and Aurora, between singing and prayer, gave the impression that the religious experience of the young women in the drug inebriation was combined with sensual-sexual feelings.

In the middle of the ceremony María Sabina asked for our request. Gordon inquired again after the health of his daughter and grandchild. He received the same good information as from the *curandera* Consuela. Mother and child were in fact well when he returned home to New York. Obviously, however, this still represents no proof of the prophetic abilities of both *curanderas*.

Evidently as an effect of the *hojas,* I found myself for some time in a state of mental sensitivity and intense experience, which, however, was not accompanied by hallucinations. Anita, Irmgard, and Gordon experienced a euphoric condition of inebriation that was influenced by the strange, mystical atmosphere. My wife was impressed by the vision of very distinct strange line patterns.

She was astonished and perplexed, later, on discovering precisely the same images in the rich ornamentation over the altar in an old church near Puebla. That was on the return trip to Mexico City, when we visited churches from colonial times. These admirable churches offer great cultural and historical interest because the Indian artists and workmen who assisted in their construction smuggled in elements of Indian style. Klaus Thomas, in his book *Die künstlich gesteuerte Seele* [*The Artificially Steered Mind*] (Ferdinand Enke Verlag, Stuttgart, 1970), writes about the possible influence of visions from psilocybin inebriation on Mesoamerican Indian art: "Surely a cultural-historical comparison of the old and new creations of Indian art . . . must convince the unbiased spectator of the harmony with the images, forms and colors of a psilocybin inebriation." The Mexican character of the visions seen in my first experience with dried *Psilocybe mexicana* mushrooms and the drawing of Li Gelpke after a psilocybin inebriation could also point to such an association.

As we took leave of María Sabina and her clan at the crack of dawn, the *curandera* said that the pills had the same power as the mushrooms, that there was no difference. This was a confirmation from the most competent authority that the synthetic psilocybin is identical with the natural product. As a parting gift I let María Sabina have a vial of psilocybin pills. She radiantly explained to our interpreter Herlinda that she could now give consultations even in the season when no mushrooms grow.

How should we judge the conduct of María Sabina, the fact that she allowed strangers, white people, access to the secret ceremony, and let them try the sacred mushrooms?

To her credit it can be said that she had thereby opened the door to the exploration of the Mexican mushroom cult in its present form, and to the scientific, botanical and chemical investigation of the sacred mushrooms. Valuable active substances, psilocybin and psilocin, resulted. Without this assistance, the ancient knowledge and experience that was concealed in these secret practices would possibly, even probably, have disappeared without a trace, without having borne fruit, in the advancement of civilization.

From another standpoint, the conduct of this *curandera* can be regarded as a profanation of a sacred custom—even as a betrayal. Some of her countrymen were of this opinion, which was expressed in acts of revenge, including the burning of her house.

The profanation of the mushroom cult did not stop with the scientific investigations. The publication about the magic mushrooms unleashed an invasion of hippies and drug seekers into the Mazatec country, many of whom behaved badly, some even criminally. Another undesirable consequence was the beginning of true tourism in Huautla de Jiménez, whereby the originality of the place was eradicated.

Such statements and considerations are, for the most part, the concern of ethnographical research. Wherever researchers and scientists trace and elucidate the remains of ancient customs that are becoming rarer, their primitiveness is lost. This loss is only more or less counterbalanced when the outcome of the research results in a lasting cultural gain.

Addendum in Tribute to Gordon[4]

There is no question that a lasting gain resulted from Gordon's discoveries in Mexico. Benefits accrued to a number of scientific fields, though from Gordon's point of view the most important was ethnomycology. Of this he has written:

> I am repeatedly asked what the future holds for our entheogenic mushrooms and the other entheogens. My interest being in proto-history and Early Man, this question is beyond the scope of my inquiry. Whether chemists and therapists will find some lasting use for the extraordinary substances that we have discovered in these strange plants, I will not predict.[5]

No prediction is necessary. Several lasting uses have already been found. Without Gordon's investigations of the Mexican mushroom cult there probably would not have followed research on the botany and later on the chemistry of the mushrooms employed sacramentally by these cults. As a result of the chemical investigations, new psychotropic substances—psilocybin and psilocin—were isolated and synthesized.

These compounds have enriched the pharmacological inventarium. Their practical significance is clear and multidisciplinary. For example, because they are related in chemical structure to natural brain hormones that function as neurophysiological transmitters, psilocybin and psilocin have proven to be useful tools for brain research.

Such compounds furthermore have a pronounced psychological effect— psychedelic activity—that involves a deep influence on human consciousness.

Psilocybin (CY 39)

o-phosphoryl-4-hydroxy-N-dimethyl-tryptamine

Physical and Chemical Data:

Empirical formula: $C_{12}H_{17}O_4N_2P \cdot CH_3OH$

Molecular Weight: $284.3 \cdot 32$

Melting Point: 200 - 210° with decomposition

Solubility: In 120 parts of water at 20°C.

pH: 5.3

Keller Color Reaction:

    0.1 mg. of Psilocybin dissolve in 1 cc. 0.01% ferric

    chloride - containing acetic acid and stratify with 1 cc.

    concentrated sulfuric acid. After mixing a permanent

    violet color is obtained.

Sandoz Pharmaceuticals
Medical Department
Hanover, New Jersey
   October, 1958

Molecular structure and other scientific data on psilocybin as published by Hofmann in 1958. *Courtesy Wasson Collection.*

This can be used as a pharmacological aid in psychoanalysis and other forms of psychotherapy. Unfortunately, this application has been interrupted by the draconian laws of the health authorities, passed in response to misuse of these substances by recreational users in the "drug scene." It can be expected, however, that the psychopharmaceutical constituents of psychedelic mushrooms will again become available to medical professionals in the future.

On a personal note, a friend's doctor has prescribed for her an ethical drug called "Visken" that is connected with the hallucinogenic *Psilocybe* mushrooms. Its main use is to treat hypertension. One of my former co-workers, Dr. Franz Troxler, who had participated in the synthesis of psilocybin, helped develop this drug while involved in a subsequent project aimed at synthesizing β-receptor-inhibitors. These medicaments, sometimes called "beta blockers," are used to regulate cardiac function. Dr. Troxler found that 4-*hydroxy-indol,* which is the main component of psilocybin (4-phosphoryloxy-dimethyltryptamine), was the most appropriate starting material for a new β-receptor-inhibitor that now is marketed as Visken: isopropylamino-2-hydroxy-propyl-4-*hydroxy-indol.* Without the research on the hallucinogenic mushrooms, where 4-*hydroxy-indol* was discovered, Visken would not have originated.

A final and related question focuses specifically on Gordon's published findings: will the books on which he lavished so much care have a lasting significance along with the practical uses of the mushrooms he investigated? Gordon himself is on record expressing concern that his life and his accomplishments might prove ephemeral. Consider these lines from the memoir "postscriptum" he wrote for *That Gettysburg Address,* an essay by his father that Gordon had privately printed as a book and sent to his friends in 1965:

> Now that mother and father and brother are all gone, I alone remain to conjure up those memories, and in a few years, when I too will be gone, there will be but the pallid evocation of a memory among those few whose eyes chance then to fall on these lines. Here in a nutshell lies the philosophic problem of reality: nothing could have been more real to us than our family circle, yet this reality is on its way to become unreality.[6]

It is true that personal events, being the only true reality at the time for the person experiencing them, disappear when that person has died. I have elsewhere described my "transmitter-receiver" concept of reality that makes this inevitable.[7] But scientific findings published in books such as Gordon's or in scientific journals can be studied for thousands of years. True discoveries—and Gordon's are undoubtedly a case in point—may endure even longer. Because they are perennially valid, such findings will be "wirklich" = wirkend = having effects—that is, *real*—for as long as a human civilization exists. Gordon's writings are therefore assured of provoking much more than just the "pallid evocation of a memory." They are likely instead to continue exciting the interest, awe and abiding admiration of all future generations as they do even now among Gordon's contemporaries.

Notes

1. *Editor's note:* Dr. Hofmann's contribution begins with an excerpt from his book *LSD: My Problem Child* (New York: McGraw-Hill Book Co., 1980, pp. 127–144), reprinted here with his permission. The final paragraphs are rearranged, with some recast as endnotes and the rest introducing a special addendum he has written for this anthology. The addendum considers the practical significance of Gordon Wasson's discoveries for modern medicine, for Dr. Hofmann personally and for posterity.
2. The herbarium samples of *hoja de la Pastora* that we brought back with us were later identified by Carl Epling and Carlos D. Játiva at the Botanical Institute of Harvard University in Cambridge, Massachusetts. They found that this plant was a hitherto

undescribed species of *Salvia,* which was named *Salvia divinorum* by these authors.

3. The chemical investigation of the juice of the magic sage in the laboratory in Basel was unsuccessful. The psychoactive principle of this drug seems to be a rather unstable substance, since the juice prepared in Mexico and preserved with alcohol proved in self-experiments to be no longer active. Where the chemical nature of the active principle is concerned, the problem of the magic plant *ska María Pastora* still awaits a solution.

4. Specially written for this 1990 anthology.

5. From R. G. Wasson's *The Wondrous Mushroom: Mycolatry in Mesoamerica* (New York: McGraw-Hill Book Co., 1980, p. xxiii).

6. From p. 102 of R. G. Wasson's postscriptum in E. A. Wasson's *That Gettysburg Address* (privately published; printed in Verona by the Officina Bodoni, 1965).

7. See A. Hofmann's *LSD: My Problem Child,* pp. 195–209; *Einsichten Ausblicke: Essays,* Sphinx-Verlag, Basel, 1986; and "The Transmitter-Receiver Concept of Reality," in *ReVision* 1988; 10(4):5–11.

Martino Mardersteig (*left*) and Gordon inspect typeset proofs of one of Gordon's books at the Stamperia Valdonega in Verona (c. late 1970s or early '80s). *Masha Wasson Britten.*

Collecting Wasson

MICHAEL HOROWITZ

For students and collectors of psychoactive drug literature, R. Gordon Wasson's name is synonymous with thoroughly researched, boldly stated and eloquently written books. Many of these works were beautifully printed by his renowned friend Giovanni Mardersteig at the Stamperia Valdonega in Verona, then distributed in the United States and other countries as weighty, expensively-bound, deluxe limited editions that became instant collector's items.

Literary works concerning psychoactive substances are typically published in mass-market formats ranging from trade books to textbooks to pulp editions, or as "underground" street publications—but seldom in high-quality, limited editions such as Wasson's. The latter, one suspects, will eventually appear in ordinary, mass-market editions, enabling far more readers to possess these extraordinarily important works. Until then, first editions of Wasson's books—and even some of the subsequent trade editions—will be virtually unobtainable or priced beyond the reach of many collectors.

For example, the price of Wasson's initial "monograph of ethnomycology," *Mushrooms, Russia and History,* has hovered between $1,250 and $2,750 in the rare book market for the past 15 years. Co-authored with his wife Valentina Pavlovna, it was published in 1957 in an edition of 512 buckram-bound, boxed copies. When issued, the two-volume set cost $125, a very steep price for that year.

His collaborative two-part work with Roger Heim, *Les Champignons Hallucinogènes du Mexique* and *Nouvelles Investigations sur les Champignons Hallucinogènes,* published in 1958 and 1967 respectively, was for a long time available only from the Muséum National d'Histoire Naturelle in Paris and never in retail book stores. Antiquarian book dealers now sell the two books together for $300 to $400.

Wasson's most revolutionary work, *Soma: Divine Mushroom of Immortality,* with a section by Wendy Doniger O'Flaherty, was published with a half blue morocco binding in an edition of 680 boxed copies in 1968, at a cost of $200. (Such was the expense of this exquisite book that Alan Watts and Timothy Leary could afford to buy only one copy between them when it was first issued.) Today, this edition is worth about $750. A trade hardback and a trade paperback, both with fewer color plates, were subsequently published within a few years; the first of these, issued in a thin cardboard box, is now worth about $100.

Wasson's next book, *María Sabina and her Mazatec Mushroom Velada* (1974), lists George and Florence Cowan and Willard Rhodes as co-authors. The deluxe edition, limited to 250 copies initially costing $275, has a binding of quarter blue morocco with a Mazatec Indian motif on the cloth-covered boards. Four long-playing phonograph records of a complete *velada,* or magic mushroom ceremony, and a 79-page printed musical score were included in a separate cloth-covered box that came inserted in the book's double slipcase. The trade edition, issued at $82.50, comprises a clothbound book imprinted with the Mazatec motif, a separate cloth-covered box containing four cassette tapes in place of the records, and the same printed musical score—but no slipcase. Prices for the limited edition have ranged between $300 and $500 in recent years; for the trade edition, up to $150.

Wasson's next printed work, *The Road to Eleusis: Unveiling the Secret of the Mysteries* (1978), was co-authored with Albert Hofmann and Carl A. P. Ruck. The book does not state a limitation and cost only $12.95 when new, but the great difficulty of finding it in its full linen hardcover issue suggests a print run of about 1,000 copies.[1] It currently sells for about $75. The paperback issue is not nearly so difficult to acquire.

The Wondrous Mushroom: Mycolatry in Mesoamerica (1980), Wasson's next work, was issued in a quarter niger-bound edition limited to 501 boxed copies, all signed by the author on the colophon leaf. Initially costing about $200, it has since been offered by book sellers for as much as $525. The paperback edition is somewhat easier to find than his other paperback editions.

A posthumous work, *Persephone's Quest: Entheogens and the Origins of Religion,* was published in 1986. The limited edition lists the authors alphabetically: Stella Kramrisch, Jonathan Ott, Carl A. P. Ruck, R. Gordon Wasson. The trade edition, available through Yale University Press, lists Wasson first, followed by Kramrisch, Ott and Ruck. Release of the limited edition of 326 copies was delayed by Wasson's death in December 1986; it had not been widely distributed in the United States as of February 1989. The cost is therefore unknown at the time of this writing.

It is interesting to note that long before he started publishing his writings on ethnomycology, Wasson published a book about the history of the famous carbine rifle. *The Hall Carbine Affair* appeared in 1941. Only 100 copies were published, making this book not only his first but his rarest as well. Two subsequent editions came out in 1948 and 1971, the latter with a new introduction.

Wasson also arranged to have privately printed in 1965 a book by his father, E. A. Wasson, called *That Gettysburg Address.* This polemic against Lincoln's famous speech was limited to 225 copies and distributed only to friends. It includes a poignant and revealing autobiographical essay by Gordon in which he reflects on his childhood upbringing.

Having taken brief note of Wasson's remarkable series of lavishly-produced limited editions which he issued with painstaking scholarly and aesthetic care over a period of 30 years, it must be said with all due irony that the single most significant publication of his long and prolific career as writer-researcher-explorer-mind adventurer was an article published in a mass-market weekly magazine.

On the cover of the May 13, 1957 issue of *Life* magazine is a photograph of the actor Bert Lahr, who had recently starred in a production of Beckett's great existentialist play *Waiting for Godot*. Inside the magazine is Wasson's 17-page article "Seeking the Magic Mushroom," a riveting account of how he and his photographer Allan Richardson became the first white outsiders to eat hallucinogenic mushrooms in a Mexican Mazatec Indian ritual. It is accompanied with photographs of Wasson waiting for what we might call Godot's "little sisters"—the mushrooms—to take effect and provide an existential imperative. Wasson and Western Man had been waiting a long time to find out that the "magic mushrooms" of which 16th Century Spanish historians wrote did indeed exist, and that the universe was not absurd—or rather, that the universe is really so hyper-absurd as to render the concept of absurdity both profound and useless.

Wasson's Godot was the charismatic Mazatec Indian shaman María Sabina. She was the second of two extraordinary women he met from cultures far different than his own. (The first, of course, was his wife Valentina Pavlovna, a Russian emigre, pediatrician and mushroom-fancier who first interested Gordon in mycology. Any collection of her husband's works would be incomplete without a copy of her much-harder-to-find account, "I Ate the Sacred Mushroom," which was published in *This Week* magazine about the same time as his *Life* article.) The dramatic confrontations that lay at the heart of the widely-read books of Castenada and Lynn Andrews owe everything to the historic meeting of the American banker and María Sabina, the dirt-poor "wise-in-the-way-of-the-plants" woman, on the night of June 29–30, 1955. The infatuation of Western psychedelicists with Native American shamans from the '60s to the present can be traced to this meeting.

Published in the year of Sputnik, of Kerouac's *On the Road* and Leary's *Interpersonal Diagnosis of Personality*, Wasson's lengthy description in *Life* of his ingestion of six pairs of psilocybin mushrooms under the guidance of the *curandera*, enhanced with Richardson's stunning photographs, is arguably the best-written and surely the most widely-broadcast single drug experiment of the modern era. Hundreds of beat and hippie mushroom-seekers (including some leading rock stars, such as Mick Jagger, John Lennon and Peter Townshend) made pilgrimages to María Sabina's remote Oaxacan village to track her down and trip under her guidance. This proved disturbing to both her and Wasson. Possibly for this reason, among others, Wasson remained aloof from the youth counterculture. In a rare interview published in *High Times* in 1976, he described himself as "a loner," adding: "These things are delicate and I've wished to plow my own furrow."

There are a number of shorter monographs in the Harvard Botanical Series written by Wasson, including an interesting rejoinder to Professor Brough, who had disputed some of the assertions made in *Soma*, and a fine bibliography of works pertaining to hallucinogenic mushrooms. For many years these were obtainable for a modest price from the Harvard Botanical Museum, but some of them by now are out of print and bring $25 to $50 on the antiquarian market.

Wasson also contributed the first feature article in the seminal issue of *The Harvard Review* on "Drugs and the Mind," which appeared in the summer of 1963. This issue of the journal, which also contains the first printing of Leary's and Alpert's "Politics of Consciousness Expansion," is worth about $50; the second printing a bit less.

Among Wasson's least-known works is his foreword to the landmark *Phantastica* catalog of rare books (now worth about $25), compiled by this author in 1979 for the antiquarian book dealers William and Victoria Dailey. Here, Wasson referred to his theories regarding the *Amanita muscaria* mushroom and to his more recently-published collaborative work in which he, along with Hofmann and Ruck, proposed that a naturally-occurring form of LSD was the mind-altering potion used at the Eleusinian Mystery Rites. He also took the opportunity to wonder at the lack of scholarly research on the introduction of ethyl alcohol to the general population of Europe during the mid-16th Century, and he briefly discusses some of the titles in the catalog—notably, John Uri Lloyd's *Etidorhpa*.

Of great interest is a never-before-published paragraph written by Wasson for that foreword, on the subject of drug control laws. It was deemed inappropriate for publication in a book seller's catalog and suppressed. A reading of this paragraph now shows its remarkable timeliness. In today's confused and absurdist atmosphere of manufactured and blatantly political drug hysteria, voices are beginning to be raised for the heretofore unthinkable position of legalizing all drugs—a position Wasson stated with his usual aplomb in 1979. Here is the deleted paragraph from the *Phantastica* catalog, published in its entirety and for the first time anywhere:

If I had my way, I would make the psychoactive drugs (except alcoholic beverages) as cheap as possible. I would make them available in every drug store without prescription to anyone. (There could well be an excise tax on them, to help our public officials pay themselves the salaries that they think they deserve.) This would mean destroying overnight the incentives for the street pushers to push their stuff; the incentive for the middle men in the present traffic to be middle men; the incentives for the big shots who now play for heavy stakes with small risk; the incentives for the elaborate grids that now operate on the coasts of Colombia and Venezuela; the incentives for the fleets of vessels and planes that now ply between those countries and the Bahamas and our eastern seaboard; the incentives for the police and FBI agents that now pursue all the foregoing people, occasionally making big captures of smuggled plants and also of the smugglers, with exciting publicity for the 'good guys;' the incentives for the army of prosecutors and their staffs in federal and state employ who now give themselves over to hunting the 'bad guys;' the incentives to our adolescents to engage in a 'romantic' adventure for the kick. My 'reform' would make our newspapers that much duller. There would be no more appeal to the young to enlist in the game for big stakes and at high risk. I would have the Surgeon General issue a further statement to the youth of the country that the ingestion of the list of psychoactive products on the market is damaging to the health. I would make clear to those who will succumb to the drugs that not one cent of public money would be spent on their rehabilitation: our society has no time to busy themselves with these weaklings. The victims of these 'drugs' would have only themselves to deal with as best they could, with such help as their families or friends might make available to them. But my program would of course face the combined opposition of our Government enforcement agents, the lawyers and the men on the firing line, and the vested interests intent on pushing the drugs and importing fresh supplies, buying cheap and selling dear. Doubtless there is no formal alliance but the efforts of these disparate elements are buttressing a big and thoroughly corrupt industry just as in Prohibition days.

The foregoing is a rare instance of Wasson addressing in print the socio-political implications of drugs.[2] More typically, he pursued the spiritual and metaphysical importance of the ritual use of mind-altering plants throughout history to a greater extent than anyone before him.

An edition of Wasson's collected writings in an affordable, uniform set of volumes will someday be available to students of his work. But even then, the original editions of his books, because of their intellectual importance, beauty and rarity, will remain among the most coveted by collectors of 20th Century psychoactive drug literature.

Notes

1. *Editor's note:* In a letter to the editor dated 10 February 1989, Martino Mardersteig of the Stamperia Valdonega in Verona reports that the size of the first edition press run of *The Road to Eleusis* was 1,500 copies.
2. *Editor's note:* When informed by Michael Horowitz that the book sellers felt his paragraph would be somewhat out of place in their catalog, Gordon replied, in a letter dated 14 October 1979: "I am sorry. I was re-writing the paragraph to make it more biting." A copy of this letter and the first draft of his foreword to the catalog are on file in the Tina and Gordon Wasson Ethnomycological Collection at Harvard's Botanical Museum.

Irmgard Weitlaner Johnson (*right*) and Gordon pick through a mound of dried mushrooms in Mexico during 1959 expedition with Roger Heim. *Courtesy Wasson Collection.*

Remembrances of Things Past

IRMGARD WEITLANER JOHNSON

The Weitlaner-Johnson family's friendship and association with R. Gordon Wasson go back many, many years.

Long before we first learned about Mr. Wasson's quest in Mesoamerica for *teonanácatl*, the divine mushroom, we had visited and carried out ethnographic and linguistic studies in the Mazatec region. My first acquaintance with the practice of curing by witchcraft took place in the Mazatec town of Huautla de Jiménez, in the northeastern portion of the State of Oaxaca. There, in July 1938, a party consisting of Louise Lacaud, Bernard Bevan, Jean Bassett Johnson (my husband-to-be) and I witnessed for the first time a curing ceremony during which the *curandero* (healer), while under the influence of the hallucinogenic mushroom, divined the patient's illness; and it was the mushroom that gave instructions on how the sick person should be cured. I shall not here provide details of these events, since our data on witchcraft and the use of the hallucinogenic mushroom were subsequently published by Jean Bassett Johnson (1939a; 1939b) in Sweden and Mexico. Although our party witnessed this unusual mushroom rite, we did not take an active part in the ceremony itself.

Two years earlier, during the Easter week of 1936, my father Robert J. Weitlaner had spent several days in Huautla de Jiménez engaged in linguistic investigations. While there, he learned through his good friend Don José Dorantes, a well-known Mazatec merchant, of the existence and use of a mushroom in curative witchcraft and divination. My father obtained and sent some of these mushrooms to Dr. Blas Pablo Reko, who some time later mailed them to botanists in the United States for identification. This, then, was one of the reasons for our party's trip to the Mazatec region in 1938—namely, to continue the investigation started by Robert J. Weitlaner.

Unfortunately, my husband's promising career in anthropology and linguistics was cut short by his untimely death in 1944. Mr. Wasson's book *María Sabina and her Mazatec Mushroom Velada* (1974) is dedicated "To the Memory of Jean Bassett Johnson 1915–1944."

Little did the Weitlaners know that in another 15 years events would take a dramatic turn. Dr. Gordon F. Ekholm, of the American Museum of Natural History, an old friend of the family's, wrote to my father in 1953 about Gordon's great interest in the hallucinogenic mushrooms, and asked him to help the Wassons get started on their Mexican venture. Due to my father's experience and extensive knowledge of Mexican Indian cultures, he was invited to accompany the Wasson party to the Mazatec region of Huautla de Jiménez.

Thus began our long years of friendship, cooperation and travel to many Indian communities. From the very beginning there was a steady and continuous correspondence. This correspondence, which covered the period 1953 to 1986, the year of Gordon's death, is an impressive record of his diverse activities, his careful planning and organization of field trips, the preparation of data gathered in the field for publication, and his trips to Verona, Italy to supervise the printing of his beautiful books.

He kept in touch with a variety of well-known anthropologists, linguists, mycologists and botanists, writers, poets, artists and bank directors from many parts of the world—not forgetting all the crackpots who wished to meet Wasson—and he left succinct impressions regarding their personalities and activities.

Gordon's letters also describe his frequent trips to the Orient, Southeast Asia and India—always in search of information on the divine mushroom. There is mention, too, of his participation in congresses and symposiums, and he gives us news of the *homenajes* and honors received. He was greatly interested in a variety of seeds and plants having narcotic properties, such as the morning glory seeds (*Turbina corymbosa*), the *Salvia divinorum,* the *Quararibea funebris* or "Flor de cacao," and others.

A Memorable Trip to the Mixeria

Of the many field trips I took with Gordon, most of which have been amply documented in his books, I would like to describe some of the experiences we had in the Mixeria. They do not include such spectacular events as the Mazatec *velada* in Huautla de Jiménez, but they do show culture traits of a different ethnic group, having its own knowledge and usage of the hallucinogenic mushrooms.

On July 25, 1959 a party of friends, scientists and colleagues of Gordon's arrived in Zacatepec Mixes, Oaxaca. The group accompanying him included: Prof. Roger Heim, renowned French mycologist and director of the Muséum National d'Histoire Naturelle, Paris; Mr. Roger Cailleux, assistant to Prof. Heim; Prof. Guy Stresser-Péan, French anthropologist and specialist in the ethnography and archeology of the Huaxteca; Mr. Walter Miller, of the Summer Institute of Linguistics, a specialist in Mixe language and customs; Francisco ("Chico") Ortega, our faithful Zapotec guide from Tehuantepec, Oaxaca; Gordon's daughter Masha Wasson Britten; and myself.

This was quite a large group of "outsiders" to descend on the town of Zacatepec, located in the northwestern part of the Mixe region, in the District of Choapan. In those days it was an isolated, remote and conservative country. It is at an elevation of approximately 1400 m, and its mountainous terrain is largely covered with rain or cloud forests. According to Dr. Ralph Beals (1945, p. 5), the western Mixe live on the flanks of the enormous mountain knot of Zempoaltepetl, "The Twenty Mountains," nearly 5000 m high; this sacred mountain of the Mixe rises to the west of Zacatepec. To this day, Indians climb the mountain to perform ancient rites and to leave offerings in a cave near the summit. Gordon wished to obtain evidence that the hallucinogenic mushrooms were known and used in this part of the Mixe region.

The 25th of July is the Patron Saint's day in Zacatepec. As we arrived, great preparations were being made for the festival dedicated to Santiago. Indians from the surrounding villages were arriving in numbers. We were able to recognize Mixe women from Espíritu Santo Tamazulapan by their traditional dress, composed of a white *huipil* (a tunic-like garment), dark blue skirt and red *faja* (belt). There were women from San Cristóbal Chichicaxtepec dressed in white *huipiles* tucked into red skirts, held by red *fajas*. Women from Santa Maria Mixistlan came wearing their strange blue-green costumes and red *rodetes* (headdress). We heard musicians from San Juan Metaltepec playing the drum and the flute. The band of Zacatepec performed admirably; several Mixe towns are famous for their fine bands of musicians and are frequently invited to perform in the State Capital of Oaxaca.

Toward noon the great plaza of Zacatepec was full of Indians arranged in an orderly manner around the public square, to watch dances, and musical and sports events. Gordon moved around from one place to another, photographing people and activities. One very amusing shot illustrates a group of Chichicaxtepec women holding printed "programs" (handed out by the authorities) in their hands; amusing because they, being monolingual, certainly could not read Spanish.

Of considerable interest are the unique photographs taken of VIPs sitting at the *Mesa de Honor,* watching the festival. The most important personage was, without doubt, the famous (or infamous) cacique Luis Rodríguez—our host. It was quite an achievement for Gordon to have taken a rare photograph of Don Luis, sitting at the table wearing his dark glasses and keeping a sharp lookout at the people and at the things going on around him. The result was a perfect image of a ruthless personality. Joining the group around Don Luis were a smiling priest and the Presidente Municipal, who was related to Don Luis. Our party was also present; in fact, we were the guests of honor. In addition, Gordon was able to record for posterity some stupendous "portraits" of other members of Don Luis' entourage: attendants and bodyguards. There is one incredible photograph of a round-faced, dark-skinned man with the scar of a deep *machete* cut running clear across his left cheek; he, another of Don Luis' relatives, seemed the perfect example of a *matón* (killer).

When the dances and games were over, we were invited to a banquet held in the open-air space next to the church. An amazing amount of food and drink was served. Long lines of Indian women from Metaltepec, dressed in flowing white *huipiles* hanging over dark blue skirts, came bearing a variety of food on platters decorated with multi-colored tissue paper. Years later, a letter from Gordon dated May 21, 1983 reminisces: "Do you remember how fine the orchestra was? And how lavish the open-air lunch was with which he [Don Luis] honored us? Do you remember how he introduced the *comodidades* to the ladies of our group? And how the menu included tinned elvers, which I had never had before, which were delicious, but which should be everywhere illegal?" Gordon forgot to mention

Watching festivities during Patron Saint's day in Zapotec are (*left to right*) Guy Stresser-Péan, Irmgard Weitlaner Johnson and Roger Cailleaux (1959). R. Gordon Wasson.

Musicians from San Juan Metaltepec play drum and flute at the festival in Zapotec (1959). R. Gordon Wasson.

that, in addition to all this, we were served imported Scotch whiskey! The evening ended with marvelous fireworks.

A few more words about Luis Rodríguez. By all accounts, he was a ruthless cacique who ruled the entire northern Mixe region with an iron fist. Don Luis wanted to "modernize" the Mixe. Furthermore, it goes without saying, he wished to suppress all ancient beliefs and superstitions: witchcraft, divination and curing ceremonies, including knowledge and usage of the hallucinogenic mushrooms.

This situation posed some problems for Gordon. Even though he presented official letters of recommendation from the Governor of the State of Oaxaca to the authorities of Zacatepec, it was not certain that we would be allowed to pursue our investigations or, more importantly, receive help from Don Luis. Gordon had to employ all his diplomatic skills, stressing over and over that our mission was a scientific one. Fortunately, science won the day and we were able to proceed with our plans.

It was decided that our party should be divided into two groups. Gordon, Masha and I were to ride to the Mixe town of San Juan Cotzocón (1380 m), there to investigate the presence and usage of the narcotic mushroom. Prof. Heim and the remainder of the party were to search the surrounding area of Zacatepec for mushrooms.

Our trip to Cotzocón turned out to be a long day of riding up and down mountain trails, through rain forests and across a river, occasionally coming out of the mist for views of beautiful scenery and the mountain of La Malinche. Chico Ortega went along to take care of our horses and luggage.

Don Luis had given orders that the trail between Zacatepec and Cotzocón (approximately 30 km) should be cleared of fallen branches, weeds and undergrowth, stones, etc. He had also given orders that the "Municipal Building" in Cotzocón be swept and cleaned for our use. (There were no hotels in Zacatepec or Cotzocón.) Furthermore, he sent a supply of food and gave orders to the Indian women to prepare our meals. In short, we were well taken care of; we were under the "protection" of Don Luis. (However, Gordon insisted that we pay the women for their work for us.)

The cacique Don Luis Rodríguez sits between Roger Heim (*left*) and a Zapotec village priest (1959). *R. Gordon Wasson.*

A relative of Don Luis Rodríguez bore the scar of a deep machete cut across his left cheek (1959). *R. Gordon Wasson.*

Cotzocón is a smaller and more conservative village than Zacatepec. Women wear handwoven costumes different from those of other Mixe towns—i.e., white *huipiles* with red brocaded designs worn over red skirts, and brown head cloths.

The day after our arrival, we were able to contact two informants who were well-versed in traditional customs. One of them, Ladislao Reyes, a well-known *curandero*, turned out to be most helpful in supplying us with important information on the *tonalpohualli* (ancient Mesoamerican calendar), telling us about the pagan rites and showing us how to keep the count of the ritual calendar. His wife, Maria Antonia, was an expert on the use of the hallucinogenic mushroom. Gordon acquired much information from her on the use of mushrooms for divination but we were not invited to participate in a *velada*.

Toward evening of our last day in Cotzocón, we were surprised to discover that Ladislao Reyes possessed an old booklet containing the hand-written names of the 260-day Mixe ritual calendar. The *curandero* owned a second, more recent copy of this old document, which he gave to me. Since the older document included additional annotations made by Reyes, Gordon decided that we had to make a complete photographic documentation of this valuable manuscript. We spent our last evening photographing it, and Masha spent hours copying by hand many pages of the manuscript. It contains precious data concerning specific instructions as to how pagan rites were to be performed, and exactly what amount and kind of offerings were necessary. Mixe Indians are accustomed to consult the *curandero*, who knows how to cure and divine and how to conduct these rituals.

A few months after our return from the Mixe country, I received word that Don Luis was dead (perhaps killed). A letter from Gordon dated October 5, 1959, says: "Your news is hardly surprising. . . . I had heard from Walter [Miller] that fighting had broken out in Alotepec [a Mixe village near Zacatepec], but not that Don Luis had died."

On November 9, 1959, Gordon writes: "The clippings arrived. The goings-on in Alotepec are exciting and leave the reader wondering why the folk in that village hate the authorities so. I hear from Walter [that] Don Luis was reported by radio the first day to have been transported to Oaxaca with a wound!"

Every page of Ladislao Reyes' old booklet containing the hand-written names of the 260-day Mixe ritual calendar was photographed by Gordon in the field (1959). *R. Gordon Wasson.*

Years later (March 2, 1978) Gordon comments again on the matter: "[P]lease buy the book on the caciques of the Mixera by Inigo Laviada and send it to me. What a story! But there is one conspicuous fault in his story which you have pointed out: Don Luis did not die in his bed during the Second World War! How do you account for that? In the seven installments that you sent me ... the writer tells of his activities in the late '40s and the fifties, and has him die a natural death in 1959, and also at the very end during World War II. Do you know who Laviada is?"

Five years later (May 21, 1983), Gordon is still interested in the matter: "I have now finished Laviada's book. I am delighted that it will be in my library at Harvard: It has most interesting things to say. Have you read it? And to think that our host in Zacatepec was Luis Rodríguez! ... We thought he was shot when he left a plane (or helicopter) at Zacatepec, and took the same plane down to Oaxaca where he died two days later. Laviada says he died in his bed in Zacatepec uneventfully. How strange! Is Laviada right or wrong? Do you remember our trip down to Cotzocón and our stay in that little community? And the jail there?"

Gordon shared many other such memories in his voluminous correspondence with the Weitlaner-Johnson family over a period of more than 30 years. In these letters, Gordon also freely discussed his most significant achievements and interests, his important triumphs and disappointments, adding much to the record already preserved in his papers and books. As he himself would no doubt say: "What a story it makes!"

References

Johnson, J. B. 1939a. The elements of Mazatec witchcraft. *Ethnological Studies* (Gothenburg) 1939; (9):128–150.

Johnson, J. B. 1939b. Some notes on the Mazatec. *Revista Mexicana de Antropología* (Mexico) 1939; 3(2):142–156.

Beals, R. 1945. *Ethnology of the Western Mixe.* Berkeley and Los Angeles: University of California Press.

The People of Miniss Kitigan
Who Were and Are
Honor the Spirit of WaussungNaabe
Who Was and Is

KEEWAYDINOQUAY ET AL.

Introduction

This contribution was prepared by Keewaydinoquay and some members of the Miniss Kitigan band.[1] Like R. Gordon Wasson, many persons involved in the events described here have since "gone West" into a new cycle. When you speak peace upon the name of Wasson, say the same for Kenwabakesi, Wabakesi, Okiiyasin and Anungoday.

Keewaydinoquay notes: "One really needs to have lived Indian to appreciate just how it was at the time of the events described below. Dr. Wasson had indeed indicated his intended visit, but so had other scholars from 'Out There.' Once they discovered the travel difficulties in reaching us and/or the primitiveness of the living accommodations on Miniss Kitigan, most of them simply didn't arrive. We had also reasoned that, if this man were as important as he and others said he was, surely he would not be bothering with a band as insignificant as ourselves. For these and other reasons, we did not expect that Dr. Wasson would actually show up. At the time, none of us had much understanding at all of the nature of academic research. Certainly, we had never heard anyone speak with a vocabulary or diction such as his."

* * * * * *

We knew that Man-from-the-East was coming. Kaum of our island, making the weekly mail/grocery trip to a larger nearby island, heard about the arrival of someone who was quite different from the usual resorter-tourist. That arrival was witnessed by Kaum's aunt and cousin, who were cleaning the windows of the tiny waiting room at Peaine airport. (We all take much pride in that airport being

Keewaydinoquay and Gordon at a conference in San Francisco (1978). *Courtesy Keewaydinoquay.*

named for the last great chief of the Islands.) They noted the man's easy affluence, tailored clothes, leather luggage and distinguished accent—along with the decisive and concise manner with which he arranged for accommodation and transportation.

Most of our native folk speak/understand both English and "necessity" Ahnishinaabem,[2] so Kaum's aunt and cousin, being proud of their family ties as Framboises, a French heritage, have developed the habit of using French in conversations they wish to keep private.

"That one," Kaum's cousin had commented to his mother, "is accustomed to command."

"And to have his commands fulfilled," his mother added with a chuckle. Imagine their embarrassment when Man-from-the-East stepped up to the waiting room area and addressed them in flawless French! He inquired politely about possible boat conveyance to Miniss Kitigan. He then initiated casual conversation concerning the effect of weather on boat travel time, asked if they were acquainted with Keewaydinoquay of the Crane clan and inquired about the degree of their familiarity with local mushroom species. From this, they deduced something of what kind of person he might be and where he might be headed, and surmised possibilities of why he had come.

When Aunt and Cousin Framboise went home that evening, they found Kaum preparing to overnight in their backyard, and they told him what had happened. Kaum promptly re-rolled his sleeping bag, squeezed into a grocery-laden canoe, and paddled off for Indian Harbor on our island. He finished the trek across our island on foot in the moonlight.

Weather then and early the next morning was mild, so we calculated that one of the Kellys for hire would be bringing us company in a motor launch. We cleaned up the family encampment area and ourselves, banked the outdoor hearthfire to simmer under venison stew, set blueberry stackcakes to rise in the clay oven, and prepared a pot of coffee with maple sap—a treat we like to offer guests.

Our calculations were correct. At first, we were disappointed not to see the elegant clothing and luggage described to us. (Among our people, fine dressing is a mode of honoring; and *we* had anointed ourselves, parted and braided our children's hair, and put on our best shirts to welcome *him*.) This man wore thrift shop clothing and carried his belongings in a rope-tied cardboard box, just as we do. Even before he spoke, however, we came to realize he had done this to make us feel at ease, and our hearts warmed toward him, whatever might be his intention. There was no mistaking the demeanor Aunt and Cousin Framboise had noted, and Kaum, watching from behind the red osier bushes, declared it was the same man for sure.

His actual words of greeting were lost in the sound of breaking waves, but his body language was impeccable. He motioned to the earth of our homeland, then offered cigarettes and wrapped sweets, shaking hands all around, from the elders to the children. Our Elders were impressed; this man knew the proprieties and performed them gracefully. Comfortably ensconced beside the encampment hearthside, Man-from-the-East accepted and held the Speaking Stick easily, as if he might have been a leading tribal speaker for many moons—as indeed he had been, though we had not heard of either the Morgan or Ethnomycology tribes before!

He told us that his name was GUERDON WAUSSUNG (so it sounded to us), which made a considerable ripple in camp both then and among some Ojibwe later, for a good name with cryptic meanings is much respected. Someone rushed off to surreptitiously check out his first name in the community's only large dictionary, but the family name needed no such investigation, as names derived from WAUSSEINE are highly honored among us. We were amazed that he should entrust us at the very beginning with the significance of his true name. We told ourselves this showed the importance he placed upon our relationship. (R. Gordon Wasson is still spoken of by us as WAUSSUNG NAABE, meaning "Shining-from-Afar-Man.")

WaussungNaabe's next declaration was certainly not an expected approach. After looking about the group to consolidate everyone's attention, he intoned importantly, "I AM A FRIEND OF CLAUDE LEVI-STRAUSS!" There followed a brief silence as we struggled to understand the meaning of this announcement.

One of us inquired, "This man of whom you speak . . . He is your *Very Good Friend?*"

"Yes, indeed," replied WaussungNaabe, very definitely and very happily. We could tell, from the manner in which he had produced this statement, that he believed this should just about take care of anything and everything. We had thought the Shining One's own name beautiful, but *this* name . . . this name was almost ugly, and furthermore highly unlikely. It wouldn't do to say so, however,

143

and some kind of response was obviously being expected. There had been, in the olden days of the Ahnishinaabeg,[3] a custom of friendship contract between men which the WhiteEyes had translated as "blood brother." It must have been just such a friendship of which WaussungNaabe spoke, we decided. That he should reveal at the outset even the name of his blood brother seemed to eloquently demonstrate his true intent and intellectual honesty. Time later proved that our experiences prior to his visit had not enabled us to comprehend Waussung-Naabe's intent at all, but his intellectual honesty we had assessed most exactly.

One of our elders responded to the unspoken expectation that hung in the air. He gave a short homily on the responsibilities of true friendship, then congratulated our guest on having such a friend and commended him for speaking of such friendship as a primary asset. Now it was WaussungNaabe's turn to cogitate for a moment in stunned silence!

During the subsequent years of our association with WaussungNaabe, we were slowly made aware of the academic world: its welter of purposes, stresses of competition and difficulties of achievement. Previously, we had not even imagined such a world. Our own wisdom is not confined in volumes. It is parcelled out ceremoniously to reliable "keepers" who are expected both to commit it to memory and to exemplify its use when needed by The People. But what if the Keepers should be careless or pass over before the knowledge is given to successors? We are, unfortunately, acutely aware of such losses.

The exquisitely created books WaussungNaabe showed us were obviously superior to any library books we had seen. We greatly admired them, for they reminded us of our own hieroglyphic scrolls and woodland quill designs. We thought they were, like our own Atiisokanek literature, spirits of beauty and wisdom in their own right. It was the academic process by which they came into being that frightened us; the danger of their content being slanted, or even misinterpreted completely, seemed formidable. WaussungNaabe emphasized this with weird tales from his own experiences. Thus we pondered the whiteman's process of preserving knowledge, glad that its responsibilities were so removed from us. (For so we still believed at the time we first encountered Waussung-Naabe, being completely unable to stretch our minds to the possibility of the Miniss Kitigan Press as it is today.) Eventually, our limited acquaintance with academia turned out to be very useful—especially to Grandmother Kee, who had been considering further formal education and went on to pursue it. Then, in rapid succession, it became valuable as well to Miniss Kitigan Press and to our young people, who discovered direct relationships to their revelations in the Vision Quest. We thank WaussungNaabe for the introduction.

While visiting us, WaussungNaabe expressed boyish exuberance over blueberry stackcakes and the band's storytelling sessions. (Later, when he could no longer come to us, we tried sending these to him. In a very real sense, he became "our" anthropologist.) The Shining-from-Afar-Man found as much delight in retelling his own life's adventures as he did in collecting a new Ahnishinaabeg yarn. His account of stopping an ocean-going freighter near the Peloponnesus and hitchhiking a ride as star passenger bids fair to becoming one of our favorite oral tales!

There is no doubt that Dr. Wasson's appreciation of antiquities engendered or re-enhanced our appreciation of our own antiquities. He would stop the whole working of a project to demand the recording of a particular chant or story that took his fancy.

"But that isn't about mushrooms or the Miskwedo (fly-agaric) at all!," someone—who really did want to help—would object.

"True, true," he would mollify the complainer. "But it is sheer poetry of life."

In this manner such items as Honor March of Bears and the biography and spirit song of Hispahkwe have been preserved.

We feel mention must be made of one very special trait of this man. There is a word for it in our language, but we search in vain for one adequate in English. He possessed a true zest for life, BIMADISIWIN, finely modulated by a discernment of the appropriate, developed beyond an art. It was almost spiritual. We have deeply appreciated that sensitive quality. If ever he needed to press beyond this point, it was with full awareness of benefit for the cause to which he had dedicated himself. Often we shared the same dedication, but not always. Our diverse cultures brought out differing attitudes toward certain matters, but we could still understand each other very well.

The warmth of kindly respect continues to flow between our spirits.

Notes

1. *Editor's note:* Keewaydinoquay is described in R. Gordon Wasson's book *The Wondrous Mushroom: Mycolatry in Mesoamerica* (New York: McGraw-Hill, 1980, p. 228) as a "remarkable Ojibway herbalist and shaman" who lives on a lovely island in one of the Great Lakes, an island that had once belonged to her ancestors. Miniss Kitigan is the native language name for this island.

2. *Editor's note:* Ahnishinaabem is the native language of the Ahnishinaabeg people; see following note.

3. *Editor's note:* In her monograph *Puhpohwee for the People: A Narrative Account of Some Uses of Fungi Among the Ahnishinaubeg* (Cambridge, Massachusetts: Botanical Museum of Harvard University, February 1978) Keewaydinoquay asserts that "a people has a right to be called by the name that they themselves have always used." For this reason she prefers the native language term *Ahnishinaabeg* (var. *Ahnishinaubeg*), meaning "The-People-Who-Came-from-the-Place-Beyond-Where-the-Sun-Rises," instead of the French *Ojibwé* and its English adaptation *Ojibway*.

Left to right: Richard Evans Schultes, Gordon, Albert Hofmann and Weston La Barre at a conference in San Francisco (1978). *Courtesy Weston La Barre.*

My Friend Gordon

WESTON LA BARRE

Gordon Wasson was one of the most remarkable and admirable men I have known. Our friendship began with a common interest in psychotropic plants. But during my visits to his home in Danbury, Connecticut, and when he visited me at my Duke Forest country place, our conversations developed into long and learned discussions concerning many subjects.

We sometimes disagreed on picayune philological points, such as whether the root *teo-* in the Nahuatl word *teonanácatl* means "god." I also contested his neologism *entheogen,* which Gordon proposed as a substitute for words such as *hallucinogen.* He was a very good writer and taught me, in the course of our discussions, to avoid the common misuse of "prestigious." I am not sure, but he may also have helped me discriminate between "exquisite" and "exquisitive."

A past master in the art of making friends, he was also a practiced and easy host who was always solicitous of his guests. Yet he made them feel more at home by working them into his daily routine. I remember one particular Sunday morning when he took me to a neighbor's swimming pool, there to continue our discussion intermittently at poolside as he read *The New York Times.* On other occasions he repaired to his garden library to work, while I read happily in the main house.

Gordon seemed to have an endless stream of interesting visitors. He also employed a superb cook. An interesting anecdote involves her indirectly. Back in 1944, there came to Lord Mountbatten's South East Asia Command in Kandy, Ceylon (now Sri Lanka), a group of young *maquis* from France. They were by then too well known by the occupying forces in their country to be of any further use back home as underground resistance fighters. However, it was felt that they might serve a useful role in former French Indochina (later Viet Nam). For this reason they came to Ceylon, where I was stationed. We young Naval officers who spoke French were given time off to entertain the *maquis,* whom we took to the Pointe de Galle for a swim in the India Ocean. All of us had a great time. Many years later, while chatting with the French husband of the cook at Gordon's house, I learned that he too had been in Ceylon for a time during the war—and, sure enough, he remembered some American Navy officers who had taken him swimming! This only goes to show the kind of interesting people Gordon gathered

147

around him, when even the houseman had experienced a colorful career!

Gordon was a broadly cultivated man, interested in an amazing variety of subjects, even recondite scholarly issues. He had many really brilliant *aperçus*—for example, his use of the well-known Aztec statue of Xochipilli, "Prince of Flowers," as a "cultural Rosetta Stone" with which he was able to demonstrate that each one of the carvings on the god's body represents a known hallucinogenic plant of the Aztec. "Flower" is a metaphor for hallucinogens in Nahuatl poetry, and of course Wasson knew this fact too and could put the poetry and sculpture together into a penetrating insight. He was one of the first to identify *teonanácatl,* "flesh of the gods," with various Mexican psilocybin mushrooms, and the first to follow up this lead with field work involving the Mazatec shamaness María Sabina, who then was still using these substances. Equally brilliant was his surmise that the wild transport in the ancient Greek Mysteries came from an LSD-like substance produced by a fungus infecting the barley or rye that was ingested, as a sacramental brew, in the Eleusinian ritual. Surely he deserved the coinage of the term "ethnomycologist" for himself!

My relation to this fine mind began in an unalloyed admiration for his identifying of the lost Vedic hallucinogen, Soma. As an anthropologist long interested in Orientalist studies, I am concerned that non-specialists recognize the true magnitude of Gordon's achievement. Sanskrit has been known to Western scholars since the 18th Century. The Rig Veda, one of the four sacred texts of Brahmins, is in fact the oldest written record of any Indo-European language, dating back perhaps to 1500 B.C. and some portions in oral form to 1800 B.C. It consists principally of invocations to Soma, a powerful hallucinogen evidently of plant origin. But when the Aryans entered India through the Hindu Kush, they lost identifying knowledge of the sacred plant, which does not grow south of the Himalayas.

Native Hindu and European scholars alike long wrestled in vain with the problem of the identity of Soma. Thus an apparently insoluble riddle lay at the heart of Vedantic studies. The entire ascetic mysticism of India and practice of yoga was a somewhat mechanical attempt to recapture the visions excited by Soma, which consequently underlies the whole of Hindu religion. Soma is ultimately the cognate of the ambrosia that the Greek gods ate to maintain their immortality. Wasson's willingness and ability to pursue the fascinating riddle through the wide range of linguistics, archeology, folklore, philology, prehistory, ethnobotany, plant ecology, human physiology and ethnography was the basis of his success. Wasson straightforwardly took the many allusions to Soma literally and found that the composite description exactly fitted *Amanita muscaria* (Fr. ex L.) Quel., a hallucinogenic mushroom still in use among Siberian tribes in modern times. That his solution is unquestionably correct—despite the miffed reluctance of some professional Sanskritists to follow him—is for me incontestably proven by the many doors to understanding opened by his discovery.[1] In my opinion, relevant American Indian data show Mesolithic horizons for the practice. Small wonder that Wasson's discovery has exhilarated some anthropologists! I draw from my earlier characterization in a professional review:[2]

The cautious scholarship and careful method of Mr. Wasson—a retired banker, formerly a partner in J. P. Morgan & Co.—may cause us to forget that as an ethnomycologist he is also an autodidact. This leads to an interesting professional question. Mr. Wasson is fully professional in scholarly behavior, and moreover in several diverse fields. An amateur affection for data, pursued with such zeal and acumen, achieves here results really as good as the best professional work.

But there are evidently autodidacts and autodidacts. One at least, L. M. Klauber (professionally an engineer), in his magnificent two-volume monograph on *Rattlesnakes,* becomes manifestly the world authority on his subject; Klauber, indeed, is no less than masterly in his handling of the voluminous folkloristic and complex ethnographic data on snakes. Others, like Robert Ardrey (a dramatist and freelance journalist), cut no such professional swathe on the nature of man. Professionals are quite unimpressed with the democratic vote of laymen as a criterion of 'success.' But professionals constitute no impenetrable Establishment. They merely respond with approval (or dismay) to the relative mastery of information and to astuteness in assessing evidence. On this score some experts may even disdain professionals like Thor Heyerdahl, Desmond Morris, and J. B. Rhine.

There is also the matter of didactic preparation, autodidact or not, for the intellectual tasks one sets himself. Benjamin Whorf, for example, was an insurance man, but he studied with the world's peer linguist, Edward Sapir, in earning his laurels as a linguist. Gordon Wasson, who later in life learned from such distinguished teachers as the mycologist Heim, the linguist Jacobson, and from Harvard's Schultes, the world's ranking authority on the ethnobotany of psychotropic plants, evidently belongs to the high company of the Klaubers and the Whorfs as a distinguished scholar in his own right. Wasson's research on Soma is a thoroughly remarkable scholarly achievement, judged even by the most stringent standards.

This book is on all accounts a magnificent one. The volume is in blue half-leather stamped in gold, and slip-covered in fine blue linen cloth. It is designed by Giovanni Mardersteig and printed in Dante type by the Stamperia Valdonega in Verona, on paper handmade by the Fratelli Magnani, with numerous color plates tipped in by hand, in an edition of 680 copies, of which 250 are available for sale in this country. This connoisseur elegance in bookmaking is an appropriate setting for the stature of the problem attacked and for the scholarly quality of its solution.

One might add that this perfectionism of Gordon Wasson was manifest in all things—in his intellect, his friendships, his way of life. This is the way we remember him.

Notes

1. For some of these related matters, see Weston La Barre's "Soma: The Three-and-One-Half Millennia Mystery," in *Culture in Context: Selected Writings of Weston La Barre* (Durham: Duke University Press, 1980, pp. 108–115).
2. La Barre, W., review of Wasson's *Soma,* in *American Anthropologist* 1970(72):368–373. Also see pp. 114–115 in the article cited above from *Culture in Context.*

Frank Lipp with Gordon at Gordon's home in Danbury. *Courtesy Frank Lipp.*

Mixe Concepts and Uses of Entheogenic Mushrooms

FRANK J. LIPP

Although R. Gordon Wasson's ethnomycological studies encompassed a variety of cultures in India, Siberia, Greece and other areas, his most intensive and prolonged research pertained to the sacred mushrooms of Mesoamerica. This endeavor culminated in the 1974 publication of his *María Sabina and her Mazatec Mushroom Velada*, a work which stands out as the most meticulous, complete and aesthetically pleasing rendition of a shamanic curing ceremony yet recorded. In this, as in his Mesoamerican research generally, his primary focus was on the Mazatecs of Oaxaca. However, Wasson conducted ethnobotanical research on the use of entheogenic plants among several other Mesoamerican Indian groups as well. In their pioneering work, *Mushrooms, Russia and History* (1957), he and his wife Valentina described the use of sacred mushrooms by the Mixe. However, since the two trips he ultimately made to the Mixe were brief, Wasson's research on these Indians and their practices using the sacred mushrooms was not as thorough as he would have liked. I therefore here present, in memory to a consummate scholar and friend, a more detailed study of this subject. It is based on my ethnographic research in the Mixe region from April 1978 through October 1979, and during brief periods in 1980, 1984 and 1987. Most of this research was centered in four villages, but primarily in Mazatlán de los Mixes.

Mixe terms presented in the following discussion are pronounced with three degrees of vowel length: short (a), mid (a·) and long (a:). There are also three types of glottalization: checked (v'), interrupted (v'') and aspirated (vH).

Entheogenic Plants and the Mixe

The Mixe, numbering 76,000, occupy the southeastern highlands of the State of Oaxaca.[1] They sustain themselves primarily by slash-and-burn agriculture, by raising and selling coffee and by trading.

Sacred plants utilized by the Mixe for divinatory and medicinal purposes include the *Na·shwi·ñ mush* (*Psilocybe* spp.), *ma''zhuṇ paHk* (*Turbina corym-*

bosa/Ipomoea violacea), ?ama'y mushtak (Datura stramonium), po:b pɨH (Brugmansia candida) and koba·c pɨH (Tagetes erecta).[2] Ipomoea violacea, also referred to as pɨH pu'ucte·sh, "broken plate flower," is considered the more powerful of the two con-volvulaceous plants used. Both are used for divinatory purposes throughout the Mixe region, as are the mushrooms. The use of Tagetes and Brugmansia is limited to two or three villages each. And D. stramonium, which has as one of its names ?ama'y ?u·'c, "dangerous plant," is restricted in use throughout the region because of its toxicity and propensity to produce violent mental states.

The Na·shwi·ñ mush, "mushrooms of the Earth," are pi·'tpa, "spindle whorl," or 'Ene· tɨ'ɨc, "Thunder's teeth" (Psilocybe mexicana, P. cordispora); atka:t, "judge" (P. hoogshagensis); and ko:ng, "lord, governor" (P. caerulescens). Since size and smell are major distinguishing criteria, another species utilized, Psilocybe yungensis, is classified as either pi·'tpa or atka:t, depending upon the plant's size.

Other than different methods of preparation, these plants have identical ritual proscriptions, attendant religious beliefs and physical effects. Therefore, in the following discussion which primarily concerns the sacred mushrooms, it should be kept in mind that the same holds true for the convolvulaceous seeds.

The Mixe consider the mushrooms extremely wise. This is because they spring from the Earth, which is all-knowing of the past and present affairs of man, and furthermore are said to be "born" from the bones of ancient sages and prophet-kings. Related to the latter notion is the belief that only persons with hollow bones are capable of becoming diviners or obtaining successful results when taking the mushrooms.

Additionally, the Earth mushrooms are considered to be soothsayers, being equated with the blood of Christ. The Mixe believe that when Jesus was on the cross, blood flowed from His heart to the ground. From it issued numerous flowers and many kinds of edible mushrooms. All but the last of these miraculous plants then disappeared. Those that remained are the Na·shwi·ñ mush.[3]

Some individuals consult the sacred plants for divinatory purposes such as divining the cause of an illness, death or affliction of a family member; the loca-tion of a lost object; the identity of a thief, sorcerer or vandal; and the resolution of a personal or family problem. The sacred plants also disclose the location of hidden treasures, ruins and ritual knowledge. They normally speak in Mixe. But sometimes they speak in Zapotec, in which case, although they might provide the desired information, the Mixe are unable to understand it. This is said to occur with the pi·'tpa but not with the ko:ng, and also with ma''zhuṇ paHk seeds obtained from outside the Mixe region.

The mushrooms and morning glory seeds are ingested for a number of ail-ments, including gastrointestinal disorders, migraine headaches, trauma, swellings, bone fractures and seizures, as well as for chronic and acute illnesses. As a poultice, the seeds of Datura stramonium are applied externally to relieve the pain of injuries, headaches, toothaches, bone fractures, burns and fevers. Datura and Tagetes seeds are also reputedly placed covertly in peoples' food in order to harm them.

It is said among the Mixe that Western medicine is able to alleviate only the symptoms of illness, since it goes into the flesh and blood, whereas the sacred plants are thought to penetrate to the source of illness, the bones. Before the advent of Western medicine, these plants were the only potent medicines available to the Mixe, and a few still assert that they have no need for Western medicine since they possess these plants.

Although they do not use them for every illness and readily admit that these plants are unable to cure all illnesses, many Mixe hold them in high esteem. Nonetheless, it is safe to say that a majority of the Mixe population do not utilize these plants. This is largely due to the adverse reactions associated with them and the plants' ability to "speak," which is regarded by some as uncanny and diabolic. Moreover, these plants are reputed to leave adults and children who take them with poor mental performance or permanently deranged. Since the seasonal appearance of the mushrooms coincides with a number of annual village festivals, some Mixe have never taken them: the combination of alcohol ingested at the festivals and the mushrooms results in acute adverse reactions.

Villagers deny that the "curers" who use these plants for healing have more knowledge about them than the general populace. Such knowledge is widespread, and to solicit a shaman to consume one of the entheogenic plants on behalf of an individual is infrequent. Personal use is preferred, since the mental condition and thought processes of each individual are considered distinctive. Some feel this is better than to seek a cure from a shaman.

Some curers obtain knowledge concerning medicinal plants and plants designated for curing rituals heuristically, by taking *Psilocybe* mushrooms or *Turbina corymbosa* seeds. These curers, primarily women, maintain a close, affective relationship with the mind-altering plants and diagnose illnesses by ingesting, or having their clients ingest, these plants. Once the plant drug has taken effect, the individual begins a lively dialogue with the spirits of the plant as to the cause of and cure for the illness. However, most Mixe shamans do not use hallucinogenic plants, except sporadically and personally, to obtain "knowledge" or in their medical practices. One curer, after receiving initial instructions from the mushrooms, never had recourse to them again.

Since the mushrooms grow only during the summer rainy season, the convolvulaceous seeds are taken during the rest of the year. The growth of the mushrooms is closely associated with the growth of maize, since the mushrooms, ostensibly "planted" by termites (*we·c* or *wa·yñ*), grow when the maize ears are ripening in June, becoming "mute" and dangerous when the maize is tasseling during a dry, hot spell called *?ambɨshɨ:k* (22 July to 28 August).

Varying cultural practices within the Mixe region designate different days as propitious for taking the sacred plants. Some ritual specialists and a calendar priest said they could be taken any time, while other curers advised taking them only during good calendar days.[4]

The mushrooms are said to grow only in sacred soil. When they are encountered in the countryside, the devout light three candles inserted in the soil, kneel and recite a prayer such as the following:

> *Tu·m ?uh* ["one world"]. Thou who art the queen of all there is and who was placed here as the healer of all sicknesses. I say to you that I will carry you from this place to heal the sickness I have in my house, for you were named as a great being of the Earth. Forgive this molestation, for I am carrying you to the place where the sick person is, so that you make clear what the suffering is that has

come to pass. I respect you. You are the Master of all and you reveal all to the sick.

The mushrooms are handled very carefully, with reverence and respect. They are placed in a gourd cup and put on the house altar where copal incense smoke is passed over them, or taken to the church and left there for three days. Since not all know what they look like or where they grow, the mushrooms are also purchased in the village and neighboring communities. They may be sun-dried for later use, to be revived in water before consuming.

During a three-day period before the mushrooms are taken, the individual abstains from sexual intercourse and is not to consume any fowl or pig meat, mescal, eggs or vegetables. Although eggs and fowl meat play an important role in Mixe religious behavior, they are proscribed as dietary items, not for symbolic reasons. Any kind of drug or pill is also prohibited, since they interact negatively with the sacred plant. The person should rest and avoid any agricultural labor. On the morning of the fourth day, he/she takes a bath and consumes a breakfast of some maize bread and gruel; after noon, nothing is eaten. The day after taking the mushrooms, a good amount of chili peppers are eaten. Meat or alcohol is proscribed for one month.

Mushroom dosage for women is seven pair; for men, nine pair. Certain individuals take 12 or 13 pair of *pi·'tpa*, six of *atka:t* and two of *ko:ng*. Dosage is also approximated by the amount of alcohol an individual can drink before becoming inebriated. The *ko:ng* can be taken by only the Mayor and Elders, since it "castigates" others too much by causing adverse reactions when they take it. The *atka:t* is taken by adults, while the *pi·'tpa* is given to sick children. (Children get three pair, with proportionately more allotted as they grow older.)

The dosage for *ma"zhuṇ paHk*, "bones of the children," is 26 seeds; for *Tagetes erecta*, nine flowers. For *Brugmansia* the initial dosage is three flowers, augmented to six if there is no reaction. The dosage for *Datura stramonium* is nine seeds for a male taken three times, for a total of 27, and seven seeds taken three times (21) for a female.

Except for the mushrooms, which are taken "crude," all plants are specially processed. The seeds of *ma"zhuṇ paHk* must be ground by a 10- to 15-year-old virgin; otherwise, the plant spirits will not speak. The ground seeds are then stirred in a cup of water and strained. The flowers of *Tagetes erecta* and *Brugmansia candida* are macerated in hot water and then squeezed with a cloth strainer. The juice of the Tree *Datura* is poured into water, whereas that of *Tagetes* is poured into a gourd cup of spiced maize gruel. Both are taken to uncover pending misfortune and all that is hidden.

All psychotropic plants, except for the *Datura*, are taken at night after 8:00 p.m. They "work" until a cock's crow stops them from speaking, normally at about 3:30 a.m.

When taking the mushrooms or other sacred plants, two eggs are placed next to them, one or two candles are lighted, and prayers for help are said before the house altar. A supplication, such as the following, is recited as burning copal incense is wafted over the mushrooms or the liquid draught:

Thou who art blessed. I am now going to swallow you so that you heal me of the illness I have. Please give me the knowledge I need, thou, who knows all of what I need and of what I have; of my problem. I ask of you the favor that you only tell me and divine what I need to know, but do nothing bad to me. I do not wish an evil heart and wickedness. I only wish to know of my problems and illness and other things which you can do for me. But I ask you, please do not frighten me, do not show me evil things but only tell all. This is for the person with a pure heart. You can do many things and I ask you to do them for me. I now ask your forgiveness for being in my stomach this night.

The mushrooms are consumed whole, with two sips of water if there is difficulty swallowing them.

In order to prevent anyone from overhearing the conversation and to exclude any undue, sudden noises or disturbances, the sacred plants are taken in an isolated hut, or a curtain is hung to partition a section of the room from the rest of the family. The door is tightly shut to keep out noise and to prevent annoying dogs or chickens from entering. The sacred plants are said to dislike noise and will stop "speaking" when it occurs, resuming when the sound has ceased. Usually, one or two relatives or trusted friends attend the imbiber. They are not to make any brusque movements or talk to the imbiber during the trance. Instead, they only listen to the imbiber so that he/she can later be informed of his/her utterances. If any problems arise, the companion blows copal smoke on the person and prays to the saints so that the complication is removed and the individual is "able to continue on his (or her) path" and receive his/her consultation.

Some take the sacred plants and then go to sleep. A young girl or some other individual listens to what the person says in his/her sleep. However, this method is not widely used because of the fear of going insane and the uncertainty that the imbiber will speak while asleep.

The initial reaction upon taking the sacred plants is comparable to alcohol inebriation. The first visions to appear are serpents and jaguars. Although they may frighten some, they merely set the stage for what occurs next. These animals disappear and after a while the imbiber sees one of the following visions: a boy and a girl (the mythical Sun and Moon); diminutive adults who are the "children of the Winds;" a mature woman (the Earth goddess); or a neighbor or deceased relative. In a convolvulaceous plant-induced trance, an angel or juvenile twins appear. Quite often there are no apparitions. Instead, a voice informs the seeker of his problem or illness. The voice or visionary figure asks why the person took the sacred plants. The person answers; whereupon, in the case of illness, the visionary figure proceeds to heal the person using traditional methods, then informs him/her of the illness and identifies a ritual means to alleviate it. To the observer, the person appears to be conducting an extended, two-way conversation in which he/she is both interrogator and respondent. This process is illustrated in the following interview with an imbiber conducted by me:

At that time when my mother took the mushrooms, she laid down on the bed and covered herself with a blanket. Now her co-mother, who was treating her, was at her side looking after her. "Now let's see whether it takes effect." Well, she was there with two other women and then when the mushrooms took effect, mother said she first saw little snakes, numerous ones, moving, small

155

and big ones, and later she saw animals like jaguars, like cats. Later there appeared a woman who was her co-mother, and my mother greeted her co-mother. She said, "Good day, co-mother. How are you, co-mother?" "You are sick here," the woman said, and my mother responded, "Yes, co-mother, I am here sick." There was no one else talking, nothing more than a revelation of the thing that took effect. Well, she then saw that her co-mother sat down near the bed. Her co-mother said, "Well, co-mother, for this illness which you have, I am going to blow and massage your stomach. You have an illness there inside." But my poor mother—she, herself—was massaging her chest. "Like this, there!," she said. She saw that the woman was massaging. The co-mother presented herself in revelation. Well, when her co-mother was finished massaging, she took leave. "I am going. We will see each other again. Let's see whether the pain you have is removed," she said. "Yes, co-mother," responded my mother, as if she was speaking with her co-mother. But she alone was speaking, not the other woman who was looking after her and listening. The effects gradually went away, like when a person becomes sober, and then disappeared, she said. She felt a lot of pain in her chest and stomach; the whole area, as if she had been vigorously massaged. After three, four days, the pain she [had] had disappeared and she was cured, she said. She was a lot better, not like when she was ill.

The imbibers experience an initial intensified conflict of opposing ideas and later are heard to speak with a change in voice when the spirit of the plant "speaks" through him or her. This suggests that the imbiber undergoes a process of mental dissociation, where the psychic complexes of memories and ideas associated with the relatively autonomous subpersonalities of the mind become conscious (Figge 1973; Hilgard 1977).

The symbolic and semantic content of these visions is closely related to the Mixe worldview and theory of disease-causation. Sorcery, for example, is a prevalent theme as a cause of illness revealed in the trance. The nature of such a vision is depicted in the following account recorded by me:

Once I took the *ma''zhuṇ paHk* when I was about to die. A child, a kind of angel with wings came down. "What do you wish? My Father has sent me." "Look, angel, I put this in my heart, in my stomach to ask a favor of you. I am sick. Tell me why I am sick." "Ah," said the angel, "you are sick. The people bewitched you." "Oh?" "The people, yes. Do you want to know who?" "Yes, I want to know." "But you know perfectly well that your co-father asked you for a hat. Then he got someone to burn pine[5] and a piece of your clothing inside of it. Four bundles they burnt with four eggs. They killed a chicken with four candles. That is harming you, no other thing," said the angel. "But you are of good character," he said. "That is why God has not abandoned you. He who wants to kill you will never succeed. This illness of yours will not last long." The next day I went to defend myself [against the bewitchment]. I counted pine and went to burn on the mountain. I got a little better and then my father-in-law went to burn and sacrificed chickens for me. I got well and they never did get me.

On a different occasion, another man sick with chills and fever took the morning glory seeds to determine the cause of his illness. A boy and girl appeared in his vision and told him, "We are going to blow and make you well. You are ill since your wife went to church, lighted a candle and then prayed to a saint so that you would get sick and die. What you have to do now is go to the same saint, light a votive candle and pray; then she will die." A month later, according to his account, the woman, who was strong and healthy, became pallid and ill, then died as foretold in the vision.

The sacred plants are not considered dangerous if properly taken, and the incidence of adverse reactions is minimal. Nevertheless, the hazards of inimical effects play a significant role in native discussions of these plants. Untoward reactions do occur, either during the trance or sometime after. During the trance, for example, the sacred plants may not "speak" at all, but rather produce visions of phantoms, "another world," hell, or numerous snakes crawling over the imbiber's body or emerging from its orifices. There is no remedy for such a situation other than letting it subside naturally. Fright illness rituals are ineffective. The best antidote is to wash one's head with soap, eat chili peppers and salt, and then go to sleep. The remedy for bad effects of *Datura* is to drink a broth of very hot chili peppers.

These adverse effects are viewed by the Mixe as "punishment" inflicted by the sacred plants. The reasons given for such occurrences are numerous, but most involve infractions of the rules prescribed for using the sacred plants. For example, the user may not have been cleaned ceremoniously; had previously eaten eggs or fowl meat; or had taken the sacred plants during a bad time period or calendar day. If the sacred plants are not taken with a "clean heart," or if they are taken with bad intent (such as asking to acquire skill in sorcery), they will produce visions of attacking snakes and tigers. Although the spirits of the plants will reveal a modicum of sorcery techniques, they do not approve of their use. If the user lacks confidence, is fearful of the sacred plants or is very aggressive, they will frighten him or her with visions. Or, since only one class of mushroom should, by tradition, be used at a time, complications are alleged to result when mixed species are ingested. Visions of serpents are also said to occur to people with compact bones, or are interpreted as an omen that one should not leave the house.

Adverse effects which occur or persist after the trance should have ended are considered far more serious than those that occur only during the trance. Such aftereffects include decreased abstract reasoning abilities and chronic psychotic reactions. Three representative cases in Chiltepec involved: a person said to have gone permanently insane after taking morning glory seeds; another, similarly afflicted, who recovered after five years; and a third who recovered after two years. In addition to shouting and talking to themselves, these unfortunates initially exhibited immodest behavior, roaming the streets naked and recklessly wandering at night from village to village. According to observers and the subjects themselves, derangements of this kind result from having taken beer, mescal or fatty pork subsequent to a sacred plant trance. The condition may occur from a day to a month after the trance. In one scenario, a man may be invited by a cofather or friend, or a woman by her sweetheart, to consume an alcoholic drink, and on so doing is suddenly taken with a seizure. The trance may have been successful, but, on the spur of the moment, the individual forgets that he or she had recently consumed the sacred plant and drinks alcohol. In another example, a woman who had eaten the mushrooms and then pork a few days later became so deranged she was jailed to protect her from harm.

When adverse reactions occur, the services of even several shamans may be of no avail; they are rarely successful in curing the distraught individual. In one case, however, a curer was able to heal a man who was delirious as a result of having taken *Brugmansia* flowers. On the seventh day of his delirium, the shaman prayed before the same number of *Brugmansia* flowers as were taken by the man, and then offered pine, eggs and fowl for his recovery.

Acknowledgments

The research on which this study is based was made possible through grants awarded by the National Science Foundation (BNS 77–19159), the Wenner-Gren Foundation for Anthropological Research and the Marstrand Foundation. I am also most grateful to Searle Hoogshagen, Irmgard Weitlaner Johnson, Richard Evans Schultes and the former director of the Escuela Nacional de Ciencias Biológicas, I.P.N., Dr. Armand Lemos Pastrana. Dr. Gastón Guzmán, I.N.I.RE.B., identified the *Psilocybe* species. I am also beholden to Drs. Rupert Barneby and Jacquelyn A. Kallunki of the New York Botanical Garden for identifying other botanical species collected.

Notes

1. See Lipp (1983; 1985) for a description of Mixe culture and society.
2. The mushrooms contain strong hallucinogens—psilocybin and/or psilocin (Hofmann 1964). *Turbina corymbosa* and *Ipomoea violacea* contain trance-inducing ergoline alkaloids (Hofmann 1966). *Datura* and *Brugmansia* contain tropane alkaloids, such as scopolamine, which may cause delirium, excitement and hallucinations (Fodor 1967; Bristol et al., 1969; Hall et al., 1977). *Tagetes erecta* contains lactones and terpenes which induce hilarity and vivid dream imagery (Díaz 1979, p. 93). Additional ethnographic data on the use of *Psilocybe* mushrooms by the Mixe may be found in the works of Heim and Wasson (1958, pp. 84–91); Hoogshagen (1959); and Miller (1966). The therapeutic effects of ritualistic hallucinogens are explored by Joralemon (1984).
3. The mythological motif of a beneficial plant emerging from the dead body of a culture hero—Jensen's Hainuwele mythologem (1973, p. 110)—is cosmopolitan in distribution and typically part of the worldview of early planters and hunters (Frobenius 1938; Zerries 1952).
4. The calendar days 5, 9, 13, 18 and 29, when they fall on a Saturday or Sunday, were cited for the convolvulaceous plants as auspicious for imbibing the ceremonial draught. Other than the fact that the Earth-deity is addressed by her calendar name when the mushrooms are picked, there is no relationship between the calendar day *tu·m ?uh* (1 June) and one of the names for the sacred mushrooms, *tu·m ?uŋ*, "one son," according to Miller (1966, p. 319).
5. Mixe curing rituals, rites of passage, village festivals, and rituals relating to agriculture, hunting and other economic pursuits include, as part of the offering, bundles of grass, pine needles, wood splints of pine or other plant materials. These rituals usually consist of two parts. The first part, *ni no'k tut* ("to burn in order to free") or *nayni wa'ac* ("to open" or "cleanse"), signifies that the offering or "payment" is to protect the petitionary's life and to remove the afflicting illness or misfortune. The second part, *?oy gishpa* ("on behalf of the good"), asks forgiveness of the *Na·shwi·ñ,* the Earth and other deities, for transgressions.

In the first part, bundles of resinous pine splints are burned with cornmeal and eggs, and sometimes a pullet is sacrificed. In the second part, two turkeys are sacrificed—first a male, then a female—over a separate pine needle bundle or series of bundles with emplaced candles, eggs and tamales. The number of needles and sticks in each bundle is determined by the nature and function of the ritual.

References

Bristol, M. L., W. C. Evans, and J. F. Lampard. 1969. The alkaloids of the genus *Datura*, section *Brugmansia* part VI. Tree *Datura* drugs (*Datura candida* Cus.) of the Colombian Sibundoy. *Lloydia* 1969; 32:123–30.

Díaz, J. L. 1979. Ethnopharmacology and taxonomy of Mexican psychodysleptic plants. *J Psychedelic Drugs* 1979; 11:71–101.

Figge, H. H. 1973. Zur Entwicklung und Stabilisierung von Sekundaerpersoenlichkeit im Rahmen von Besessenheits Kulten. *Confinia Psychiatrica* 1973; 16:18–37.

Fodor, G. 1967. The tropane alkaloids. In *The Alkaloids*, Vol. 9, edited by R. H. F. Mansje. New York: Academic Press, 1967, pp. 269–304.

Frobenius, L. 1938. Das archiv fuer folkloristik. *Paideuma* 1938; 1(1):1–18.

Hall, R. C. W., M. K. Popkin, and L. E. McHenry. 1977. Angel's Trumpet psychosis: a central nervous system anticholinergic syndrome. *Am J Psychiatry* 1977; 134:312–14.

Heim, R., and R. G. Wasson. 1958. *Les Champignons Hallucinogènes du Mexique: Etude Ethnologique, Taxinomiques, Biologiques, Physiologiques et Chimiques.* Paris: Editions du Muséum National d'Histoire Naturelle.

Hilgard, E. R. 1977. *Divided Consciousness: Multiple Controls in Human Thought and Action.* New York: Wiley.

Hofmann, A. 1964. Mexican witchcraft drugs and their active principles. *Plant Medica* 1964; 12:341–52.

Hofmann, A. 1966. The active principles of the seeds of *Rivea corymbosa* (L.) Hal. F. (Ololiuhqui, Badoh) and *Ipomoea tricolor* Cav. (Badoh Negro). In *Summa Antropológica en Homenaje a Roberto J. Weitlaner*, edited by A. Pompa y Pompa. Mexico City: Instituto Nacional de Antropología e Historia, S.E.P.

Hoogshagen, S. 1959. Notes on the sacred (narcotic) mushroom from Coatlán, Oaxaca, Mexico. *Oklahoma Anthropological Soc Bull* 1959; 7:71–74.

Jensen, A. E. 1973. *Myth and Cult among Primitive Peoples.* Chicago: University of Chicago Press.

Joralemon, D. 1984. The role of hallucinogenic drugs and sensory stimuli in Peruvian ritual healing. *Culture, Medicine and Psychiatry* 1984; 8:399–430.

Lipp, F. J. 1983. *The Mixe Calendrical System: Concepts and Behavior.* Ann Arbor, Michigan: University Microfilms (No. KKA830353).

Lipp, F. J. 1985. Mixe ritual: an ethnographic and epigraphical comparison. *Mexicon* (Berlin) 1985; 7:83–87.

Miller, W. S. 1966. El Tonalamatl Mixe y los Hongos Sagrados. In *Summa Antropológica en Homenaje a Roberto J. Weitlaner*, edited by A. Pompa y Pompa. Mexico City: Instituto Nacional de Antropología e Historia, S.E.P., pp. 317–28.

Wasson, R. G., G. Cowan, F. Cowan, and W. Rhodes. 1974. *María Sabina and her Mazatec Mushroom Velada.* New York and London: Harcourt Brace Jovanovich.

Wasson, V. P., and R. G. Wasson. 1957. *Mushrooms, Russia and History.* New York: Pantheon Books.

Zerries, O. 1952. Die Kulturgeschichtliche Bedeutung einiger Mythen aus Suedamerika ueber den Ursprung der Pflanzen. *Zeitschrift fuer Ethnologie* 1952; 77:62–82.

Bernard Lowy (*center*) and Richard Evans Schultes listen to Gordon's presentation at the formal dedication of the Wasson Collection at Harvard (1983). *Masha Wasson Britten.*

'The Banquet of his Interests'

BERNARD LOWY

Remarks at the dedication of the Tina and Gordon Wasson
Ethnomycological Collection, Harvard University,
February 12, 1983

Dean Rosovsky, Dr. Erickson, Prof. Schultes, distinguished guests, fellow bibliophiles and mycophiles, I am honored to participate in the dedication of the Tina and R. Gordon Wasson Ethnomycological Collection, which now becomes part of Harvard University. Innovative research has been the hallmark of the Wassons' contributions to ethnomycology for decades past, and to them belongs the distinction of having forged this new discipline in the life sciences. The beginnings were humble, as is often the case with potentially influential, even world-shaking phenomena. Here is Gordon Wasson's revealing account of his conversion from mycophobia to mycophilia, recorded in a statement made at the New York Academy of Sciences in 1959:

> My wife and I began to gather our material long ago, in 1927, on an August afternoon in the Catskills, as we strolled along a mountain path on the edge of a forest. She was of Russian birth and I of Anglo-Saxon ancestry. I knew nothing of mushrooms and cared less. They were for me rather repellent, and likely as not, deadly poisonous. In the years that we had known each other, I had never discussed mushrooms with my wife. Suddenly she darted away from my side: she had seen a forest floor carpeted with mushrooms of many kinds. She knelt before them, called them by endearing Russian names, and over my protests insisted on gathering them in her dress and taking them back to the lodge, where she went so far as to cook them and eat them—alone . . . This episode made so deep an impression on us that from then on, as circumstances permitted, we gathered all the information that we could about the attitude of various peoples toward mushrooms—what kinds they know, their names for them, the etymology of those names, the folklore and legends in which mushrooms figure, references to them in proverbs and literature and mythology.

This incident eventually led Wasson to undertake extensive travels to distant parts of the earth in search of clues to explain certain perplexing magico-religious phenomena; to the consultation and collaboration with specialists in disciplines as diverse as linguistics, musicology, chemistry and mycology; to the collection and identification of numerous narcotic plants and fungi in their far-flung natural habitats; to the publication of books as splendid in their attire as they are significant and memorable in their content; and in general, to the shedding of light where previously there was darkness.

The pursuit of learning and the extension of the horizons of knowledge are among the primary functions and attributes of a university. Harvard, through the centuries, has been a bulwark of learning, and with her have been associated many formidable scholars whose studies have enriched some facet of human life. Gordon Wasson and his late wife are among them. His published works over the past 25 years are a monument to his careful research, the result of a disciplined yet imaginative and creative mind. He and his collaborators have produced a corpus of works upon which future scholarly investigations in this field will rest for generations to come. Testimony to his influence upon our contemporaries may be clearly discerned from the fact that his books *Mushrooms, Russia and History, Soma: Divine Mushroom of Immortality, María Sabina and her Mazatec Mushroom Velada* and *The Wondrous Mushroom: Mycolatry in Mesoamerica* have all become classics in our lifetime.

I first met Gordon Wasson on August 30, 1960, in Stillwater, Oklahoma, where he gave an address on ethnomycology before the Mycological Society of America.[1] It was the first time that the Society had invited a non-mycologist to present the annual address, and it was a memorable one. Until this evening we have not had the occasion to meet again. So I am highly gratified that I have that opportunity tonight. This is not to say that we have been incommunicado all that time. We have, in fact, been avid correspondents for many years, have frequently consulted one another about ethnomycological problems of mutual interest, and quite recently have collaborated on some aspects of mycological lore pertaining to the Guatemalan highlands. Currently, the interpretation now generally accepted concerning the significance of mushroom stones has come under scrutiny, a normal healthy procedure, since theories in science are born only to be further criticized and evaluated. With Gordon's indispensable collaboration I am presently reviewing some assumptions pertaining to this controversial subject.

Gordon Wasson's research has concentrated on the solution of major problems that have confronted and challenged students of the fungi, among others, for centuries. His endeavors have frequently resulted in novel discoveries that have offered penetrating insights. His work is a notable example of what can be done in interdisciplinary research, and he has somehow managed to integrate anthropological, linguistic, archeological and mycological studies into that complex organic fabric called ethnomycology. Often in brilliant and unexpected ways Wasson has utilized his deep fund of knowledge to make audacious proposals and present new interpretations. A fragmentary list of the surprising diversity of subject matter that he has at various times considered, dissected, interpreted, synthesized or incorporated into his works is illuminating. Included are: mushroom stones; Soma; Aztec statuary; ancient frescoed walls; Florentine mosaics; terra cotta figurines; Colonial American, European and Asiatic paintings; pre-Columbian archeological sites; Cherrugeresque extravagances; Mesoamerican flora and fauna; beliefs of New Guinea tribesmen; Mazatec *veladas* and their accompanying music, clapping and thumping; unmentionable Siberian shamanistic pastimes; mentionable but occult North American Indian rituals; toads and toadstools; arcane amulets; devils and assorted demons; fly killers; bedbugs; sacred figs; the Buddha; sorcerers cum apprentices; Mesoamerican codices; cabalistic glyphs; bark-paper maps; diving gods and floating cherubs. We also encounter choice references to Chinese and pre-Columbian textiles; Zoroastrianism; German children's songs; curandero's chants; shamanistic vigils; marvelous herbs, divine inebriants and other entheogens (this term being one of several Wassonian neologisms that have already entered our language); the Rig Veda; Eleusinian Mysteries; mycophagy, mycophilia and mycophobia, and mycolatry (here again the fine hand of Wasson is seen); ololiuqui, piciete, teonanácatl and *Psilocybe* (these are considerably pre-Wassonian); Nahuatl poetry, Shakespeare, Keats, Shelley . . . The list is very long. This must suffice as a kind of *hors d'oeuvre* (which, incidentally, Gordon's late and cherished mycological colleague and mentor, Roger Heim, would have savored), a kind of *hors d'oeuvre*, as I have mentioned, to the banquet of his interests, and to his ardent, unswerving pursuit of ethnomycological research.

I can think of no more fitting accolade that applies to a man of Gordon's talents than Chaucer's succinct evaluation of the Clerk of Oxford: "Gladly would he learn, and gladly teach."

Notes

1. *Editor's note:* In a letter to the editor dated 1 February 1988, Prof. Lowy adds the following anecdote:

> At Stillwater, I asked Gordon whether he was acquainted with a bit of Hungarian folklore that referred to mushrooms, and since I have a command of Hungarian, I thought I would test his knowledge. I told him that, as a child, I had heard my Hungarian mother use the expression "bolond gomba," and he immediately seized upon this with enthusiasm, correctly translating it as "crazy mushroom." I was astonished that he was able to do this, since to most of the world Hungarian is a totally obscure language. I should add that the expression is still in current use, as I found years later when I made inquiry in Hungary. It is generally used by a vexed parent indignantly addressing a child who has said something outlandish. The sentence in which the words occur is: "Mi a baj, ettél talán bolond gombát?," or some variation thereof, meaning, "What's the matter, have you eaten crazy mushrooms?"

Wasson's Literary Precursors

TERENCE McKENNA

There can be no doubt that the modern era of ethnomycology begins with the work of Gordon and Valentina Wasson. The late Mr. Wasson is the Abraham of the reborn awareness in Western civilization of the presence of the shamanically empowering mushroom. Yet, like all great innovative thinkers, the Wassons had their precursors. Others before them had stumbled into an awareness of the visionary potential of fungi, though their experiences, their findings, did not become a *cause celebre* or an academic discipline. Many simply chose to keep secret what they had discovered—a sensible response to Western society's longstanding bias against these mushrooms, which was reinforced by frightening reports on "mushroom intoxication" that never acknowledged anything salutary about the experience. A good example, set down by A. E. Merrill in a 1914 issue of *Science,* concerned an accidental ingestion of *Panaeolus papilionaceus*[1] in Oxford, Maine. It was later recounted by Robert Graves (1960):

> Mr. W. gathered about a pound of *Panaeolus papilionaceus* mushrooms, and fried them in butter for himself and his niece. The immediate effect was that both felt a bit tipsy, and soon their surroundings seemed to take on bright colours, in which a vivid green predominated. Next both experienced an irresistible impulse to run and jump, which they did hilariously, laughing almost to the point of hysteria at the witty remarks they exchanged . . . When they left the house to take a walk, they lost all sense of time—a long period seemed short and contrariwise; the same with distances.
>
> . . . [W]allpaper patterns appeared to creep and crawl about, though at first remaining two dimensional; then began to grow out toward him from the walls with uncanny motions. He looked at a bunch of large red roses, all of one kind, which lay on the table; and at another on a writing-desk. At once the room seemed to fill with roses of various red colours and many sizes in lavish bunches, wreaths and chains.
>
> Feeling a sudden rush of blood to his head, he lay down. Then followed an illusion of countless hideous faces of every sort and extending in multitudes over long distances, all grimacing at him vapidly and horribly, and coloured like fireworks—intense reds, purples, greens, and yellows.

It would have been difficult indeed for any voluntary user of hallucinogenic fungi to openly defend such effects as desirable, even for artistic inspiration.

Instead, the inclination was for mushroom cognoscenti to keep silent.

Yet some, it appears, found a way to safely publicize their personal familiarity with psychoactive mushrooms, by disguising it as literary fiction. It is useful for us now, in the expanded intellectual arena that ethnomycology has created for itself, to examine these brave futurists of the past whose works implied what Gordon Wasson made explicit: the presence of an awesome spiritual power resident in visionary fungi.

The Wassons acknowledged a few of these "literary precursors" in *Mushrooms, Russia and History*. One was Lewis Carroll, whose 1865 masterpiece *Alice in Wonderland* includes an interesting section in which Alice, the protagonist, eats pieces of a mushroom that cause her to alternately shrink and grow. The Wassons observed that:

> All of Alice's subsequent distortions, softened by the loving irony of Lewis Carroll's imagination, retain the flavor of mushroomic hallucinations. Is there not something uncanny about the injection of this mushroom into Alice's story? What led the quiet Oxford don to hit on a device so felicitous, but at the same time sinister for the initiated readers, when he launched his maiden on her way? Did he dredge up this curious specimen of wondrous and even fearsome lore from some deep well of half-conscious folk-knowledge? (Wasson & Wasson 1957, pp. 194–195)

The possibility that Carroll may have drawn on a personal experience with psychoactive mushrooms is not acknowledged by the Wassons. Instead, they proceed to develop the thesis that he got his inspiration from a different source: Mordecai Cooke's *A Plain and Easy Account of British Fungi* (1862). This book included what the Wassons call "horrifying accounts of the amanita-eating Korjaks" of Siberia, who, upon eating this mushroom (*Amanita muscaria*), experienced "erroneous impressions of size and distance" among other psychoactive effects.

Interestingly, the Wassons ignored a significant piece of evidence that strengthens the case for their thesis considerably. It is found in the beginning of the scene in which Alice encounters the magical mushroom:

> There was a large mushroom growing near her, about the same height as herself: and, when she had looked under it, and on both sides of it, and behind it, it occurred to her that she might as well look and see what was on top of it.
>
> She stretched herself up on tiptoe, and peeped over the edge of the mushroom, and her eyes immediately met those of a large blue caterpillar, that was sitting on the top, with its arms folded, quietly smoking a long hookah, and taking not the smallest notice of her or of anything else. (Carroll 1960, p. 66)

At the time the Wassons wrote *Mushrooms, Russia and History* they did not know of another, more relevant book by Mordecai Cooke, *The Seven Sisters of Sleep* (1860), which discussed seven major varieties of psychoactive substances. Included among them are both *Amanita muscaria* and cannabis, united by Carroll in one striking image of a hookah-smoking caterpillar perched on a mushroom with magical properties.

Also in *Mushrooms, Russia and History*, the Wassons acknowledged an inter-

Peering over the edge of a magical mushroom, Alice finds a hookah-smoking caterpillar. *Drawing by J. Tenniel.*

esting psychoactive mushroom story written by H. G. Wells in the late 19th Century. "The Purple Pileus" tells the ostensibly fictional tale of one Mr. Coombes, a meek, henpecked man who tries to kill himself by eating what he thinks to be a poisonous mushroom he finds in the forest. This mushroom, writes Wells (1966, pp. 191–200), is "a peculiarly poisonous-looking purple: slimy, shiny, and emitting a sour odor." When broken by Coombes, its creamy white inner flesh changes "like magic in the space of ten seconds to a yellowish-green colour." Its taste is so pungent he almost spits it out. Within minutes, his pulse starts to race and he feels a tingling sensation in his fingertips and toes. Then, before he can pick more purple pilei from a cluster he sees in the distance, Coombes is distracted by the mushroom's full effect. It induces a powerful change in his psychology for several hours, transforming him into a veritable lion. He rushes home, gaily singing and dancing, to confront his wife. His eyes as he enters the house are described as "unnaturally large and bright." After frightening off his wife's boyfriend and earning her lasting respect, he falls into a "deep and healing sleep."

The Wassons (1957, pp. 50–51) make it clear, in their analysis of Wells' story, that Coombes had not eaten *Amanita muscaria*. Instead, they conclude, Wells had "filled out the necessities of a given plot by inventing the needed mushroom." They do not suggest, and apparently never considered, that Wells' purple pileus may have been a thinly-disguised *Psilocybe* mushroom. Like *Psilocybe,* it changes color when broken; has a markedly pungent taste when eaten fresh; often grows in clusters; quickly causes profound psychological and somatic effects, including dilation of the pupils; and induces deep sleep as an after-effect. Also worth noting is that Wasson (1978) later made much of a psychoactive fungus' purple color in *The Road to Eleusis: Unveiling the Secret of the Mysteries,* where he and co-authors Albert Hofmann and Carl A. P. Ruck argued convincingly that the sacramental drink imbibed at Eleusis contained the psychoactive fungus *Claviceps purpurea.* According to them, the purple color of the vestments of the priests who conducted the Mysteries was identical to, and therefore emblematic of, this fungus, which grows throughout Europe.

Could Wells have been personally familiar with the effects of *Psilocybe, Claviceps* or some other species of hallucinogenic mushroom? Certain other of his stories seem to resonate with insights that may well have been derived from such experiences. In "The Plattner Story," for example, the title character finds himself transported by a chemical explosion to an eerie, hallucinatory "Other-World" with a green sun, where the left and right sides of his body are transposed. The green illumination is consistent with the earlier cited experience of Mr. W, whose "surroundings seemed to take on bright colors, in which a vivid green predominated" when he ate what were reported to be *Panaeolus papilionaceus* mushrooms. The transposition of Plattner's body reminds us both of Alice's adventures "through the looking glass" (since mirrors cause a similar transposition) and modern theories that hallucinogens shift emphasis from left- to right-brain thinking. Of equal interest, this "Other-World" co-exists with ours and is accessible to us when our perceptions are enhanced (or altered). "It seems quite possible," wrote Wells (1966, pp. 141–157), "that people with unusually keen eyesight may occasionally catch a glimpse of this strange Other-World about us."

Another Wells story, "The New Accelerator," tells of a man who takes a drug that speeds his metabolism to such a degree that the world around him appears to be standing still. The impression of "stopping the world" is another effect that occurs with hallucinogens, though Wells (1966, pp. 165–176) compares it instead to the effect of nitrous oxide: "You know that blank non-existence into which one drops when one has taken 'gas,'" says his protagonist. "For an indefinite interval it was like that."

The possibility that Wells experienced psychoactive substances is therefore compelling. An even stronger case can be made for Wells' contemporary John Uri Lloyd (1895; 1976; 1978), who almost certainly had personal awareness of the psychoactivity of psilocybin-containing mushrooms. The first publication date of Lloyd's crypto-discourse on psilocybin, *Etidorhpa,* was 1895—nearly 60 years before Gordon Wasson's first trip to Huautla.

The evidence, both circumstantial and prima facie, that Lloyd had experienced intoxification by psilocybin is ample. Lloyd was a *fin de siecle* character, both a competent pharmaceutical chemist and a man with a passion for occult literature and speculation. According to Neal Wilgas, author of the introduction to both later editions of *Etidorhpa* (1966; 1976), Lloyd was born in West Bloomfield, New York, on April 19, 1849, the eldest son of a civil engineer and a descendent of Governor John Webster of Massachusetts. His family moved to Kentucky and then to Cincinnati. It was there, at the age of 15, that John Uri Lloyd began to learn the drug trade. He became the laboratory manager of a drug firm and later a partner in the company. Lloyd and his brother published a quarterly journal, *Drugs and Medicines of North America*. Later, he was to participate in the establishment of the Lloyd Library of Botany and Pharmacy. To this day, in the field of phytochemistry, the preeminence of the journal *Lloydia* is a testament to the Lloyd brothers' passion for pharmacology and pharmacognosy.

John Uri's brother, Curtis Gates Lloyd, has been described as one of the leading fungi botanists in the world of his time. C. G. Lloyd made extensive collections of fungi in the gulf states and deep south. There can be little doubt that if a mushroom species such as *Stropharia cubensis* were present in those places at that time with even a fraction of their current frequency, he would have collected and been familiar with it. C. G. Lloyd's specimen collections deposited with the Smithsonian Institute number several thousands. Perhaps an examination of those collections would yield specimens of psychoactive fungi and field notes concerning them.

In any case, it seems clear that John Uri Lloyd's bizarre hollow-earth novel *Etidorhpa* was for him a kind of labyrinth at whose center he wished to place the apotheosis he had personally experienced in his own peregrinations in the realm of gigantic fungi. In it, he describes not only his encounter with the anagrammatically scrambled mother-goddess Etidorhpa (*Aphrodite* spelled backwards), but also a theory of time that bears the unmistakable imprint of the mushroom *philosophe*. At the end of seven chapters devoted to a classic psychopompic initiation via visionary fungi, Lloyd places a footnote that lets the cat out of the bag:

> If, in the course of experimentation, a chemist should strike upon a compound that in traces only would subject his mind and drive his pen to record such seemingly extravagant ideas as are found in the hallucinations herein pictured, or to frame word-sentences foreign to normal conditions, and beyond his natural ability, and yet could he not know the end of such a drug, would it not be his duty to bury the discovery from others, to cover from mankind the existence of such noxious fruit of the chemist's or pharmaceutist's art? To sip once or twice from such a potent liquid and then to write lines that tell the story of its power may do no harm to an individual on his guard, but all mankind in common should never possess such a penetrating essence. Introduce such an intoxicant, and start it to ferment in humanity's blood, and it may spread from soul to soul, until, before the world is advised of its possible results, the ever increasing potency will gain such headway as to destroy, or debase, our civilization, and even to exterminate mankind. (Lloyd 1895, p. 276)

And what are the extravagant ideas and hallucinations of which John Uri Lloyd wishes to speak? At the close of Chapter XXXIII, the hero of *Etidorhpa* is told to drink the juice of "a peculiar fungus." Our hero's guide minces no words: "[H]e spoke the single word, 'Drink,' and I did as directed." The three chapters which follow are a virtual monologue on the methods of intoxication known to humanity world-wide and throughout history. The horror of inebriation and addiction is graphically depicted and reaches a climax in Chapter XXXIX, "Among the Drunkards." If these chapters are the obligatory hell experience of 19th Century drug reportage, then Chapter XL is the paradisiacal apotheosis. It is also the climax of the book, and contains the incident in which the hero confronts Etidorhpa.

169

The protagonist in J. U. Lloyd's novel *Etidorhpa* is conducted through a forest of gigantic, mind-altering fungi. *Illustration by J A. Knapp.*

Indeed, J. Augustus Knapp's beautiful etching of her is tipped into this chapter. Her appearance and retinue set off a cascade of florid (and psychedelic-sounding) Victorian prose:

> Could any man from the data of my past experiences have predicted such a scene? Never before had the semblance of a woman appeared, never before had an intimation been given that the gentle sex existed in these silent chambers. Now, from the grotesque figures and horrible cries of the former occupants of this same cavern, the scene had changed to a conception of the

beautiful and artistic, such as a poetic spirit might evolve in an extravagant dream of higher fairy land. I glanced above; the great hall was clothed in brilliant colors, the bare rocks had disappeared, the dome of that vast arch, reaching to an immeasurable height, was decorated in all colors of the rainbow. Flags and streamers fluttered in breezes that also moved the garments of the angelic throng about me, but which I could not sense.

The band of spirits or fairy forms reached the rock at my feet, but I did not know how long a time they consumed in doing this; it may have been a second, and it may have been an eternity. Neither did I care. A single moment of existence such as I had experienced, seemed worth an age of any other pleasure. (p. 253)

The appearance of the goddess is quickly followed by reestablishment of the theme of suffering and terror as the hero imagines himself lost and wandering for days in an arid wasteland, at first tormented by the sun, later frozen by its absence. As this hallucination fades:

. . . the ice scene dissolved, the enveloped frozen form of myself faded from view, the sand shrunk into nothingness, and with my natural body, and in normal condition, I found myself back in the earth cavern, on my knees, beside the curious inverted fungus, of which fruit I had eaten in obedience to my guide's directions. (p. 270)

At the beginning of Chapter XLII, the hero argues with his guide concerning the nature of what he has just experienced. The psychopomp speaks first:

"You ate of the narcotic fungus; you have been intoxicated." "I have not," I retorted. "I have been through your accursed caverns and into hell beyond. I have been consumed by eternal damnation in the journey, have experienced a heaven of delight, and also an eternity of misery." "Upon the contrary, the time that has passed since you drank the liquid contents of that fungus fruit has only been that which permitted you to fall upon your knees. You swallowed the liquor when I handed you the shell cup; you dropped upon your knees and then instantly awoke. See," he said, "in corroboration of my assertion the shell of the fungus fruit at your feet is still dripping with the liquid you did not drink. Time has been annihilated. Under the influence of this potent earth-bread narcoto-intoxicant, your dream began inside of eternity; you did not pass into it." (pp. 272–273)

These passages are more than sufficient to convince the open-minded reader that John Uri Lloyd, 19th Century savant, pharmacist, occultist and author, had discovered the consciousness-expanding properties of psilocybin mushrooms, experienced them and then decided to suppress his discovery. Why he should have done so is less obvious. It may be that he feared public censure. Or perhaps he was deeply ambivalent about the visionary state, as evidenced by his diatribes against intoxication. Another possibility is that Lloyd was an elitist. He may have felt, with Aldous Huxley, that only "superior" people should use hallucinogens. From this perspective, *Etidorhpa* serves a twofold purpose. It shares the secret of the mushroom with those "on their guard," who are sufficiently insightful to perceive the ostensible fantasy's real meaning. And it frightens away the unworthy who fail to delve beneath its surface, thus assuring that "all mankind in common should never possess such a penetrating essence."

Wasson's interest in Lloyd is a matter of record, though he never, to my knowledge, wrote about him in his books. There is a file at the Tina and Gordon Wasson Ethnomycological Collection at Harvard's Botanical Museum where Wasson saved newspaper clippings, letters, notes and other information for a second edition of *Mushrooms, Russia and History* that never materialized. In this file is a letter from a Mr. Bernard Lentz dated May 16, 1957, recommending that Wasson read *Etidorhpa*. A copy of Wasson's reply, dated June 4, 1957, is also on file. It reads in part: "I shall try to look the matter up, when I have time. John Uri Lloyd—a well known name." Years later, in his foreword to a book seller's catalog, Wasson (1979) wrote the following:

> There is one item in it [the catalog] that interests me especially: *Etidorhpa* ("Aphrodite" spelled backwards), by J. U. Lloyd, a strange novel, or better a fantasy, first published in 1895, a novel that Michael Horowitz [the cataloguer] rightly says was a psychoactive mushroom tale. Where did Lloyd's ideas come from? He must have read carefully Captain John G. Bourke's *Scatalogic* [sic] *Rites of All Nations,* published in 1891, that immense and amazing collection of scatological materials. Lloyd's mushrooms are clearly not *A. muscaria.* Did he possess a copy of that rarest of all entheogenic books, by the famous English mycologist M. C. Cooke, the *Seven Sisters of Sleep,* a book that Cooke himself never referred to in his later mycological writings? Here is indeed a manual of psychoactive drugs, published almost 120 years ago! Nor was it included in the bibliography of his writings published on his death. How did Lloyd hit on this mushroom fantasy? Is there latent in our society a memory of mushroom use, long long ago, a subliminal memory that crops out in Lloyd's tale, also in *Alice in Wonderland?* The suggestive shapes and delicate changing colors of mushrooms, their sudden appearance and disappearance, the endless diversity in their odors, one for each species—all support a mushroom mythology that is backed up, when one knows about it, by the compelling entheogenic potency residing in some of them.

Again, as with Carroll and Wells, Wasson failed to admit the possibility that Lloyd's insights were based on personal experiences. But neither did he rule it out. The matter thus remains to be resolved by a new generation of ethnomycologists—or literary archaeologists.

Finally, we turn to a work of the 20th Century overlooked by Wasson. In the May 1915 issue of the *Irish Ecclesiastical Record* a piece appeared, written by A. Newman, titled "Monsieur Among the Mushrooms." It purported to be a nonfiction recollection of a person known to the author. "Monsieur Among the Mushrooms" was reprinted in 1917, one piece among many in a book titled *Unknown Immortals—In the Northern City of Success,* by one Herbert Moore Pim, a minor essayist and journalist of the time. In his preface, Pim thanks the editors of the *Irish Ecclesiastical Record* for permission to reprint "Monsieur"—a possible hint that Pim was himself A. Newman, or vice-versa.

While the mushrooms described in "Monsieur" are not overtly psychoactive (except, perhaps, as a fetish), the story is relevant to our discussion for being (1) the record of a person with an appreciation of mushrooms on a cosmic scale and (2) the earliest known instance of a modern cult of mushroom users. Pim writes in his preface:

The original of *Monsieur Among the Mushrooms* is alive and prosperous. He is a perfectly amazing person, a man of considerable fortune who has, I believe, been detained in an asylum on several occasions. He conducts a large business during the day-time; but he may be discovered at four or five o'clock in the morning, pouring forth a stream of brilliance, and holding men in the cold street against their will. His brain works with such rapidity that he has constructed a language of his own, by means of which only the absolutely essential thought is presented to the hearer. I have seen calm men whipped into fury when they found themselves simply swept off their feet in argument with my model.

In *Monsieur* I have drawn him exactly as he exists, save in the matter of the physical description. Apart from that fact, there is nothing exaggerated; and the debate between Monsieur and the members of the committee is almost as true as a description of such a debate could be. There you see my model and his method. (Pim 1915; 1917)

Ostensibly, "Monsieur Among the Mushrooms" comprises the recollections of a remarkable, even obsessed, personality, a man who has been committed to an asylum because of his unusual ideas concerning fungi. In it, Monsieur argues before the asylum's release committee that he is sane. But first, Pim relates the remarkable philosophy and history of his "model," as he calls Monsieur:

Here it was that Monsieur learned how the mushroom might be persuaded to grow; and here it was that for many days he toiled unobserved, appropriately attired in black, with a light heart and a somewhat lightened purse. And in those first, fresh, active days he found time even to press his theory upon others as a physic to be received in small measures, while the giver retains something, if it even be the bottle. And so it came that, in a little while, there arose a respectful company of believers.

"How great, indeed," Monsieur would exclaim, "is the mushroom! It has claimed the world round for its habitation; and when man rears his cities of stone it demands of him that even in the heart of cities it shall be given space to express itself in silence."

There was the world itself to be considered. For presently it put its claw into its own stomach, where Monsieur and his disciples were digesting wisdom, and demanded to know the reason for an aesthetic appreciation of a commodity interesting only for its commercial value. (Pim 1915, p. 588)

Needless to say, a hearing was held, its conclusions foregone:

And who can forget the genial and superior smile which rested upon the faces of his judges? The mushroom was the all-powerful exception! Just so. Who could doubt it? But for such as believed it there had been made complete provision.

"But how," exclaimed Monsieur, "shall I make progress in my investigation?"

He was assured with gentleness that, even though placed *extra muros,* he should have "every facility," "ample scope," and that, above all, he might hope to be well again.

"But to what end?" he interrupted.

"In order," it was explained to him, "that you may be in harmony with the majority."

"But the majority here are mushrooms! Man, their toy, is nowhere. It is he who is *extra muros!*" (p. 590)

A virtual prisoner, Monsieur spends his days in the asylum contemplating the irony of his situation and ultimately hatches a plan of escape. His asylum musings revolve around only one theme:

> With his knowledge of the mushroom he was all-powerful. Behind the material which witnessed to a supremely strong exception, there was the energy of mind that drove and guided, swept and conquered. And in the mushroom itself there was unity without contact. The mushroom was, indeed, a giant body torn and strewn over the earth. There was the fungus of the hair, there was that which, by its shape, clearly proved the existence of the brain. There was a form which made certain that the egg was the origin of that which it contained. There was the manifestation of that which generates. And there was a growth which appertained to the lower animals. There were many things besides: the star-like eyes, from which the sun and moon derived their radiance; the great masses of body and limb; the fingers and the features; the mouth that devoured. There was the warrior from whose wounds blood could flow. There was that which indicated the cellular structure of the human body, and indeed of all living things. And yet all this was incalculably strong, and all this was inexplicably united. (p. 592)

Eventually Monsieur effects a daring, mushroom-assisted escape. Then we are told in a footnote that:

> He lived for some time under the protection of the keeper of a plant nursery, who had become so enthusiastic a believer in the doctrine of the Mushroom that he painted his glass-houses with a black light-excluding fluid, and cultivated the mushroom reverently. A primitive worship had already developed when Monsieur was restored to his followers. I have reason to believe that he prepared to encourage this, and, in some respects to modify it. But the world interfered. There is a journal before me which records frequent attacks upon the glass-houses; and there are references to search parties from the asylum. I am enabled to trace the purchase of a sailing ship by the keeper of the plant nursery, and the embarkation of Monsieur and his followers upon this ship, the hold of which contained mushroom-spore bricks. After that I have no reliable evidence. (p. 605)

Aside from the early date of its composition, what makes "Monsieur Among the Mushrooms" so interesting is that it purports to be a factual account of a group of people, informed as to the transformative power of the mushroom and united around a leader and a set of cult practices. It is difficult to believe that Pim would have given such a central philosophical role to mushrooms had he not been aware of the visionary experience imparted by psilocybin-containing species.[2]

Ultimately, unless Pim was facetious in claiming that his story was veridical, we are left with a number of unanswered questions. Who was the mysterious Monsieur, and where and how did he discover the psychoactive properties of mushrooms? Was Monsieur actually Herbert Moore Pim himself?[3] Are there other written records concerning Monsieur's remarkable career? Into what asylum was Monsieur placed? In what "journals" did Pim find the references he mentioned concerning attacks on the glass-houses of Monsieur's benefactor? And finally, what became of Monsieur, his writings and his disciples?

All fascinating questions, answers to which might illuminate the status of awareness of psilocybin in the pre-Wasson era.

* * * * *

I have included in my essay's bibliography, below, all known editions of Lloyd's *Etidorhpa*. An introduction by Neal Wilgas appears in both of the later editions. Titled "The Pharmaceutical Alchemist," this introduction is scholarly and informative and discusses the psychedelic interests of John Uri Lloyd. I have also included citations for both instances in which "Monsieur Among the Mushrooms" was printed, as both are obscure and difficult to obtain. The 1915 version has extensive footnotes which were dropped from the 1917 reprinting.

Special deep thanks to Michael Horowitz of Flashback Books, Petaluma, California, for calling my attention to the work of Herbert Moore Pim.

Notes

1. This mushroom may have been misidentified, though its effects are very likely due to psilocybin.
2. Perhaps Pim had read *Etidorhpa*. This book was very popular in its time, having influenced Howard Phillips Lovecraft, the inventor of "cosmic horror" science fiction and the Cthulhu mythos. Lovecraft makes reference to *Etidorhpa* in material contained in his *Selected Letters* and *Marginalia,* noting, for instance, that his visit to the Endless Caverns in Virginia made him think "above all else, of that strange old novel *Etidorhpa* once pass'd around our Kleicomolo circle."
3. A respectable precedent for this was set by William James (1985, pp. 134–135 and endnote, p. 447), when he published a description of his own severe attacks of melancholia attributed to someone else in *The Varieties of Religious Experience* (1902).

References

Carroll, L. 1960. *The Annotated Alice: Alice's Adventures in Wonderland & Through the Looking Glass,* with an introduction and notes by Martin Gardner. New York: Clarkson N. Potter, Inc.

Graves, R. 1960. *Food for Centaurs.* New York: Doubleday & Co. Inc.

James, W. 1985. *The Varieties of Religious Experience.* Cambridge and London: Harvard University Press.

Lloyd, J. U. 1895. *Etidorhpa or The End of Earth.* Cincinnati: Author's Limited Edition.

Lloyd, J. U. 1976. *Etidorhpa or The End of Earth.* Albuquerque: Sunn Publishing Co.

Lloyd, J. U. 1978. *Etidorhpa or The End of Earth.* New York: Pocket Books.

Merrill, A. E. 1914. The narrative of Mr. W. *Science* 1914;40(1029).

Newman, A. [apparently a pseudonym for H. M. Pim] 1915. Monsieur among the mushrooms. *Irish Ecclesiastical Record* January–June 1915; 5:586–603.

Pim, H. M. 1917. Monsieur among the mushrooms. Reprinted, from *Irish Ecclesiastical Record* (1915), in *Unknown Immortals—In the Northern City of Success.* Dublin: The Talbot Press Ltd.

Wasson, V. P., and R. G. Wasson. 1957. *Mushrooms, Russia and History.* New York: Pantheon Books.

Wasson, R. G., A. Hofmann, and C. A. P. Ruck. 1978. *The Road to Eleusis: Unveiling the Secret of the Mysteries.* New York and London: Harcourt Brace Jovanovich, Inc.

Wasson, R. G. 1979. Foreword. In *Phantastica: Rare and Important Psychoactive Drug Literature from 1700 to the Present.* Privately published by William and Victoria Dailey Antiquarian Books and Fine Prints, 8216 Melrose Avenue, Los Angeles, California 90046.

Wells, H. G. 1966. *Best Science Fiction Stories of H. G. Wells.* New York: Dover Publications, Inc.

Left to right: Roberto Weitlaner, Gordon and Walter Miller negotiate rough trail during 1954 expedition to Mazatlán.

Searle Hoogshagen (*right*) and Gordon speak with two "informants" from the village of Santa María Nativitas Coatlán who vividly describe the effects of hallucinogenic mushrooms (1954).

Unless otherwise credited, the photographs are by Allan B. Richardson, through the courtesy of the Wasson Collection.

Plate 1

Huautla de Jiménez from the air (1956).

The landing strip at San Andrés (1955).

The trail to Huautla (1955).

The house on the outskirts of Huautla where Gordon and Allan Richardson first took the sacred mushrooms (1955).
R. Gordon Wasson. Courtesy Wasson Collection.

Plate 3

Gordon, camera at the ready, watches María Sabina prepare for the sacred mushroom ceremony on the evening of 29 June 1955. Photos were not permitted once the mushrooms took effect.

Plate 4

Allan Richardson eats the mushrooms prepared by María Sabina. No photo of Gordon doing likewise at this historic session survives (1955). *R. Gordon Wasson.*

Opposite: Psilocybe caerulescens Murrill var. *Mazetecorum* Heim, also known as the "Landslide Mushroom," was the species ingested by Gordon and Allan in 1955.

Plate 5

The marketplace of Huautla (1955).

Wares sold by an Indian woman in the marketplace include wax candles, copal incense and goose eggs—all used in religious ceremonies (1955).

Plate 6

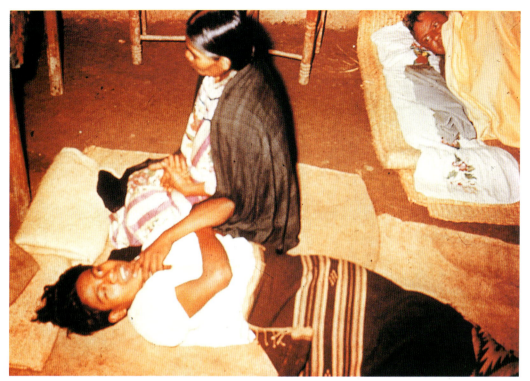

María Sabina tends to her mentally retarded son during a second mushroom ceremony witnessed by Gordon (seen watching in the background) during the 1955 expedition.

Gordon and Valentina in Huautla (1955). The jars contain samples of hallucinogenic mushrooms later sent to Roger Heim for identification.

Plate 7

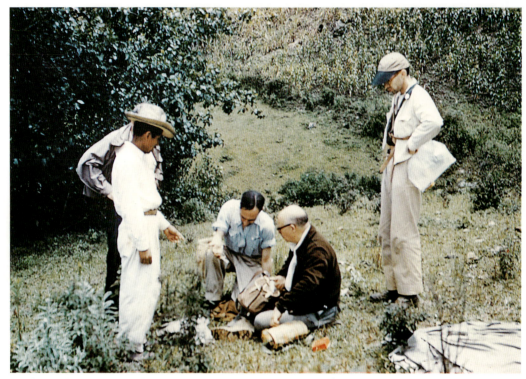

Left to right: Gordon (behind Indian guide), Guy Stresser-Péan, Roger Heim and James Moore gather mushrooms during 1956 expedition to Mexico.

The watercolor paintings of hallucinogenic mushrooms that illustrated Gordon's 1957 article in *Life* were all prepared by Roger Heim, here photographed at work during the 1956 expedition. Seven of these paintings are shown on the following pages.

Opposite: Unknown to Gordon at the time, James Moore was a CIA operative who joined the 1956 expedition to get samples of hallucinogenic mushrooms for possible use in developing mind-control drugs.

Plate 9

Psilocybe zapotecorum Heim

Psilocybe cubensis (Earle) Singer

Psilocybe caerulescens Murrill var. *mazetecorum* Heim

Roger Heim. Courtesy Wasson Collection.

Conocybe siligineoides Heim

Psilocybe caerulescens Murrill var. *nigripes* Heim

Psilocybe mexicana Heim

Psilocybe aztecorum Heim

Roger Heim. Courtesy Wasson Collection.

Plate 11

María Sabina prepares for a *velada* using sacred mushrooms gathered in a paperboard box by Gordon's team in 1956.

Plate 12

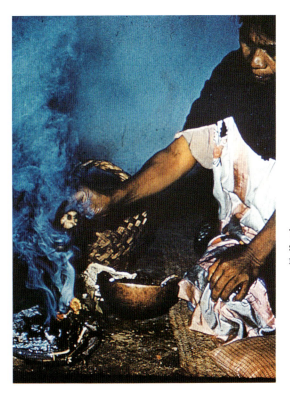

The mushrooms are passed through the smoke of burning copal as part of an ancient ritual (1956).

Gordon accepts his portion from María (1956).

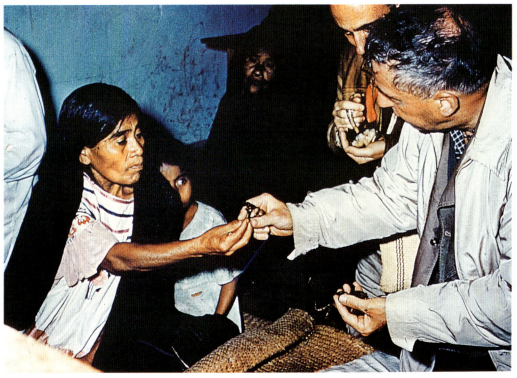

Plate 13

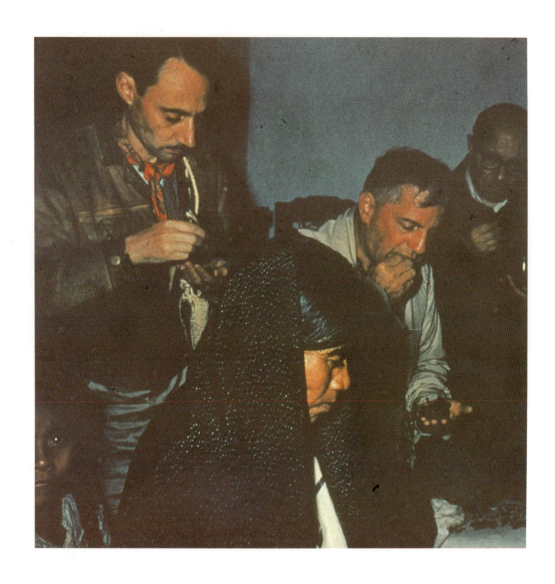

Plate 14

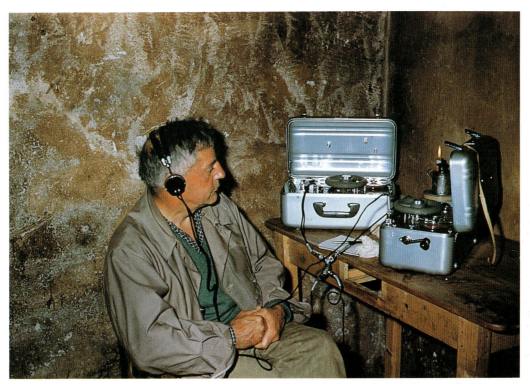

Also during the 1956 expedition, Gordon recorded a mushroom *velada* with this portable tape recording system. It was published in the form of a vinyl record with accompanying text by Folkway Records in 1957.

Herlinda Cid, Aurelio Carreras and Gordon pick through a pile of hallucinogenic mushrooms (*Psilocybe caerulescens* var. *Mazetecorum* Heim) in July 1958.

Plate 15

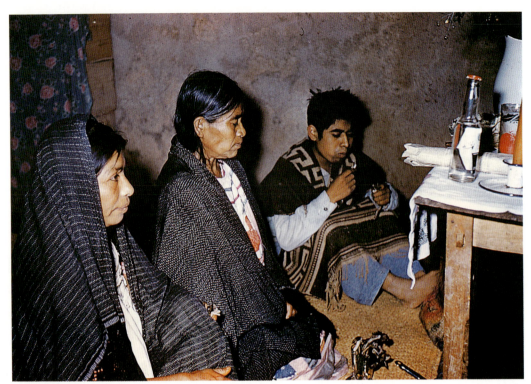

María Sabina prepares for the 1958 mushroom ceremony Gordon immortalized in the book of which he felt most proud, *María Sabina and her Mazatec Mushroom Velada*.

At the beginning of the ceremony, held to ask the mushrooms what will become of the ailing boy seated to her left, María claps in a rhythmic pattern (1958).

Plate 16

As the mushrooms take effect and she starts to see visions, María prays for enlightenment (1958).

Plate 17

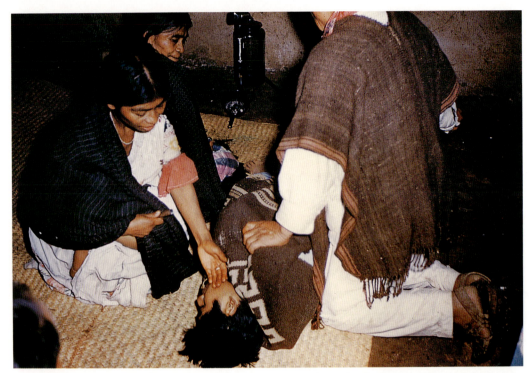

Upon learning from María that the mushrooms prognosticate death, the boy falls to the ground in despair. He did in fact die a few days later of undiagnosed, but apparently natural, causes. Note canister-type microphone under table, with which Gordon recorded the ceremony unobtrusively (1958).

A Mexican Indian girl crushes leaves of the *Salvia divinorum* plant to prepare a psychoactive powder (1962). *Courtesy Wasson Collection.*

Plate 18

Gordon, in the foreground, and a white-clad Albert Hofmann hike a rocky trail through the Sierra Mazateca mountains in October 1962. *Courtesy Wasson Collection.*

Plate 19

Researchers and students such as Rick Mason have found the Wasson Collection at Harvard Botanical Museum an invaluable resource for ethnomycological studies (1988). *Thomas Riedlinger.*

First editions of R. Gordon Wasson's books on ethnomycology include, *left to right:* the two-volume *Mushrooms, Russia and History* (1957); the companion books *Les Champignons Hallucinogènes du Mexique* (1958) and, lying before it, *Nouvelles Investigations sur les Champignons Hallucinogènes* (1967); *Soma* (1968); *María Sabina and her Mazatec Mushroom Velada* (1974), with the contents of its double slipcase displayed in the foreground; *The Road to Eleusis* (1978); *The Wondrous Mushroom* (1982), seen here outside its slipcase to expose the "disembodied eyes" cover motif; and *Persephone's Quest* (1986). *George Tarbay*

Plate 20

A Posthumous 'Encounter' with R. Gordon Wasson

CLAUDIO NARANJO

I am in a peculiar position as I set out to dictate the following in response to the invitation to participate in this volume. In spite of warm feelings for Gordon Wasson, and despite the encomiastic quality that characterizes contributions to a memorial publication, the best I can offer is the experience of my relationship with the man and his ideas, and this is a story of ongoing disagreement. I therefore trust that others will honor Wasson the fine gentleman with a twinkle in his eye, Wasson the scholar and Wasson the philanthropist, while I tell rather where I stand in regard to his beliefs and contentions.

I first encountered Gordon Wasson through his writings: a copy of his *Life* magazine article on the Mexican magic mushrooms came into my hands while I still lived in Chile. Only later, in 1967, did we meet at the San Francisco conference on the "Ethnopharmacologic Search for Psychoactive Drugs," chaired by Bo Holmstedt of the Karolinska Institute in Sweden. There I heard him relate how he, a mycophobic Anglo-Saxon, learned to love mushrooms thanks to "contagion" by his Russian mycophilic wife. I had previously read this story in one of Wasson's articles, published perhaps in *The Psychedelic Review,* which Ralph Metzner had shown me during my stay at Harvard in 1962. It taught me that those curious pieces of umbrella-shaped sculpture that archaeologists, for want of a better name, had simply called "mushrooms" did in fact represent mushrooms. From these and Wasson's other accounts I had formed an image of him as one who was able to win the confidence of the Indians in secret matters. Furthermore, by the time of the conference I had experienced my first psychedelic session, as a volunteer for an experiment in isolation at the University of Chile. As a consequence, I deeply sympathized with Wasson's penetration of contemporary mushroom use. It seemed obvious to me that his sincerity and the personal, rather than academic, nature of his interest had opened the door for him.

Though my principal role in the conference was a presentation concerning my harmaline research and, perhaps most eventfully, a report on joint research with Dr. Alexander Shulgin (which had the unforeseen result of bringing MDA—methylenedioxyamphetamine—to the streets), I was also invited to join a panel

177

under the chairmanship of Dr. Daniel H. Efron, co-sponsor of the conference through the U.S. Department of Health, Education and Welfare. It was on this occasion that I first "crossed words" with Wasson on the question of the origins and identity of Soma.

The panel discussion[1] opened with Dr. Efron's question: might *Banisteriopsis* have been used by Tibetan gurus? I said I was practically sure that *Banisteriopsis* did not grow in that part of the world due to unfavorable climatic conditions, but pointed out that *Peganum harmala,* containing harmaline (which, I had recently verified, is a more powerful hallucinogen than the harmine of *Banisteriopsis*), does grow in central Asia. I was of the opinion that *Peganum,* rather than *Amanita muscaria* or other mushrooms, was a more likely plant source candidate for Soma, because harmala alkaloids bring about a state much more comparable to Yogic trance, the specialty of the Indian tradition. Psychoactive mushrooms, by comparison, stimulate the nervous system; *Amanita's* effects include agitation, hallucinations and even delirium. Wasson responded to my hypothesis by suggesting that if there were any *Peganum* connected with Soma it would have to be a plant called "green *Peganum*—though not the plant known today by that name." The chairman then moved to the next item on his agenda.

Recently, however, another scholar, David Flattery, has argued persuasively that *Peganum,* not Wasson's *Amanita,* is the source of the ancient Soma. Flattery, a research associate in the Department of Near Eastern Studies at the University of California and an expert on both the Avestan language and Near Eastern ethnobotany, has written a book on the origins of Soma in collaboration with Martin Schwartz, which is soon to be published by the University of California Press. Recognizing Wasson's profound interest in this subject, and his willingness to clash with those who contested his theory, I offer the following summary of Flattery's major arguments:

1. **Geographical distribution.** No known psychotropic plant is as abundant over the Iranian area as is *Peganum harmala,* long known in Iran to have psychoactive properties.
2. **Psychopharmacology.** The suitability of *Peganum harmala* for use as Soma is demonstrated by the "parallel roles" played by Soma in ancient Iran and by *Banisteriopsis* extracts containing the same alkaloids in certain South American cultures.
3. **Name.** Modern Iranian names for *Peganum harmala* indicate that it was named "sacred" in ancient Iran. There is no plausible explanation for the present names for *Peganum harmala* unless it was Soma.
4. **Ethnobotanic attributes.** In pre-Islamic Avestan and Pahlavi texts *haoma* (i.e., Soma) is invoked in verses which: (1) attribute the origin of the use of *haoma* to the founding figures of Zoroastrianism; and (2) assert that *haoma* brings healing, victory, salvation and protection, originates in mountains, promotes childbirth (as an aphrodisiac?) and is chief of drugs. In post-Islamic Persian, Mandaic and Turkish texts *Peganum harmala* is invoked in verses which: (1) attribute the origin of *Peganum harmala* to the founding figures of Shi'a Islam; and (2) assert that *Peganum harmala* brings healing, victory, salvation and protection, originates in mountains, promotes childbirth (as an aphrodisiac) and is king of drugs.

My path crossed Wasson's again at two subsequent conferences, in the late 1970s and early 1980s. Though he knew I did not share his theory that Soma was *Amanita* or indeed any other mushroom, the time we spent together was cordial. Only later, too late to discuss it with him, did I fully understand his views, their implications and the effect they had on his many admirers. Only later, in fact, did my own thoughts on these matters finally crystalize. I now address them, in effect to Wasson posthumously, in the following paragraphs.

Let us begin with the claim, set forth in Wasson's last book, *Persephone's Quest: Entheogens and the Origins of Religion* (1986), that a psychedelic mushroom could have been the forbidden fruit in the Garden of Eden. My view is that the "forbidden fruit" idea is one of such rich and multi-faceted mythic significance that the suggestion it alludes to a real event or events in historical time—such as psychoactive plant usage—is unlikely. The very suggestion seems based, in my opinion, on a familiar manifestation of enthusiasm for the psychedelic experience: a wish to see psychedelics behind a variety of different events in the past.

Perhaps I can convey most clearly what I mean by discussing my response to the theory that psychoactive mushrooms helped stimulate evolutionary growth of the human brain, a theory explicitly stated by a scientist at an Esalen conference I once attended. Behind his hypothesis one could discern an implicit view that certain psychoactive mushrooms embody a spirit sent to this planet to assist fallen humanity. When he finished speaking and feedback was invited, I laughed a little and said I hoped he was being humorous or putting us to a test. But my comment was received with general seriousness by the speaker and his audience, and was not followed up. Nor was my subsequent observation that the state of mind induced by the mushrooms is typically less conducive to verbalization than is ordinary consciousness, and therefore could hardly have been, as the speaker had claimed, a stimulus for the development of human language. It seemed to me that I was in the midst of a group of enthusiasts who embraced the idea of the mushroom as messiah, an idea that had earlier been endorsed by John Allegro, using arguments from his translation of the Dead Sea scrolls, in *The Sacred Mushroom and the Cross* (1970).

I think it was through the "crypto-messianism" of Wasson's admirers that I first became aware of Wasson's own streak of mushroom messianism in his later years. I had failed to notice it sooner in his carefully-chosen words and fine scholarship. Even a non-exhibitionist and a person full of common sense like Wasson may succumb to emotional factors that generate what psychologists call "overvalued ideas." Just as a person's hidden sense of insecurity may compel grandiosity, I think a scholar's insecurity caused by society's disapproval of experiences he or she perceives as valuable may compel exaggerated claims for his or her research findings—claims that are usually based on an analysis of historical matters, rather than on the results of scientific investigation. Such a claim, I believe, was Wasson's theory of the origin of religion. It is stated most explicitly in Robert Forte's introduction to his recently published interview with Wasson:[2] "To him, we owe the rediscovery of the sacred mushroom and of its role *in generating* the ancient religions of Mexico, Greece, and India" (italics mine). I think it would be more correct simply to speak of the mushroom's role *in* the ancient religions of

Mexico, for we cannot be sure about the psychedelic theory of religious origins either in Mexico or in other cultures.

We still do not know whether human beings first gained access to altered states of consciousness by eating psychoactive plants or by spontaneous mysticism caused by confrontation with death, religious vocation and grace. In the present state of insufficient information we can only declare our intuition. And my own, based on the fact of both spontaneous and ascetic mysticism through the ages, is that spiritual perception was available to human beings before they first ate the "forbidden fruit," going back—to stay within the language of this metaphor— even so far as the time of Adam's insufflation by the breath of God (i.e., the emergence of true humanity from animal life).

Throughout most of his life, Mircea Eliade believed that the use of hallucinogens was characteristic of a late and relatively degenerative form of shamanism, while original mysticism relied solely on vocation and the "techniques of ecstasy." He changed his mind about this near the end of his life through the influence of Robert Graves, who in turn had been stimulated by Wasson to demonstrate mushroom use in the Dionysian religion of ancient Greece. My own view somewhat echoes that which Eliade relinquished. To me it seems likely that mysticism has existed at all times independently of plant use, but that psychedelics such as *Amanita, Peganum, Cannabis, Iboga* and others contributed much to the early diffusion of religious experience and played a part in the origin of religious *movements.* I am thus in agreement with the *wording* of Wasson's hypothesis on the "origins of religion," though not with his intended meaning: I believe in a possible role for psychedelics in the origin of early religious institutions, if not in the origin of the religious experience.

Had my thoughts on these matters been clear to me sooner, I would have told Wasson the next time we met that his research, while surely something to be praised and welcomed, seemed to me somewhat affected by his intense mycophilia and a consequent desire to see the shadow of the mushroom in every verbal and pictorial echo of the past. I would have challenged him to recognize to what extent his theory of religion represents an act of faith; for while I honor Gordon Wasson the collector, the field worker and the scholar, I disagree with and question Gordon Wasson the believer.

And yet I must admit to harboring not only sympathy toward Wasson's "will to believe," but also an appreciation of the value of his intellectual position in a society inimical to both magic mushrooms and experiential depth. However biased and simplistic, his mushroom hypothesis may help to counteract the prohibitionist attitude of the American ethos. If so, it may be that his contention embodies a kind of truth proper to the oracles and different from scientific truth: a truth that does not lie in the fact that something *is* the case, but rather in the fact that when something is *taken* as truth it furthers "lived truth," with salutary consequences.

Notes

1. Published in the proceedings: 1967. *Ethnopharmacologic Search for Psychoactive Drugs,* edited by D. H. Efron, B. Holmstedt, and N. S. Kline. NIMH Workshop Series No. 2. Dept. of Health, Education and Welfare. Washington, D.C.: U.S. Government Printing Office. The relevant section appears on pp. 445–446.
2. Forte, R. 1988. A conversation with R. Gordon Wasson (1898–1986). *ReVision* 1988; 10(4):13–30.

A Twentieth Century Darwin

JONATHAN OTT

I met R. Gordon Wasson in the summer of 1974. My undergraduate studies of natural products chemistry, which led me to work on the toxicology of the psychoactive *Amanita muscaria* mushroom species, combined with my interest in Indian religion, had led me to a reading of *Soma: Divine Mushroom of Immortality*. For me this was a revelation, as two quite different fields of study of compelling interest to me, fields which I had thought to be quite distinct and separate, suddenly merged into one. I had had the good fortune to meet Dick Schultes in Seattle earlier that year, and he had invited me to visit him in Cambridge and avail myself of the resources of the Oakes Ames Library of economic botany. Discussing topics of mutual interest with Schultes in his office that summer 14 years ago, he had suddenly said, "You must meet Gordon Wasson!" Imagine my delight as Schultes promptly telephoned Wasson in Danbury and told him, "There is a young man here you should meet," and then passed me the telephone! Wasson cordially invited me to have lunch with him a few days later and gave me the somewhat complicated directions to his country home. He was a bit surprised when I told him I would be hitch-hiking, and he said I should call him when I arrived at a nearby intersection, so he could pick me up. As it happened, my uncle loaned me his car on the appointed day, but I got lost and had to phone Gordon anyway for more detailed directions.

While I was somewhat in awe at first meeting Gordon, owing to my admiration for his work, his friendliness and evident delight in the interest I took in ethnomycology soon put me at ease. Though I only stayed for lunch and an evening of conversation, it seemed much longer, and I left Danbury reluctantly. Gordon and I began to correspond. Our letters at first dealt exclusively with ethnopharmacological topics. At my request he began to send me offprints of his early papers, and I kept him apprised of my *Amanita* research. In early 1975 I undertook a research fellowship in Mexico City, and Gordon informed me he would be visiting Mexico that summer to see friends and conduct a bit of field work. This was an opportunity not to be missed, so I eagerly offered my assistance.

We had dinner together in Mexico City, and Gordon invited me to accompany him on a weekend sojourn in Puebla with his old friend Irmgard Weitlaner Johnson and the Mexican mycologists Gastón Guzmán and Teófilo Herrera. We

183

first went to the little town of Chignahuapan, in which there was a church known colloquially as Nuestro Señor del Honguito ("Our Lord of the Little Mushroom"). It seems the church had a relic consisting of a saprophytic polypore on which an image of Christ was inscribed, along with a sun and a moon and other cryptic markings. The fungal relic was enshrined behind a glass lens in the center of the little church, although the priest then in charge didn't approve of it and was contriving to move it ever further from the liturgical center; when we saw it, it stood in the aisle. We were able to persuade the caretaker of the church to let us open the cross to clean and examine the specimen, allowing us to surreptitiously take a tiny sample of tissue for identification. The fungus did not prove to be psychoactive, nor did we find any evidence suggesting that the cult of *teonanácatl* survived in the vicinity. Our field trip also took us to a village called Nanacamilpa ("Mushroom Field"), where we likewise failed to find evidence of a surviving use of the sacred mushrooms. For me, however, this little trip was like a dream come true, as I was able to observe firsthand Gordon's gentle, probing technique of investigation. He had learned, he told me, to treat the indigenous people *as* equals (rather than *as if* they were equals), and I saw ample evidence of this as he easily won their confidence—no mean feat—and patiently brought them around to ethnomycological subjects. I saw that his ability to treat *campesinos* as equals was no act, that he genuinely liked and respected these uncomplicated people; and I fancied he felt more at home among them than he would have in a room filled with university professors.

When I was invited to give a five-lecture mini-course on psychoactive mushrooms at the Aspen Mushroom Conference in the summer of 1976, Gordon wrote me an urgent letter saying I must come to Danbury to study his files on the history of the scientific discovery of the entheogenic mushrooms. It happened that Drs. Alexander Smith and Rolf Singer were also members of the faculty for this conference, and Gordon wished me to know in detail the history of their involvement with this research. Of course I accepted his offer to spend a week with him in Danbury, and so made the long journey—for me rather like a pilgrimage—by bus from Mexico City to Texas, then hitch-hiked the rest of the way in the cool weather of early spring.

I passed that week immersed in Gordon's voluminous files, and in lengthy discussions with him on ethnomycological topics. I was in heaven! Gordon began to speak of me as his "disciple," and treated me as a friend and junior colleague. He gave me full access to his files, often delving into them himself to point out areas of particular interest. He gave me a huge pile of reprints of his and others' papers, as well as copies of two of his books. This treasure trove was so bulky that there was no way I could carry it with me, so we trussed it up in packages to consign to the post.

During that week I came to know Gordon well, and to learn much about his life and work. I told him I was an aspiring writer, that I considered him to be a master of the art, and asked him to teach me. From that point onward, Gordon always corrected any verbal mistakes I might make, often discoursing at length on etymological fine points. I would often send him articles I was working on, for his comments and criticisms, which were always apt and carefully considered. We discussed his book *Soma*—which I consider to be his masterwork and a brilliant piece

184

of writing—in minute detail. On more than one occasion he commented with delight, "My, you know my book better than I do!" Of course this wasn't true, but merely reflected the fact that I had studied the book with care and had it fresh in my mind.

Gordon frequently complained that his memory was failing him, as he often hesitated a moment or two in trying to recall a name or date. But I was amazed at his powers of recall: he knew so much, and he had such command of what he knew! I suppose his memory and speed of recall had been better before his stroke in the early 1960s, but in 1976, at the age of 78, he was a formidable mnemonist. I came to Danbury that spring as a pilgrim seeking the Grail, and I left a true disciple.

I next saw Gordon that fall, when he came to visit me in Olympia, Washington, and to take part in a conference on psychoactive mushrooms I had organized. Our correspondence became more frequent and detailed, and I visited him periodically in Danbury. He came to Washington again in 1977 for another conference I had organized, as well as for one the following year, in San Francisco. During these conferences I had an opportunity to see Gordon's businessman side (we must remember that he had a long and successful career as a banker). As an adjunct to the conference I had arranged to sell his books. This proved to be a minor success, and I remember well his delight when in settling accounts at the end of the 1977 conference he pocketed a tidy sum—about $2,000. The following years saw me publishing the proceedings of one of those conferences and an extract of another, during which period I had the good fortune to work closely with Gordon.

I often visited him in Danbury, usually for about a week at a time. On one occasion I interviewed him at length about his life and work, making about five hours of tape recordings. The details he recalled of his youth and childhood, including names, dates and places, was remarkable.

After one of his trips to Europe, he presented me with a complete set of Roger Heim's publications dealing with the entheogenic mushrooms. Such was my interest in his work that I had at length acquired all of his publications, as well as his father's two books. (Significantly, his father, Edmund Atwill Wasson, wrote a book dealing with the religious aspects of drug use, *Religion and Drink,* published in 1914, in which he refuted the pseudo-religious arguments then being advanced to prohibit alcohol. As Gordon was later to do with his own books, Edmund Wasson financed and arranged the publication of his book himself, to exercise complete control.) As a consequence of my interest in acquiring Gordon's books and my efforts to aid him in selling them, I was aware of the prices they were fetching. This was a topic of keenest interest to Gordon. Although he was in no sense a book collector, he was very familiar with the rare book market, as he had purchased so many rare and old volumes in the course of his ethnomycological research. The considerable value of *Mushrooms, Russia and History* (which sold for as much as $2,600 in the 1970s, making it the most valuable book ever by a living writer) was a source of pride to him.

It was during this period that Gordon and I collaborated with Jeremy Bigwood, Carl Ruck and Danny Staples on the launching of a neologism to replace words such as *hallucinogen,* which we felt had the adverse connotation of falsity or illusion, and *psychedelic,* invested with associations with modern drug use, 1960s rock music and the like, to describe plant drugs such as *teonanácatl*. At Carl Ruck's suggestion, we settled on the word *entheogen* and summarily published it. I am pleased to say that, despite being scorned by some and falling into disfavor with one prominent expert in the field, *entheogen* has been quite widely accepted in the scientific and ethnological literature, thanks in no small part to Gordon's adoption of the term, which he used in the title of his last book, *Persephone's Quest: Entheogens and the Origins of Religion* (1986). It was of the utmost importance to Gordon to have a term which would not prejudice the sacred mushrooms, for he always winced at the use of *hallucinogenic* or *psychedelic* to describe his beloved entheogens, and hated the use of the word *drug* in connection with them. While I find the term *entheogenic drug* useful, especially to describe artificial or semi-synthetic entheogens, Gordon never accepted it, opting instead for simply *entheogens* or *entheogenic plants*. In the issue of *The Journal of Psychedelic Drugs* in which we launched the neologism on its career, I suggested wistfully that one day the journal might be called *The Journal of Entheogenic Drugs*. This hasn't come to pass, but the fact that the editors saw fit to change the title to *The Journal of Psychoactive Drugs* shortly after the publication of our article was a sort of victory, consigning *psychedelic* ever more to the obscurity it so richly deserves.

My 1976 research on the history of the modern rediscovery of the sacred mushrooms had led me to pen a sarcastic footnote in my book *Hallucinogenic Plants of North America*, suggesting that Dr. Rolf Singer had in 1957 "followed in the footsteps" of the Wassons, hastily duplicating some of their work in a "whirlwind trip" and then rushing into print with what he found, in order to gain a spurious priority. I hesitate to bring the matter up again, but this was a very sore point for Gordon, who was eager to expose what he felt, in this case, to be a fraud. Indeed, his prime motivation for inviting me Danbury in 1976, and what led to our becoming friends, were my efforts to resolve the affair satisfactorily. He furnished me detailed information, including copies of the correspondence between Dr. Singer and his patron Dr. Sam Stein. (Stein, who had originally hired Singer to collect specimens and cultures of the Wassons' mushrooms for him, eventually became so disgusted with Singer's behavior that he apologized to Gordon and sent him copies of his and Singer's correspondence.)

As I later learned from the man who had been Singer's Mexican field assistant, Singer's notebook contained copies of Dr. Roger Heim's papers (Heim was Gordon's mycological collaborator and close friend) pasted in, along with copies of Heim's watercolor illustrations which had appeared earlier that year in *Life* (May 13, 1957). So closely did Singer follow in the Wassons' footsteps that he actually overtook them, meeting Gordon for the first and only time in a remote village in Oaxaca during his brief trip. His research uncovered little new information, but in collaboration with Dr. Alexander Smith he arranged for the publication *out of order* in the journal *Mycologia* two papers, one of which gave Latin descriptions of the mushrooms. This article appeared less than a month before a paper by Heim in which Heim gave the Latin binomial *Psilocybe wassonii* Heim to one of the mushrooms. Singer and Smith named the same mushroom *Psilocybe muliercula* Singer & Smith, and their name had chronological priority. However, Singer had not collected this mushroom in the field, but rather purchased it in a market *per* information gained from a paper by Heim in which Heim announced his intention to name the mushroom after the Wassons. The fact that Singer had once been Heim's student particularly underscored the discourtesy to a senior colleague.

Nevertheless, when I published my harsh footnote, Gordon scolded me for stepping out of line. While he agreed with everything I had said and appreciated my support, he felt I had been too heavy-handed. He was too much a gentleman to, as he put it, "engage in a pissing contest with a skunk." However, an unexpected sequel to this little footnote changed his mind.

After I had published two papers in *Mycologia* dealing with psilocybin mushrooms of the Pacific Northwest, the editors of the journal, curiously, saw fit to publish a vitriolic diatribe by Dr. Smith entitled "Some Comments on Hallucinogenic Agarics and the Hallucinations of Those Who Study Them"—by which he meant me. In Smith's article, pedantry and ill-conceived mycological nitpicking very thinly disguised a personal attack on me, motivated by my offending footnote in *Hallucinogenic Plants of North America*. Although Smith never actually stated the basis of his splenetic attack, he alluded obliquely to the footnote, citing as a grave error my purported mis-citation of his papers in *Mycologia*. In fact, he had evidently read only this footnote in my book, as he mistook numbers in parentheses (which referred to my bibliography, where his papers were correctly cited) for volume numbers in *Mycologia*!

Wasson was delighted that Smith had put his foot in it, and began to counsel me at length as to how I should proceed. He helped me write my rejoinder to Smith, and advised me how to deal with the editors of *Mycologia*. There had been some protests on the part of members of the Society which publishes the journal, concerning the fact that the editors had published a personal attack with no scientific content. A few members wrote me expressing their support. The editors of *Mycologia* agreed I had been wronged and offered to print my rejoinder—but not in its entirety. They would publish factual corrections only, and refused to publish the facts relating to Singer's and Smith's mistreatment of the Wassons and Heim, facts which were fundamental to understanding the motivation of Smith's hysterical broadside. After consulting with Gordon, I told the *Mycologia* editors that it was everything or nothing, and apprised them of my intention to publish my paper elsewhere if necessary. They stuck to their guns, and, as we had arranged, Gordon published my rejoinder as a paper (No. 6) in his Ethnomycological Studies series, under the imprimatur of the Harvard Botanical Museum.

Gordon obtained the mailing list of *Mycologia* from one of his colleagues at the New York Botanical Garden, which publishes the journal, and of which Gordon was an Honorary Life Manager, and sent copies of the rejoinder to all subscribers. This smoked out Singer, who then wrote a reply to my rejoinder, which Gordon graciously and generously published as No. 8 in the Ethnomycological Studies series in the same manner, followed one week later by the final salvo of the war—his own reply, No. 9 in the series, to Singer's! Gordon told me that he had changed his mind about my footnote, that he was very pleased at the fruit it had borne, and was grateful to me for bringing out into the open something that had been an annoyance for two decades.

During the several years in which this little drama ran its course, Gordon and I were in constant and close contact. We corresponded regularly and at length, and spoke on the phone often. (So closely were we in contact that one evening when I telephoned I was surprised when he answered the phone before it rang. But his surprise exceeded mine, for he had just picked up the receiver to call me and had not yet dialed a number—only to find I was already there on the other end of the line!) And, although I was then struggling to establish my business and build a house, I always managed to scrape together funds to visit Danbury at least once a year.

One humorous incident from that era sticks in my mind. Gordon and I were to meet in New York, and he called me at my hotel to invite me to dine at the Century Club. I didn't feel at all out of place strolling through the Times Square area to the club dressed in jeans, sandals and a Guatemalan peasant shirt, but the moment I stepped into the Century Club I knew something was amiss. The butler hastened to collar me and discharge me into the street, lest any members be offended by the sight of an undesirable element. But he was stopped in his tracks by Gordon, who was waiting on the stair landing, apparently with a presentiment that something of this sort might take place. "Wait, he's my friend," he told the butler, who was flabbergasted and nonplussed. It simply couldn't be possible that the ragamuffin before him was the guest of a distinguished and elderly member! When he came to his senses, the butler sputtered something to the effect that I couldn't be admitted unless properly attired. After some negotiation he reluctantly consented to admit me, but only after I had donned one of the waiters' jackets. He said it was the first time anyone had been admitted without a tie, much less someone wearing sandals! Gordon was heartily amused by the episode and not the least embarrassed nor upset. However, on the eve of my next trip to New England he told me to buy a sports jacket, which I summarily did. I always thereafter wore it when I visited him.

My last collaboration with Gordon involved "disembodied eyes." During a trip to Mexico researching my 1985 book on chocolate, I had photographed the ruins of the Palace of Quetzalpapálotl at Teotihuacán, in which the motif of the disembodied eyes is prominent. Gordon urged me to write a small paper for publication in the Harvard *Botanical Museum Leaflets,* and became so enthusiastic about the result that he included this little paper in his final book, *Persephone's Quest.*

I learned of Gordon's death while in India, in the State of Orissa, not far from the Simlipal Hills. The latter is the site of one of his important field investigations, which figures in *Persephone's Quest* in the chapter on the death of the Buddha. Curiously, I don't remember when last I saw him in Danbury. It must have been in the spring of 1985. But I think I must have blocked that visit from my mind, as he was then debilitated and suffering, with his memory failing him and his attention span short. I suppose I prefer, unconsciously, to remember him as he had been before his decline. I do remember the visit as having been somewhat sad, and I remember thinking as we parted that I would never see him again. The news of his death was therefore not surprising; I had expected it for some time. Nor was it shocking, for I had had time to adjust to the inevitable fact.

Nevertheless, the world is not quite the same for me since his death. He was my teacher, and his tireless enthusiasm was a constant source of inspiration to me. Moreover, despite the half-century difference in our ages, we saw eye-to-eye and communicated well. Gordon was one of the best friends I've ever had; helpful and loyal, generous and kind.

But while I, with Gordon's death, lost a dear friend and revered teacher, the world lost one of the finest writers of the 20th Century and one of the greatest, most original thinkers of modern science. I had always thought that great writing was the province of novelists and poets, but in the writing of Gordon Wasson I discovered literary brilliance in purely expository writing. He commanded a stunning vocabulary and was a natural linguist. He possessed a wonderful gift for expression, trained by diligent study of English, French and Spanish literature. However, his greatest ability as a writer was his genius in economy of expression—saying more with less. In this respect he was without peer among modern writers. Yet his was no clipped, exsanguinated, scientific prose. Even ignoring its content, his book *Soma* would stand as a modern masterpiece of expository writing. His 1960 paper (originally written as a speech for a mycological society) "The Hallucinogenic Fungi of Mexico: An Inquiry into the Origins of the Religious Idea among Primitive Peoples" is likewise a brilliant piece of writing, a thrilling adventure tale masquerading as a scientific paper. In this context I might also mention the final chapter, "Vale," of *Mushrooms, Russia and History* (though Gordon wrote this book with his wife Valentina, this chapter, as well as the bulk of the second volume, bears the unmistakable mark of Gordon's hand). In the explication of this exciting and, so far as I know, entirely novel theory of linguistics, Gordon evoked, through the beauty and poetry of his words, the emotional responses that, according to the theory, are the very essence of these "secretions peculiar to the human organism." Thus Gordon achieved a convincing demonstration of his theory in the process of propounding it! I have a tape recording of Gordon reading the final pages of *Mushrooms, Russia and History*. A stunned silence follows the end of this reading, followed by Gordon's surprised statement: "Did I write that? I think that's pretty good!" I couldn't agree more!

But of course one cannot ignore the content of Gordon's writing, for it embodies one of the most important scientific theories of the 20th Century, a brilliant synthesis of linguistics, anthropology, history and pharmacognosy. Underlying the theory, which must henceforth be known as The Wasson Theory, is the greatest discovery ever made in the nascent field of ethnopharmacognosy. The Wasson Theory, which Tina and Gordon Wasson "adumbrated rather diffidently" in *Mushrooms, Russia and History*—so much so that it escaped the notice of all reviewers of this book, including the famous critic Edmund Wilson—is no less than a scientific explanation for the origin of religions in antiquity. Based on their study of the linguistics of mushroom names, the Wassons had been led to surmise that there had been a religious role for mushrooms in antiquity. Following up a few scattered clues in the ethnographic literature, and by dint of persistent field work in Mexico, Gordon Wasson and his photographer Allan Richardson became, in June 1955, the first outsiders to participate in a shamanic ritual involving the use of entheogenic mushrooms.

This experience confirmed the Wassons' surmise in a dramatic way, and Gordon at once realized its implications. He had come face to face with the "wellsprings of cultural history," with the fundamental experience underlying the religious life of preliterate humankind in remotest antiquity. Others have studied shamanism, but the Wassons were the first to discover its essence, the first to ingest the shamanic sacrament, and the first to recognize shamanism as the prototypical religion. Gordon Wasson also became the first, and so far the only, investigator to record and publish a full translation of a shamanic curing ceremony. *María Sabina and her Mazatec Mushroom Velada* is one of the most important ethnographic documents in existence, and Gordon felt that this book was the most important piece of work he ever did. The fact that all his research was conducted independently and self-financed makes it all the more remarkable in this era of expensive institutional science.

I once told Gordon that I thought he was the Darwin of the 20th Century, a comparison I find increasingly apt. Like Darwin, whose fundamental discovery was a mechanism to *explain* evolution (which already was accepted by many in his day), Gordon Wasson discovered the mechanism to explain a central aspect of cultural evolution, the genesis of religion.

In the case of The Wasson Theory, many had already come to accept that religion had a naturalistic explanation, but none before the Wassons had proposed a convincing theory. Gordon Wasson, like Darwin, came from a family steeped in religion. Like Darwin, his religious background led to some hesitancy in expounding a theory which would prove controversial and iconoclastic. Darwin sat on his theory for many years before publishing it, evidently fearful of its implications. Gordon Wasson and his wife Valentina diffidently slipped their theory between the lines of *Mushrooms, Russia and History*; Gordon then offered it speculatively as an "Inquiry into the Origins of the Religious Idea" three years later, and only after two more decades had passed and he had amassed overwhelming evidence in its favor did he begin to state it openly and plainly—at first in *The Wondrous Mushroom: Mycolatry in Mesoamerica* and finally in *Persephone's Quest*.

The similarities end there, however. While The Darwinian Theory was met with either instant acceptance or rejection, the Wasson Theory has oddly been met with silence. Perhaps when the current American fad of religious fundamentalism has run its course the world will awake to the implications of this idea. Darwin had the advantage of propounding his theory in a world in which many already accepted the historical fact of evolution and in which there was a general climate of doubt surrounding religion, owing to the findings of 19th Century science. The Wassons, on the other hand, had the disadvantage of disseminating their theory at a time when the wheel had turned full circle and when science engendered doubt, thereby setting the stage for a renaissance of fundamentalist religion.

This is unfortunate, but the silence on the part of specialists is downright disgraceful. With a few prominent exceptions, Vedic scholars chose to stigmatize Wasson as an amateur and dismiss him out of hand rather than deal with *Soma* responsibly and seriously. Similarly, classical scholars hid behind their titles in fear rather than review and discuss *The Road to Eleusis: Unveiling the Secret of the Mysteries*. Perhaps most unfortunate was the appearance of *farceurs* like Andrija Puharich and the late John Allegro, who spun absurd theories based on the Wassons' research to make a fast buck. Finally, the legal hysteria following on the

heels of widespread use of entheogenic drugs in the 1960s has made it more difficult for The Wasson Theory to get a fair hearing.

If, as I believe, The Wasson Theory is true, then the truth will eventually
emerge, and Gordon and Valentina Wasson will take their rightful places in the
company of the other giants of modern science. Perhaps, like Gregor Mendel, the
Wassons will be recognized as great pioneers only after they have lain dead many
years, by a generation that never knew them.

I never knew Valentina Wasson. She died when I was nine years old. But I
knew Gordon well, and admired and respected him for what he was: a brilliant
thinker, a superb writer, a patient teacher, a kind and decent man, a loyal friend. I
will always miss him, and the world will not soon know his equal.

Allan Richardson at Gordon's New York apartment in 1957, shortly after publication of the two-volume set *Mushrooms, Russia and History* (seen on the mantelpiece). *R. Gordon Wasson.*

Recollections of R. Gordon Wasson's 'Friend and Photographer'

ALLAN B. RICHARDSON[1]

Gordon Wasson and I were the first "white outsiders" to eat the sacred mushrooms in the Mazatec Indian ceremony. I accompanied him five times on his trips to Mexico between 1954 and 1958. In his books Gordon writes that I went as his "friend and photographer." He used my photographs to illustrate the *Life* article and several of his later books related to the subject. I wouldn't have missed the adventure for anything, but after the last trip in 1958 I'd had enough. Gordon continued his search for hallucinogenic mushrooms for several years beyond that.

Our paths first crossed in the early 1950s. At the time I was visual aids coordinator at The Brearley School in New York City, where Gordon's daughter Masha was a student. Brearley is a 13-year college preparatory school for girls. I had taught photography there for about 10 years in addition to my profession as an illustration photographer and photographing weddings for the Bachrach studio in the city. My assignments for Bachrach were usually of prominent peoples' weddings, such as the Rockefellers.

In April of 1954 Masha introduced me to her father at one of the school's functions, where he learned that I was a photographer. A few days later he came to me and said, "I'd like you to show me how to use my camera." Curious, I asked his plans. He answered that he was going to Mexico to look for mushrooms. My first reaction was to ask him, rather facetiously, "What the heck do you want to do that for when there are plenty around here?" He explained that he and his wife Valentina were going in search of hallucinogenic mushrooms. I had no idea what "hallucinogenic" meant, but the trip sounded irresistible. So I begged Gordon, "Oh, take me with you!" He hesitated, pointing out that the primitive Indians are touchy with strangers. During his contact with them the previous year he had felt their inherent mistrust. We parted on that note and I put it out of mind.

A week later Gordon phoned, asking, "Would you like to go along as my friend and photographer?" Would I! I jumped at the chance. My wife Mary, however, was less enthusiastic. She said, "If you eat those weird toadstools anything can happen to you!" I confess she had something there. But eventually her resistance subsided. I think her curiosity overcame her fear, as my curiosity overcame mine when I finally did take the mushrooms about a year later.

My arrangement with Gordon was that he would take care of all expenses and insure my life for $50,000.[2] Other than that my only compensation would be based on what *Life* might pay Gordon for using my pictures, for which he retained publication rights. Gordon eventually sent me a check for $2,500—very good pay for the time. My camera colleagues sure did envy me! I was just a struggling freelancer before the article appeared.

I made a short-lived attempt at keeping a diary during that first expedition with Gordon, so I'm able to reconstruct what took place in some detail. On Friday, 21 May, 1954, I breakfasted with the Wassons in New York before leaving with Gordon in a beautiful Cadillac limousine for the International Airport. We took off at 10:45 a.m. on a DC7, arriving in Dallas at 1:25 p.m. after an uneventful but noisy trip—our forward seats were between the propellers. In Dallas we changed planes after passing through customs. Our connecting flight to Mexico was a DC6B that got us to Mexico City at 5:40 p.m.

After passing through customs again we took a taxi to the Cortés Hotel, which had a pleasant patio where Gordon and I sipped white wine while awaiting the arrival of the missionary Eunice Pike and representatives of Banco National du Mexico. He had been quiet and untalkative during the plane trip, but opened up on the patio over his aperitif. He was characteristically intolerant of foolish conversation and foolish questions. So I always tried to make my conversations with him meaningful. Consequently we got along fine because we respected each other's personalities.

Pike and the bank representatives soon arrived. We had a nice chat. The representatives confirmed that the bank would make available its plane, a De Havilland Dove. They left shortly after. Eunice Pike stayed with us for dinner, although she had already eaten.

We retired about 11:00 p.m. Restless from being overtired, I woke at 2:30 a.m. and decided to check my equipment—a Zeiss Contaflex camera that I used on all my Mexican trips with Gordon and a strobe that could be detached from it, put on an extension cord and moved around for side lighting. No telephoto or wide angle lens. From experience I had learned that the simpler the equipment the better for quick candids. Instead of a telephoto I just moved in closer; for a wide-angle effect I backed up. Plenty of Kodachrome film completed the list. I also brought some topographic maps of the area we were going to explore, which I had obtained from the Air Force through my Reserve unit.

The next day, Saturday, 22 May, we got an early start. We went to a radio station suggested by Eunice Pike, where Gordon communicated by radio with Walter Miller. Walter was a minister with the Summer Institute of Linguistics who had spent a lot of time with the Mixe Indians and knew their language. He was reluctant to join us at first because he had just returned from a stateside absence to put his daughter in school. But finally he consented to go, and planned to meet us at the airstrip in Oaxaca.

The copilot of the Bank of Mexico plane which was put at our service came to meet us at the hotel. He drove us to the airport. There we met the captain, a very nice young man named Carlos Borja, and we took off immediately. Our beautiful little plane traveled at about 120 miles per hour past snow-capped mountains for about 45 minutes, setting down in Oaxaca. During the trip I took pictures and managed, in my haste, to spill a whole cup of coffee on myself and the seat. My apologies were profuse.

Walter Miller met us at the airport. He was just as pleasant as expected. We left immediately for Ixtepec, arriving about half an hour later. The tropical heat was oppressive. It stayed with us until the next day, when we got into the Sierras.

At Ixtepec, we were joined by Roberto Weitlaner, who had taken the train because he was afraid of flying. Roberto had hired a car for the first day's trip to Ixtepec, where we bought supplies. From there we went to picturesque Juchitán, where I got my first introduction to Indian food—beef soup; the others had chicken soup. It was good, but difficult to eat without utensils. From that point on we had no eating utensils until we whittled wooden spoons from sticks.

Next we drove to Lagunas and from there to Santo Domingo Petapa, a Zapotec village. The stretch of road between Lagunas and Santo Domingo was in terrible shape. Gordon described it as "execrable" in one of his books.

We stayed overnight in Santo Domingo and set off the next morning, on foot and by mule, toward the village of Mazatlán. I rode till my rear couldn't stand it, then I walked till my feet protested. It was pretty slow progress because we were traveling up and down mountains, through arroyos and fording streams. By night-fall we reached Plantillo and spent the night there in a schoolhouse.

The second day out of Santo Domingo was also slow going. We spent our second night in the forest. Near the end of that day, I remember the sky clouded over and looked very threatening. I thought it was going to rain. So I complained to Walter Miller, "What are we going to do? We have no protection." And he said, "Well, I'll pray about it. It'll be all right." Then he went to the edge of a field and knelt down and prayed. Fifteen minutes later the sky cleared, the stars came out and everything was fine.

We entered Mazatlán shortly after noon on the following day. It is located up in the mountains at an altitude of 3,000 feet. The view was spectacular. The population at that time was about 1,200 comprising 202 families.

A festival was going on, their annual fiesta in honor of the Virgin de la Soledad. There was a lot of music, dancing and drinking. We made our way to the municipio and met the presidente and other officials, who were supervising the festivities. They checked our credentials and then put us up in a house near the church, which had no priest. Our house was a one-room, adobe structure with a thatched roof. It had no glass in the windows. The natives kept peeking in on us, interrupting our privacy.

We stayed there six days. During that time, one of the natives was found to be fooling around with someone else's wife. An old story. The torch-light search for this guy was really dramatic. I don't know if they ever caught up with him.

As for the mushrooms, we didn't get anywhere in Mazatlán. The people there wouldn't share any worthwhile information. But they were very nice to us. Even though they wouldn't let us do the mushrooms, they staged a farewell party for us before we left. They asked Gordon to be the bartender! One of the Indians got drunk and kept demanding more drinks. And Gordon said, "No. You've had enough." The Indian got furious, produced a big knife and started for Gordon. Luckily, his friends restrained him.

Left to right: Walter Miller, Gordon and Roberto Weitlaner pose for Allan Richardson in 1954 outside the thatched adobe hut where they stayed in Mazatlán. *Allan B. Richardson. Courtesy Wasson Collection.*

We left Mazatlán on Monday, 1 June. On the return trip we ran out of food before reaching Santo Domingo. Our guide said, "There's a little village off the beaten path, if you want to take a chance on it." So we decided that we'd go there. As we approached the village, the head man came out to meet us. He warned, "There's a revolution going on, and the 'ins' have chased the 'outs' into the hills. If the outs should see you coming they're going to kill you, to show the ins that they can take care of the situation." We thought that was a very nice reception! Nevertheless, he guided us in and put us in a schoolhouse. I said to Walter, "If we're still alive in the morning let's get out of here pronto!" Again he replied, "Well, I'll pray about it. It'll be all right." So he did. I woke up about 1:00 in the morning in my sleeping bag and found that nothing had happened: I was still alive. Now I wonder how I managed to get any sleep at all that night, but I was really dog-tired from walking all that way.

By 5 June we were back in New York, a distance of 3,000 miles and what seemed to be 1,000 years from Mazatlán. It had been a fascinating trip.

Gordon asked me to go back with him in June of '55. Again, my wife Mary didn't want to give permission. I think Gordon invited her to go with us, but she didn't feel up to it. During this trip I took the hallucinogenic mushrooms with Gordon. And according to the article in *Life*, Mary made me promise before leaving that I wouldn't let "those nasty toadstools" cross my lips. But I didn't promise that. I said, "I promise to be *careful*."

Masha and Gordon's wife Valentina came with us this time. In fact, they and Gordon flew down before me, on Friday, 24 June, and rented a small villa near Mexico City. I left to join them on Sunday, 26 June. My plane took off at 10:00 a.m. and touched down in Mexico at 5:40 p.m. It took me an hour to get through customs. I was awfully tired and got to bed by 11:00.

Next day, Gordon and I left by car about 2:00 p.m., bound for Tehuacán. It took four hours. On the way we stopped at Puebla and saw a magnificent church. In Tehuacán we walked through the outdoor market in the evening. Most interesting. I went to bed early again, with a headache.

Our ultimate destination was Huautla de Jiménez. On Tuesday morning we left the Tehuacán airport at 8:00 a.m. to get there, in a Cessna airplane. That was some trip—and some landing, on a strip that looked like an aircraft carrier! We waited there two hours for our baggage to be brought from Tehuacán on the return flight. Then we trekked to the home of Herlinda Cid, a school teacher living in Huautla who had agreed to put us up during our stay. We arrived there by nightfall, met Herlinda and had a drink with her. That evening I had my shoes repaired, rested a little and met some local officials. But we got nowhere with them.

The next day, Wednesday, 28 June, we met the vice president of the town, who was immediately helpful. Gordon said to him something like, "We want to tell the people back in the United States what a wonderful thing you have here. Could we possibly see it?" And the man said, "Oh yes, certainly." Then he introduced us to the woman *curandera*, María Sabina, who agreed to hold a mushroom session for us that evening.

A couple of the men took us down into the valley where a lot of sugar cane was decaying beside a little pool. This was the sugar cane bagasse, or refuse, on which the mushrooms grow. We photographed and picked some. (Later during our stay I decided to take a little swim in the pool to get clean. When I stripped and went in there was no one in sight. Then, suddenly, I looked over to shore and saw about five Indians watching and laughing. I got out of there and covered my confusion as best I could.)

That evening, about 8:15 p.m., we went to a house on the edge of Huautla, the home of Cayetano and Guadalupe García. The *curandera* used our mushrooms to conduct an all-night ceremony. Gordon, of course, has described it in detail. I took the mushrooms when they were passed to me, in part because Gordon said they'd probably be insulted if I didn't and also because my curiosity got the better of me. The taste was a little on the bitter or acrid side, but not too much so.

I had no idea what I was in for; I could only guess. And I certainly wasn't disappointed. It took about an hour for the mushrooms to take effect after we'd eaten them. My first visions were of Chinese motifs, like *Kubla Khan*—palaces, oriental designs and so forth. After that, I saw the vision of a beautiful mantlepiece with the portrait of a Spanish caballero over it, which I happen to remember very well. Later, when we got back to Mexico City on our way home, the man who had the Buick franchise invited us to his estate, a hacienda. When we walked into the drawing room, there was the portrait I had seen in my vision. I don't yet know what to make of that.

I've read Gordon's account of that evening. Beautiful words. We should put him down as one of the poets of the age, I think. As for myself, I don't remember having had any spiritual feelings or religious experiences. I was just awed and impressed by it all. Years later, three sisters came to my studio and I posed them with expressions on their faces that reflected the emotions I experienced during that evening: alternately pensive, happy and frightened. It wasn't the Indians that frightened me. They were very nice. I didn't like losing control of myself. That's why once was enough; I never again took the mushrooms.

No photographs were permitted that first evening during the ceremony. Gordon took all the notes. We finished at 5:30 in the morning. Before leaving we had chocolate to drink and some bread. Then we rested most of the morning, went to the town square, saw more officials and presented our passports. Gordon picked more mushrooms, which he dried and bottled. We had a delicious supper and retired at 9:30 p.m.

Tina and Masha came into Huautla the next morning, on the early plane at 9:00 a.m. We settled them, ate, went to town and then rested.

By the following day, Saturday, 2 July, Gordon felt he had to arrange for another mushroom ceremony. María agreed. Gordon asked if we could take photographs this time, telling them, "It's such a wonderful ceremony. Would you mind if we took just a few?" So they consented to it.

In the early 1960s, Allan Richardson posed three sisters to represent three main emotions he felt when he took the sacred mushrooms: alternately pensive, joyful and frightened. *Allan B. Richardson.*

That was the session with María's son Aurelio, who seemed to have something wrong with him—possibly he was retarded. Gordon took the mushrooms again that night. María's attention was directed to Aurelio. "Her performance," Gordon wrote in *Mushrooms, Russia and History,* "was the dramatic expression of a mother's love for her child, a lyric to mother-love, and interpreted in this way it was profoundly moving. The tenderness in her voice as she sang and spoke, and her gestures as she leaned over Aurelio to caress him, moved us profoundly. As strangers we should have been embarrassed, had we not seen in this *curandera* possessed of the mushrooms a symbol of eternal motherhood, rather than the anguished cry of an individual parent." I like that. It adds poetry to what happened. The picture I took that appeared in *Life* pretty well justifies and explains it. Years later the boy died a violent death. He was stabbed in the throat by a mule trader and died in his mother's arms.

There were problems taking pictures of María that night. I was shooting with a strobe in total darkness, so I had to aim at the sound of her voice. The first time I set off the flash and took her picture she stopped dead in her chant. Apparently I surprised her. And I thought, "Oh my God, what have I done?" Then she started up again probably half a minute later. So I didn't dare take too many pictures for fear of upsetting her. I think I took only a dozen all night long.

We were stranded by rain in Huautla for most of the following week. Gordon and I slept in one room in Herlinda's two-room hut; Tina and Masha slept in the other. We were bored stiff with nothing to do all day long. So Tina and Masha took mushrooms one afternoon just to have the experience. I don't remember much about it. I didn't take photographs.

Gordon stayed on for a while in Mexico after we got back from Huautla. I went home. The next year, 1956, I joined him again with some others he'd invited to go back to Huautla. One of them, Jim Moore, later turned out to be working for the CIA. I didn't find out about that until something like 20 years later, when I received a call from someone doing research for a book published shortly thereafter—*The Search for the "Manchurian Candidate,"* by John Marks. Almost an entire chapter is devoted to our 1956 expedition. At the time, all we knew was that we didn't like Jim. Something was wrong with him. When he first arrived, in a jeep or whatever it was he was driving, he carelessly dropped some delicate instruments on the ground. And that started things off badly. I avoided him and we went our separate ways. He asked me to get in touch with his wife when I got back to the states. That's the last I heard or saw of him.

Roger Heim was on that expedition, too. He was very nice, quiet and unassuming. During this trip he drew and painted the pictures of mushrooms published with Gordon's *Life* article. My impression of Guy Stresser-Péan was also very favorable.

Sometime just before or soon after our return from the '56 expedition, Gordon and I were dining at the Century Club in New York. He noticed Ed Thompson, the managing editor of *Life* magazine, alone at a table nearby, and asked him to join us. We talked about the article Gordon was working on to publicize what he'd discovered in Mexico. Thompson said *Life* might be interested in publishing it, and invited us to make a presentation at his offices. So we brought the slides over there, I think a few days later, and prepared to show the magazine

Allan Richardson and Gordon deplane in Mexico City during their final trip together in 1958. *Courtesy Wasson Collection.*

executives what we had. I remember Gordon prefaced it by saying, "We're going to show you the slides of our expeditions. And afterward, if you don't like it, I don't give a damn." I couldn't believe my ears! But following our presentation the executives said, "We like it!" That's how the article wound up in *Life*.

In all I made five trips with Gordon to Mexico. But details of those we made after 1955 have passed from my memory. I became bored because they seemed repititious. Our last expedition together was in 1958. That's when I took the photographs Gordon used in *María Sabina and her Mazatec Mushroom Velada*. By then I'd had enough. When we landed in Mexico City on our way back, there wasn't a room left in the hotel in which we'd stayed previously. So we went to some fourth-rate hotel. They didn't even have hot water, and I was dying for a bath, so I went down to the desk to complain. They said, "Oh, we'll turn on the furnace." Big help! I found I could get a plane out to New York that afternoon. So I simply said good-bye to Gordon and left. I couldn't get home quick enough. Gordon asked me to go with him again the following year but I declined.

After that, Huautla was discovered by the hippies. They seemed to have a nose for those good places. It got to be commercialized. That was a shame. When we first went down, there were no roads leading into Huautla for vehicles. Later they cleared it for jeeps and brought in electricity, ending the primitive quality. I guess it's inevitable.

For a while, Gordon and I gave some lectures together about our adventures. They were somewhat successful, though some audiences weren't very interested. A few years later I tried doing the lectures alone, but they usually weren't at all good. I wanted to do one at assembly at The Brearley School, but the

Allan Richardson with the Zeiss Contaflex camera he used in the 1950s to take all his photographs of María Sabina (1989). *Thomas Riedlinger.*

headmistress there wouldn't let me. I felt that was pretty thoughtless of her because all the girls were impressed by the whole thing. I don't know why she objected. Finally, I tried to give a lecture here in Winchester several years ago. I'd forgotten so much that I made a disaster of it.

Gordon and I remained warm friends, though I didn't see much of him after Mary and I moved to Winchester. We visited him at his home in Danbury a couple of times when we were traveling through that area, and I talked with him by telephone now and then. I also processed the photographs he used in *Soma: Divine Mushroom of Immortality.*

He was awfully nice, but in some ways very strange. Once, for example, I was taking portraits of Emily Post. When I mentioned it to him he asked, "Who is Emily Post?" He didn't seem to know about some of these popular characters. It wasn't up his line. The thing he really cared about was mushrooms.

Notes

1. As told to Thomas Riedlinger at Allan's home in Winchester, Virginia, in 1988. Allan was 88 years old at the time.
2. *Editor's note:* The history of this arrangement is recorded in three letters Gordon wrote to Allan, copies of which are on file in the Tina and Gordon Wasson Ethnomycological Collection at Harvard. The first, dated April 8, 1954, reveals that Gordon originally thought of placing his article in *The National Geographic.* It reads:

> Dear Allan:
>
> If you care to join up with me on a partnership basis on my trip to Mexico, here is what I propose. We shall leave on the evening of May 21 and hope to return by the night flight from Mexico City leaving there on the evening of June 2. I shall pay your expenses down and back, beginning with the departure of the plane. All pictures that you take relating to the subject matter that interests us will be ours to use in technical and learned journals and publications of limited circulation. (The National Geographic Magazine is a popular journal of very large circulation.)
>
> All of the photos relating to the matter that interests us will be held in reserve for use when we are ready, and at that time you and we exercising good will and good faith on both sides will reach an arrangement under which we can both benefit. You will hold me harmless for any accident or illness or other mishaps that may happen to you in the course of this little adventure.
>
> Perhaps you will think of other points to cover. I shall inquire about insurance to cover you on the trip and let you know. If you wish to do this, having in mind the possibilities that the trip may be unsuccessful and that for one reason or another we might not be able to catch the June 2 plane and be delayed a day or two more, then will you countersign the carbon copy of this letter and send it back to me.
>
> This will be a great trip if it is successful and it will be fun to travel together.
>
> > Sincerely yours,
> > Gordon
>
> P.S. By the way, I must clear this with Roberto Weitlaner, my anthropologist friend in Mexico, which should not be difficult.

Allan countersigned as Gordon had requested, adding a note in his own handwriting: "I feel too that it will be a wonderful trip."

The second letter, dated March 7, 1956, followed by several months the famous *velada* of 1955 in which Gordon and Allan were allowed to take the mushrooms for the first time:

> Dear Allan:
>
> Last Saturday you and I, as we drove downtown, discussed the financial relations between us in regard to the photographic material developed on our Mexican trips. The photos are of course indispensable. But we must not forget that they would never have come into existence if Tina and I had not worked up the cultural theme by dint of years of research, that we performed this preliminary work entirely at our own considerable expense over the years including the initial trip to Mexico, that at all times you have had your expenses paid and a modest supplementary sum in lieu of compensation, such payments being out of our pockets, and that selling of the material to *Life* or the *National Geographic* has been initiated and handled entirely by me.

My understanding of our talk last Saturday is as follows:

1. We confirmed our previous arrangement, entered into orally some time ago, that you are to receive the first $1,000 derived from the first serial use of the photos made to date [and] on our forthcoming trip, presumably in *Life*. (This really means $1,000 plus the payments previously made, for under our previous arrangement, before you were to receive further payments, I was to be reimbursed for those advances out of your part in the sales.) This arrangement excludes any proceeds from the sale of photos to *Vision,* an outlet that you dug up, which are to go to you. But *Vision* is not to use mushroom photos or mention mushrooms.

2. If *Life* should use for its cover a photo made on our trips, either by you or by me, or if some other magazine uses our photos paying for second serial rights, we shall divide the proceeds from the photos fifty-fifty.

3. In our book Tina and I can make use of any of the photos without further compensation. All of the Mexican photos in fact are Tina's and my property.

4. If I receive the hoped-for grants from various foundations that will cover not only the cost of equipment but expenses, etc., then, after deducting the cost of equipment and expenses, such grants will provide compensation for you up to but not exceeding $200 a week or part thereof, for not less than three weeks nor more than five, according to the length of your trip.

If the foregoing is in accord with your understanding, would you please initial one copy and return it to me.

As ever yours,
Gordon

Apparently, one of the "various foundations" from which Gordon was hoping to obtain a grant was the Geschickter Fund in Washington, D.C. It had been mentioned to him as a possible source of funding by James Moore, the CIA operative, when he initially contacted Gordon in August 1955. Unknown to Gordon, the Fund was a front for the CIA to channel money secretly. According to John Marks' book *The Search for the "Manchurian Candidate"* (New York: Dell, 1979), it anted up $2,000 to help finance Gordon's expedition in the spring of 1956.

On November 7, 1956, Gordon wrote Allan the last of his letters concerning their arrangement:

Dear Allan:

I have suggested to Phil that he send you a check for $2,500. In conformity with what we said yesterday, I shall consider this payment in full for your services to date and for all of the pictures.

This has been a most happy relationship for me. I wish everything in life went as smoothly as our adventures have gone. Should you find a market hereafter for the pictures, you will be entitled to part of the consideration, depending entirely on all of the circumstances. I am sure that we shall have no trouble arriving at terms satisfactory to us both.

Should I find a market for the pictures, with or without text, the proceeds will go to me.

With warm regards,

Sincerely,
Gordon

A Latecomer's View of R. Gordon Wasson

THOMAS J. RIEDLINGER

I was not privileged to meet R. Gordon Wasson until late in his life. By then 87 years old, he had suffered some setbacks in his health and was debilitated. His mind, however, seemed to me quite clear and remarkably agile. Undoubtedly his friends who had known him longer than I had a different view. They must have seen in Gordon, near the end, something less than he had been, and may have found his company somewhat painful for that reason. I had the advantage of starting out fresh with him; my memory could conjure up no image of Gordon in better days with which to compare him unfavorably. To me he was an old man, yes, and obviously ill, but still a man of superior character, charm and intelligence.

The chain of events leading up to our meetings began one evening in October 1981. My wife June and I were visiting two new friends, William Cashman and his wife Joann, at their home in the town of St. James on Beaver Island.

Beaver, the largest of a lovely group of islands in Lake Michigan, is accessible only by airplane or boat. No bridges connect it to the mainland, which is 20 miles distant at the nearest point. To June and me, who live near Chicago, it seems the proverbial "middle of nowhere."

A few months earlier we had purchased some land on the island from Bill Cashman. Upon our return in October for a short vacation he and Joann had invited us for dinner. That evening, we learned they had lived in San Francisco for several years before moving to the island in the mid-1970s. Now Bill, an aspiring writer, made his living as a house builder. He was articulate and obviously well-informed on many subjects.

Nonetheless, I was surprised to hear him make a casual reference to "Wasson, the mycologist," while discussing native mushroom species. At the time, I knew of Gordon's work but had not yet met him or even considered attempting to do so. Psychology rather than mycology was my field of primary interest, especially transpersonal psychology with emphasis on mystical, or "peak," religious experiences. Gordon's writings on the ritual use of psychoactive mushrooms by religious cults in Mexico and elsewhere had captured my attention and imagination. However, I was not aware that his work was familiar to many outside academe, where I had first encountered it during my college days. So it struck me as somewhat fantastic that Bill Cashman even knew his name.

205

"You've heard of Wasson?," I asked.

"He was here," Bill answered.

My surprise at least doubled. "Here on Beaver Island?," I asked.

"He was here in this house!," Bill replied.

As Bill then explained it, a mutual friend had brought Gordon to their door one summer's day in 1976. It was an unannounced and unexpected pleasure. Gordon had come to the islands in search of wild mushrooms. Their mutual friend, unaware that the Cashmans already knew Gordon by reputation, felt that Bill would enjoy meeting him.

"It was a fortunate coincidence," said Bill. He and Joann owned a rare copy of the limited edition of *Soma: Divine Mushroom of Immortality,* which they had purchased in San Francisco a few years earlier for $200. Gordon was delighted to discover that the Cashmans were invested fans.

"I will double its value by signing it," Gordon proposed. On the flyleaf— where else more appropriate for a book about the fly-agaric mushroom?—he wrote: "This copy of my SOMA I inscribe with pleasure to Joann and Bill Cashman."

That was on 2 August 1976. A few days later Gordon visited again. He, Bill, Joann and their friend all went to visit Burt Wallace (not his real name), a college professor of literature who vacations on the island, for what Bill called "an evening of repartee." What ensued, as related by Bill, is a fine Wasson anecdote.

"Burt's a very brilliant guy, but sometimes he drinks a little too much Scotch and then becomes insulting," said Bill.

"Burt is someone who claims to have known everybody that was anybody: to have drunk Jack Kerouac under the table, to have out-fornicated this guy or that guy, to have been there at the crucial moment when certain movements were born, and so on. I had visited him with other people and he'd always been able to top their stories. If they knew *of* so-and-so, he *knew* so-and-so. If they *knew* so-and-so, he was the one that had actually advised so-and-so to produce such-and-such. So he was used to getting in the last word.

"Burt had urged me to show him some of my own work and had been very promising about what he could do for me. He held out the prospect of showing it to a good friend of his, an agent, who could take a struggling young writer such as myself and make him a success. But when Burt saw my work, on a previous visit, he started drinking and became insulting. So I was kind of glad to be returning now with Wasson as my unwitting champion.

"Almost immediately, Wasson and Burt became the two people at the table. The rest of us backed offstage, so to speak; we were down in the audience. The way things went was, Burt would tell a story, and then Wasson would tell a story about the same subject—only his was a more interesting story of greater depth and greater import. The subject was personal anecdotes about the foibles of writers.

"As the stories went back and forth the fever quickened. It soon became apparent that Burt Wallace was having an experience he didn't relish: having his 'topping stories' topped. There was a certain nonchalance about the way Wasson did it that I found very rewarding.

"And then it came down to Burt's trump card. He trotted out the name of a relatively minor English poet—I think it was Edward Fitzgerald—in such a

206

manner as to say: 'Well, that may very well be true, Mr. Wasson, but as Fitzgerald would say,' and then such-and-such. Wasson countered, without missing a step: 'Ah, yes. Well, those rather risque poems that he wrote—which were never published, of course—put a different light on that matter.' And then Burt said: 'Yes, I've heard of those poems. But they're under lock and key in the British Museum. I've never been able to see them.' To which Wasson replied: 'Well, you really should make the effort. Let me see if I can remember how they go.' And he began to quote them verbatim, line after line. It was truly amazing. Even Burt Wallace was stunned into silence.

"That was the end of a perfect evening as far as I was concerned," Bill concluded.

I also learned that Joann Cashman had spent two weeks in the Mexican village of Huautla de Jiménez in the mid-1960s. She had gone there with two friends to meet María Sabina, who prepared the sacred mushrooms for them twice.

The experience of crossing Gordon's path on Beaver Island and hearing the Cashmans tell their stories had a lingering effect on me. Upon returning from the island, I continued to wonder at what seemed a rather unlikely coincidence. There, in the middle of nowhere, I had found myself standing in the great Wasson's footprints, so to speak!

By and by I resolved to write Gordon a letter describing this turn of events, as an entree to establishing a correspondence with him. For good measure, I included a manuscript copy of "Sartre's Rite of Passage," my paper on Jean-Paul Sartre's 1935 mescaline experience that was later published in *The Journal of Transpersonal Psychology*. I addressed the large envelope to Gordon in care of his most recent publisher and mailed it on 16 June 1982.

Would the publisher forward it promptly?, I wondered. If so, would Gordon deign to respond?

Never fear. The first thing I learned about Gordon was how quickly he answered his mail. His response, dated 26 June 1982, arrived less than two weeks later. He was cordial in acknowledging the news of my meeting with Bill and Joann Cashman. Concerning my manuscript, however, he was rather abrupt. "I must make a confession to you," Gordon wrote. "I have never been able to read Sartre: nothing of his that I have come across seems worth while." Then he added: "On p. 30 [of the manuscript], 'like he' is of course wrong. You don't say that, do you? You mean 'like him.' There are several other mistakes in your essay in the writing of English."

The criticism stung a bit, especially since I make my living as an editor and writer. But then I laughed out loud, appreciatively. Gordon's comments seemed characteristic of the image I had formed of him from reading his carefully written books and articles. He was obviously a scrupulous grammarian, a dedicated wordsmith and defender of the English language. That he had tended to my manuscript was clear and complimentary. As for the errors he detected in my grammar—they were there, of course. I decided not to take it badly, writing back to thank him for pointing them out to me. Thus began several years of correspondence leading up to my first meeting with him as a guest in his lovely home at 42, Long Ridge Road, Danbury, Connecticut.

207

The date was 2 December 1985. I had arranged to visit Gordon at 10:00 that morning, after which I would travel to Baltimore on business. The day before, a Sunday, had been spent with friends in Maine. From there I had driven to Boston's Logan Airport, from which I had expected to make a short plane ride to an airport in Hartford, near Danbury. But things had not worked out as planned. Due to some confusion about my flight's schedule I had missed it and been forced to rent a car instead. Leaving Logan at about 10:30 p.m., I drove straight through to Danbury, a trip of about four hours. Upon arriving I checked into a local motel and got a restless night's sleep.

Perhaps I was haunted by nightmares of Gordon as I had begun to imagine him: extremely stiff and proper; overbearing and a little condescending; somewhat stern and entirely humorless; a curmudgeon, albeit an erudite one. Because of his background as a banker—in fact, a vice president—at J. P. Morgan & Co. (now Morgan Guaranty), I confess that I also suspected he lived in baronial splendor. It seems laughable, looking back now, that I let myself be led so far astray by bourgeois prejudices.

The sky was overcast that morning. Occasional short bursts of rain rattled down on the roof of my car as I drove into the foothills outside the town proper. Mild temperatures for the season had conspired with the rainfall to generate mist—good mushroom weather. As I drove up Long Ridge Road the mist appeared to be trapped, like a smoky blue haze, in the nearly leafless treetops of the hardwoods receding below me.

Gordon's red-trimmed, white brick house was not the mansion I had expected. Set close to the road on two acres of beautifully landscaped property, it impressed me as large but not lavishly so. The style was "antique colonial," according to a local real estate broker who provided details a few years after Gordon's death: a renovated farmhouse, two stories tall, with dormers on the second floor providing a view of the road on one side and the grassy, well-tended back yard on the other. This main building (I learned from the broker) comprised just over 1,800 square feet, not including the attached garage. An adjacent white frame bungalow in which Gordon's housekeeper stayed had about 1,200 square feet of living space. And several hundred feet behind the main building stood a more contemporary "artist's studio" providing another 1,200 square feet. The latter, I later discovered, was where Gordon spent most of his time before his failing health prevented it.

The garage, attached to the house by a short breezeway, had an empty bay facing the road. I pulled into it, turned off my car and prepared to meet Gordon. Expecting that perhaps he wore a sports coat and tie, even at home, I was similarly dressed. Also, in deference to his imagined gourmet tastes, I had brought along a bottle of fine old port. With this in hand I walked up to the door at the side of Gordon's house near the garage and knocked loudly.

From the moment he opened that door to greet me personally, my misconceptions about Gordon were quickly demolished. Instead of a sports coat and tie, he was comfortably dressed in loose trousers and a red cotton shirt of the "lumberjack" type, over which he wore, unbuttoned, a plain brown vest with wool lining. The ensemble would not have been inappropriate for a camping expedition.

208

But my strongest first impression, and the one that put me most at ease immediately, was his smile. Gordon's mouth was wide, as his eyes were large, and together these features combined to make his face quite expressive. When he gripped my hand in a welcoming gesture and smiled at me, I felt less like a stranger and more like an old friend whose visit had been eagerly anticipated.

Gordon ushered me over the threshold and into his large country kitchen. There he accepted the wine with approval and thanks, then suggested I follow him into the adjoining living room. As he led the way I noticed he was shorter than I had imagined—less than my 5 feet, 7 inches—and slightly stooped by the twin burdens of age and infirmity. These troubles affected his gait as well, which was slow and shuffling. He needlessly apologized for not being able to move about faster and urged me to go on ahead.

The living room, which overlooked the spacious back yard, was high-ceilinged, much longer than wide and naturally well-lighted by tall windows and glass doors along one wall. I had expected that it would be filled with objets d'art, but in fact it was furnished and decorated simply. The dominant artwork was a painting of his late wife Valentina, who was also represented by a drawing in a picture frame on one of the bookshelves. There were fewer books in evidence than I had anticipated; later, I learned he had shipped almost all of his personal collection to Harvard University's Botanical Museum in 1982.

A large rug on the hardwood floor created a kind of oasis or island on which were clustered chairs and a sofa. The nondescript furniture all looked old, very solid and invitingly comfortable. Gordon sat in an overstuffed chair and I settled myself on the opposite sofa. He promised that Yvonne, his cook and housekeeper, would soon arrive and make lunch for us.

First, he asked if I had heard the news that María Sabina had died a few days earlier, on 22 November. I told him I had; that her death had rated a large write-up in the obituary section of the *Chicago Tribune*. He too had seen a similar account in another newspaper, but noted disapprovingly that it had listed several celebrities, such as Bob Dylan, John Lennon and Mick Jagger, who had visited her in Huautla de Jiménez while neglecting to mention such people as his friend Albert Hofmann. He had not yet been formally notified of María's death by his friends in Mexico, Gordon said; he expected to hear from them shortly.

This prompted a discussion of his famous article, "Seeking the Magic Mushroom," which had appeared in the May 13, 1957 issue of *Life* magazine. He recalled that the magazine's publisher, Henry Luce, had promised him that not a single word of the article's text would be changed.

"That seemed to anger some of the people on his staff," observed Gordon. But Luce kept his word. *Life*'s editors contributed only captions for the photographs that ran with the piece. These, Gordon said, were "completely satisfactory."

I wondered if his Morgan bank employers had expressed reservations when told he was about to reveal to the world that he had partaken of the magic mushrooms.

"Not at all," Gordon said. "I suspect that *only* Morgan would have tolerated what I did." He had shown an advance copy of the article to his "superior" at the bank, a man named Henry Alexander, who had judged it "superb" and then said he looked forward to seeing it published.

The very idea that executives at Morgan bank, a highly respectable institution, would have tolerated, let alone applauded, Gordon's efforts in exploring hallucinogens surprised me.[1] I later phoned two of his former bank colleagues and asked them about it. Their comments, recorded in 1988, are worth briefly mentioning here.

DeWitt Peterkin, retired vice president in charge of domestic lending, joined J. P. Morgan & Co. in 1937. Gordon was already there, initially as a credit banker. He soon proved himself "a great person for putting together the background and history" of Morgan's accounts, recalled Peterkin.

It was a talent that led naturally to Gordon's first book, *The Hall Carbine Affair: A Study in Contemporary Folklore* (1941), which chronicled the development of a spurious rumor that J. P. Morgan had sold defective carbine rifles to the U.S. Army in the early days of the Civil War. Peterkin said the book did not cause a sensation in the bank's inner circle. "I don't think most of us had ever believed anything except the truth about the matter," he explained. "Gordon just proved the facts."

In subsequent years, Gordon's role as a credit banker gave way to new responsibilities. Eventually, as vice president, he wound up in charge of "communications, public relations—that sort of thing," recalled Peterkin. Personal contact with overseas clients was part of the job.

"Unbeknownst to most people, we for many years were one of the bankers for the Vatican," Peterkin said. "And Gordon used to have private audiences with the Pope." Though he could not recall which particular Pope, other sources later told me it had been Pius XII—and that Gordon had not liked him much.

According to Peterkin, no preview copy of Gordon's *Life* article had made the rounds within the bank before publication. But his colleagues reportedly took it in stride when they finally saw it.

"I was never surprised with anything he came out with," said Peterkin, explaining that he and the others found Gordon's intelligence so amazing that virtually nothing he accomplished would have surprised them—with one exception.

"He knew very little about sports," remembered Peterkin. "There used to be a quiz show in the early days of radio and TV, 'The $64,000 Question.' And we said: 'Gordon, you could go on one of those shows, but don't ever take up the subject of sports.' I think he thought it was beneath his dignity to be bothered with sports."

Asked if he thought it incongruous that Gordon had been studying hallucinogenic mushrooms while employed as a Morgan banker, Peterkin answered, sincerely:

"Oh no, because people had their avocations and their hobbies. We had mountain climbers, skiers; a little bit of everything. Mushrooms are a little more esoteric than most things, but Gordon had a very inquiring mind."

I got a similar reaction from another of Gordon's bank colleagues, James Brugger, who had started with Morgan in 1954 and retired after serving many years as public relations manager. He had not known of Gordon's interest in mushrooms until the eve of publication of the *Life* magazine article.

"Like much about Gordon," Brugger recalled, "it evoked mild wonder. And after the first mild wonder, a feeling that it shouldn't have surprised us, really, for

Gordon was a man of many parts. We just wondered: what next? You have to remember that this was an institution of people not easily bemused."

As for whether he thought it ironic that Gordon had been studying magic mushrooms as a sideline to his bank job, Brugger said: "I think his association with the project conferred a respectability on it that it might not have had if someone else had been doing it."

At any rate, Brugger emphasized, Gordon's mushroom hobby never interfered with his performance as a banker: "He wasn't growing anything at his desk, so far as I know."

In an interview with Gordon conducted by Robert Forte and published in *ReVision* (1988, Vol. 10, No. 1), Gordon revealed that he had avoided discussing his mushroom investigations with his banking colleagues. "I did not wish to get involved in trying to educate people who were totally ignorant of the subject," he explained. "Anyway, I did not. I may have been wrong. Perhaps I should have started to convert everyone. It may have been the source of a revolution on Wall Street!"

Gordon and I next talked about Timothy Leary and the so-called "psychedelic revolution" of the 1960s and early '70s. I prefaced this discussion by acknowledging that Gordon had rejected the word *psychedelic,* which he felt had acquired a negative onus, in favor of *entheogen,* meaning "god generated within." From Gordon's ethnological perspective, the latter was a clearer indication that these substances are sacred, or at least are so perceived by the indigenous peoples who use them in religious ceremonies.

However, I had to admit that I still favored *psychedelic.* Would he mind, since the word came more naturally to me, if I used it in our conversations?

Gordon graciously assented, but was curious to know why I considered *"that word"* more acceptable.

My answer was that *psychedelic* signifies "mind-manifesting." (I carelessly used the word "signifies" as if it were synonymous with "means," and this undoubtedly explains why Gordon did not challenge my error. For according to the relevant linguistic rules, the Greek roots *psyche,* meaning "soul" or "mind," and *deloun,* "to show," should rather be combined as *psychodelic.*)

Psychedelics, I explained, had been found in a number of studies to function psychologically as "non-specific amplifiers." In other words, they can project into consciousness (amplify) unconscious feelings, perceptions and other psychological effects that are different (non-specific) for each individual. Furthermore, as Aldous Huxley noted in *The Doors of Perception* (1954), these substances often induce for their users a sense of identification with a universal consciousness, or "Mind-at-Large." In either case, mind (or Mind) becomes manifest as a result of ingesting the substances, thus validating *psychedelic's* meaning. It therefore seemed wasteful to me, I told Gordon, to throw out the term just because it had suffered some bad associations in the past.

"Well," Gordon said after hearing me out, "perhaps you'll change your mind some day."

We then turned our attention to Timothy Leary, whom Gordon lit into at once, tooth and nail.

"He's vain and superficial. But the worse thing is, he doesn't know it yet—*to this day,"* declared Gordon. "He's still acting as if he had something important to say. But he's reckless. No scholarship there."

I listened quietly, not wanting to take sides in what I felt to be a private feud.[2] Gordon noted my silence and asked, in a perfectly level tone of voice:

"Are you a friend of Tim Leary's?"

"I've met him," I answered, "and we've corresponded." Gordon did not need to ask if I shared his opinion of Leary; the unspoken question hung over my head like a thundercloud.

"I think his early work is something you'd find valid," I suggested.

"What work do you mean?," Gordon asked.

I cited Leary's *Interpersonal Diagnosis of Personality* (1957), judged "[p]erhaps the most important clinical book to appear this year" by the 1958 *Annual Review of Psychology*; his psychotherapy work with criminals at the Massachusetts Correctional Institute in Concord from 1961 to 1963, using *Psilocybe* mushrooms as an adjunct; and a paper he delivered in the early 1960s to a group of Lutheran ministers, on the religious implications of the sacred mushrooms.

The latter, which was published as "The Religious Experience: Its Production and Interpretation" in a 1964 issue of *The Psychedelic Review* (Vol. 1, No. 3), piqued Gordon's interest.

"I have never seen that paper," he told me.[3] "What did he say about the mushrooms?" Rather than attempting to describe it, I promised to send him a copy.

We then briefly discussed the widespread experimentation with psychedelics by young people during the 1960s and '70s often blamed, simplistically, on Leary. (At least three recent books[4] have suggested that Gordon's *Life* article was one of the factors which triggered this social phenomenon.) We both lamented that repressive laws passed by the government to stop it had also derailed legitimate research with these substances.[5]

"The government!," Gordon declared, becoming exercised even more than he had when discussing Timothy Leary. "Have you heard about the man who was given LSD *without his knowledge,* by the CIA? And then he leapt to his death from a window. That they should do such a thing!"

I failed to ask him about his own earlier brush with the CIA, of which I knew only a little at the time of our meetings. Forte's interview quotes him as saying: "[The CIA] called me up and asked whether I would work with them, and I said, 'Give me a day to think it over.' I called them up the next day and said I would not. It would tie my hands, since I couldn't talk any more." Subsequently, Gordon was contacted by James Moore, a chemist from the University of Delaware who also, unknown to Gordon, was a CIA operative (see diary entry on facing page). Moore offered to arrange some funding ($2,000) for Gordon's next expedition to Mexico, in 1956, through one of the CIA's front groups, the Geschickter Fund, if Moore were allowed to join him. Complete details may be found in *The Search for the "Manchurian Candidate"* (1979), by John Marks.

In preparation for writing this essay, I placed a Freedom of Information Act request with the CIA for a copy of Gordon's file, to see if other covert efforts were involved. The only item it contained was a letter sent to Gordon by an admirer in the Soviet Union in the late 1950s, which the CIA had intercepted under an

Monday, Aug 15.

Dr Moore, now of Parke Davis but shortly to join the faculty of the Univ. of Delaware, called on me today and for perhaps two hours we discussed our mushroom problem. He will apparently have at his disposal certain funds of the Geschickter Fund in Washington. (Geschickter, Charles, M.D.), of Georgetown University.

*   *   *

Dr Moore would like to work with me in isolating the active agent of our mushrooms. I said I wished to think over the problem and told him I should like first to talk with Abramson, with Curnow, perhaps with Parker. I should write him at Parke Davis early in September.

Moore said that his interest was not commercial — purely scientific, to discover a new series of alcoloids! I said our interest was originally cultural, that is anthropological, but we had in mind that if any by-product of our inquiries had commercial value, we wished to be among the beneficiaries.

Entry in Gordon's diary for 15 August 1955 records his first contact with James Moore. Unknown to Gordon, Moore was a CIA operative assigned to investigate hallucinogenic mushrooms. *Courtesy Wasson Collection.*

authorization that has since been rescinded. The apparently innocuous, brief letter concerned only scholarly matters.

Before leaving the subject of the psychedelic movement, I told Gordon that I felt more good than harm had been done by the widespread exposure of millions of people to these substances, though perhaps it would take many years before this became obvious. Until then, I did not expect the government to appreciably loosen its restrictions on psychedelic research.

Gordon agreed it was only a matter of time before these substances were popular again. He blamed the hiatus on more than the government's legal restraints. "Entheogens will be appreciated in 15 or 20 years—30 years at most," he predicted. "People don't want to be awed these days."

Yvonne, Gordon's housekeeper, arrived late in the morning. She was a middle-aged, bustling, talkative woman. I liked her at once. She introduced herself, traded some small talk and retired to the kitchen to start making lunch.

"She's a very good cook," Gordon said. "I 'inherited' her from my friend Tom Cleland when I bought this house from him [in 1960 or 1961]. She was with him for 15 years and has been with me since."

Through the course of that day and my visit the following year I learned more about T. M. Cleland, an American artist/illustrator. His hands had become so arthritic when he reached his late 70s he found he could no longer handle the tools of his trade. Thus deprived of a livelihood, Cleland proposed an arrangement: he would sell his house to Gordon for the bargain price of $25,000, on the condition he could stay there rent-free for the rest of his life, with Gordon supplying his food and other necessities.

"I thought it over and agreed," Gordon said, adding matter-of-factly: "He died four years later, at the age of 84."

Yvonne later added a footnote. "Before agreeing to the terms," she told me privately, "Mr. Wasson consulted an actuary to see how long Mr. Cleland might live."

"What was the answer?," I asked.

"Four years," she replied. It was clear that Yvonne was impressed by the fact that Tom Cleland had died right on schedule, but also somewhat scandalized, it seemed, that Gordon had been so precautious. "I wouldn't do that to a friend," she said. "Would you?"

Actually, I might have, given the circumstance. For Gordon, however, it was almost *de rigeuer*, in light of his background as a credit banker. In any event, it turned out to be a very good business deal. Following Gordon's death in December 1986, the house changed hands and was put on the market again by its subsequent owners. Their asking price, in August 1988, was well over half a million dollars!

When lunch was announced, Gordon and I took our seats at the dining room table, located in an area just off the living room and adjacent to the kitchen. One of Cleland's water color paintings depicting a rustic village scene hung over a cupboard behind Gordon. To my admittedly untrained eye, its style reminded me somewhat of Rowlandson's.

We continued our talk over poached eggs with cheddar cheese and wine sauce, followed by crepes filled with blackberries and topped with sour cream. My drink of choice was Bass ale, which Gordon recommended as his favorite brew. He

214

restricted himself to a glass of milk, then joined me in having a cup of espresso with our crepes.

After lunch we poured small glasses of the port I had brought and returned to the living room. I commented that I considered his house very peaceful.

"I have no television set," he pointed out, "and I don't listen to the radio." He did, however, read newspapers daily. And sometimes (though apparently not often in his home) he listened to recorded music.

"My daughter has these little discs I listen to at her house," he said, leaning forward with a wide-eyed look of interest as he formed his hands into the shape of a small circle.

From his description and the circle's size I realized he was referring to compact discs. "They're encoded with music that's read by a laser," I told him. I then tried to describe basic laser technology.

Gordon listened attentively. "Do you see any relevance there for the field in which you and I share an interest?," he asked when I finished. By this, of course, he meant psychoactive botanical substances.

I responded by mentioning Stanislav Grof's recent writings on the holonomic theory of consciousness. But then, to explain this theory, I had to describe how a hologram works. Gordon claimed to have never before even heard of a hologram. He smiled and held out his hands, palms up—a gesture of surrender.

"You young people understand things like that," he said. "I stopped learning about them when computers began to proliferate. I haven't any use for them."

Returning to the subject of the sacred mushrooms, Gordon gladly recounted the high points of the first night he had taken them, more than 30 years past. I got the impression it gave him a genuine thrill of excitement to harken me back to the dark little hut in Huautla, though this was a tale he must have told often.

"Imagine the feeling!," he challenged me. "The darkness was total except for the glowing red embers." Now I was the one who sat wide-eyed, listening carefully, as Gordon recalled the amazing events of that evening.

When he finished, I asked how often since then he had taken the mushrooms.

"About 30 times," Gordon answered, "but not for many years now. Have you ever taken them?"

"Yes," I admitted. Then I told him of something peculiar I had experienced when doing so. Though I do not believe in conventional spiritual entities, the mushrooms had induced a strong feeling that some kind of spirit was present, an invisible, silent entity that stood at the opposite verge of my consciousness. I sensed it to be, not an angel or devil, but something connected in some way to earth and the physical realm; perhaps a tutelary spirit such as primitive societies believed to be dwelling in trees, rocks and rivers.

"Have you ever had a similar experience?," I wondered.

Gordon leaned forward again, intensely interested. "There's definitely something there," he told me. "I've written a chapter about it in my next book. It's precisely what I feel on the subject. I weighed every word."

By this, I later learned, he meant his theory that entheogens make people who ingest them aware of the "god generated within." This is not to say the substances themselves are gods, or even that the god within is nothing but a chemical

215

effect. It may be that they awaken us, if only for several hours, to the presence of a real god within or near ourselves, a god that may be one with and also transcendent of our human flesh. Gordon's book, *Persephone's Quest: Entheogens and the Origins of Religion* (1986), which includes contributions by Stella Kramrisch, Jonathan Ott and Carl A. P. Ruck, does not favor any one of these or other explanations. It seeks only to affirm that "sacred" psychoactive plants—the entheogens— have a certain spontaneous power to compel religious ideation.

"I haven't found a publisher yet," Gordon added. "But there are the page proofs." He gestured to a table between me and some bookshelves a few feet away. I got up and inspected the proofs, which included his hand-written notes and corrections. As with all his other limited editions, the typography was beautifully designed by the Stamperia Valdonega in Verona. While turning the pages I noticed a reference to "the Tim Learys and their ilk," which made me smile. Gordon Wasson would certainly never be too old to "give 'em hell," I thought.

I then wandered over to look at his small book collection. Included were the limited editions of all the books he had written on mushrooms and several books by other authors. Gordon noticed my interest and told me the bulk of his books had been donated to the Harvard Botanical Museum, where they were shipped a few years earlier to establish the Tina and Gordon Wasson Ethnomycological Collection.

"I decided I should do it now," he explained. "Later I may not be able to, because I won't be here."

Gordon suddenly shook his head slightly, annoyed at himself. "I shouldn't use that euphemism, 'won't be here,' " he said. "I mean I'll be dead, for I will be soon. It doesn't bother me especially."

The way he said it, combined with the look on his face, made the statement convincing. His acceptance of imminent death was serene, not merely resigned. The only sadness I sensed in the room was my own.

For the remainder of the afternoon we talked about mutual friends and acquaintances. I left in time to make the drive back to Boston and thence, by plane, to Baltimore.

We were not to meet again until the middle of November 1986, when I returned for an overnight visit to his home. This time I brought with me a copy of *The Psychedelic Review* containing the Timothy Leary article Gordon wanted to see, and a box of dried morels, which Yvonne prepared nicely as an accompaniment to one of our meals.

As soon as we were settled in the living room, I inquired after Gordon's health. He replied that he had contracted double pneumonia during the early part of 1986. A full recovery had taken several weeks. He had subsequently visited some friends in Europe, and now was looking forward to making a trip to Belize in January.

Gordon then told me that when he was seven years old, he had suffered a bout with rheumatic fever that had left him with a permanent heart murmur. Obviously, this had not stopped him from making his physically demanding expeditions to far-away places. And his heart had functioned well enough to make him an octogenarian. But thinking back to childhood reminded him of something much more painful to bear than his physical problems: the loss of his beloved older brother, Thomas Campbell Wasson.

In addition to their blood ties, he and Thomas (Gordon's senior by two and a half years) had forged a close bond of friendship during boyhood. This bond was annealed, so to speak, in the "warm glow" of their tightly-knit family circle.[6] Together, the boys had enjoyed their mother's legendary cooking; looked forward to, and finally saw through a telescope, the near approach of Halley's comet in 1910; played stickball on the streets outside their home in Newark, New Jersey; and made frequent excursions, without adult chaperons, to New York. Their father, E. A. Wasson, an Episcopalian minister who had fought Prohibition "root and branch," had personally taught them the extraordinary academic skills and love of learning that served Gordon so well for the rest of his life. Later, when Gordon was 16 years old, their father sent them to England to enhance their education. When World War I broke out shortly after, both enlisted to join in the fighting.

Thomas later signed up with the U.S. foreign service. In 1948 he was appointed consul general to Jerusalem and sent there to help resolve Israeli settlement problems at the time of the original Arab-Israeli conflict. On 23 May of that year, he was killed by an unknown sniper in Jerusalem. It fell to Gordon to break the hard news to his parents.

"His death," Gordon told me with genuine sadness, "has weighed heavily on me all my life." (I did not appreciate fully how much this was true until later, after Gordon's death, when I learned that his ashes had been mingled with his brother's in a single urn interred in the Washington Cathedral's columbarium.[7])

Aside from this brief, understandable pang of dejection, Gordon's spirits when I visited were generally positive. Compared to my previous visit, the only detectable change in his health was that he seemed less energetic. As I spoke with him, he sometimes fell asleep for a minute or two, slowly closing his eyes and letting his chin drop slightly toward his chest. When he did so, to spare him embarrassment, I stopped talking in mid-sentence. Then, when he opened his eyes and raised his head, I completed the sentence as if nothing had transpired. Like the Dormouse in *Alice in Wonderland*, he was able to bridge the disruption without noticing it.

Among other things, we talked about his books. Gordon said he was proudest of *María Sabina and her Mazatec Mushroom Velada* (1974). It was therefore especially difficult for him to fathom the lukewarm reception it had received from most mycologists.

However, Gordon added, in a way it did not matter. He professed to be losing his interest in material things. What concerned him most was the loss of his personal freedom and mobility. Once again he expressed (as he had on my earlier visit) disappointment that his doctor had told him to stop sleeping on the open back porch of his studio building, a practice he used to enjoy. This was partly to protect him from the elements and partly to keep him in range of quick medical help, should he need it.

"I no longer dare go there alone," he explained. "If I fell I couldn't get up."

"You slept there even in the winter?," I asked.

"Yes, indeed," Gordon answered.

Yvonne, who had been listening to our conversation, remembered that once, when a terrible cold snap had settled on Danbury during the night, she had rushed to the studio porch the next morning expecting to find Mr. Wasson frozen stiff. But he was fine.

"I slept with a pig," Gordon added, smiling mischievously. When I failed to look shocked, he asked: "Do you know what kind of 'pig' I mean?"

It happened that I did: a heated piece of iron, wrapped in cloth.

Gordon then suggested that Yvonne might be willing to take me on a tour of the studio.

"I used to do all my work there," he said. "It's a wonderful space."

That it was. Removed several hundred yards from the main house, it comprised one large, white-walled room with a very high ceiling, long skylights and a built-in stone fireplace; a small kitchen; a bedroom; and the open back porch with a narrow bed on which Gordon used to sleep. The main room's walls were lined with bookshelves still containing many books, now damp and mildewed, which Gordon must have deemed irrelevant for the collection at Harvard. It was a quiet place, roomy yet comfortable, conducive to piecing together complex thoughts. I could easily imagine Gordon working there by morning light or midnight oil, writing his *Soma, María Sabina, The Wondrous Mushroom* and part of *Persephone's Quest*. How difficult it must have been for him to give it up!

On returning to the main house, I found Gordon reading Timothy Leary's long article. He told me he worried he might not have time to finish it before I left. So I urged him to keep that particular copy of the journal as a donation to the Wasson Collection at Harvard. Before shipping it there, he assured me, he would finish Leary's article and tell me his opinion of it.

All too soon, my second visit with him ended. As we said our good-byes, he invited me back in the spring and insisted that next time I should bring my wife June. I promised him another visit "sometime soon."

One week later, in a letter dated 18 November 1986, Gordon reported:

> I have now read what Timothy Leary wrote about the sacred mushrooms. It is well done. He has organized his material well and has said about the mushrooms what is obvious to those who know them. I was somewhat bored by the length to which he goes to convert his readers. Of course [in my case] he was writing for one who knows the mushrooms and [who] did not need to read what he said. He was writing for the converted in my case.

In my subsequent letter to Gordon dated 28 November 1986, I asked him if he had considered that his work, as well as Leary's, may have served a greater purpose by "unleashing" psychedelics on a world in need of enlightenment. In other words, was it not possibly good that his article in *Life* had helped democratize awareness of and access to these sacred substances?

Gordon's answer, in a letter dated December 17, 1986, was written less than a week before he died:

> I have never dealt with that question. When I attended the opening ceremony of Maria Sabina in June of 1955, I came away feeling quite dismayed by what I

had seen. I felt a responsibility to render [a report of these experiences] to American readers—and the world—in a light worthy of the performance. I did so to my own satisfaction and whether my article accomplished the ends you suggest has never preoccupied me and never will.

Of course, though he was not "preoccupied" with it, he knew his article had this effect. It would not have been seemly for Gordon to suppose that his work had set in motion great events. He preferred the less exalted role of scholar.

I did not forget, after Gordon died, that I had promised him another visit. This compelled me to make an excursion in March 1988 to the site of his interment at the Washington Cathedral. Appropriately, this was during a trip to attend the dedication of a book (*Where the Gods Reign*) by Gordon's long-time friend and colleague Richard Evans Schultes, at the World Wildlife Fund headquarters in Washington, D.C.

Before the dedication I drove up to the cathedral, a majestic gray stone structure designed, with flying buttresses and pinnacles, to emulate the great European cathedrals of the late medieval period.

Within its sturdy walls quiet dignity reigns. Numerous chapels and bays, exquisitely decorated, open from corridors skirting the nave. I wandered from one to another, passing alternately through the glow of candles in amber glass cylinders and pools of colored light cast down from magnificent stained-glass windows. At last I located the Chapel of St. Joseph of Arimathea, from which access is gained to the columbarium.

Defined by the circular piers supporting the cathedral's central tower, this chapel is formed in the shape of a Greek cross. Its namesake was the saint who gave his sepulchre as a tomb for the crucified Christ. He was also the first to take possession of the Holy Grail following the Last Supper, and is reputed to have been the first Christian missionary to England, "by his travels far antedating the later arrivals" (according to the guidebook). Even had his brother's ashes not been interred beside this chapel, Gordon might have considered St. Joseph of Arimathea a worthwhile spiritual patron in light of his own pioneering efforts to establish a different mission—the field of ethnomycology.

But then I discovered that the entrance to the columbarium is barred by huge wrought-iron gates, denying access to all would-be pilgrims who might come to visit Gordon in his final resting place. Disappointed, I paid my respects in the chapel and left the cathedral.

A cool, light rain was falling as I stepped outside. It put me in mind of the day I had first met Gordon. Again I remembered his smile and welcoming handshake, his kindness and great generosity. I thought about his powerful intelligence. And suddenly, I realized what should have been obvious to me all along:

I had not needed to come here, to this cathedral, to keep the promise I had made to him. It would have been sufficient to have visited his books in my own home. For Gordon Wasson continues to live in the words he so carefully crafted and the concepts they embody. Iron gates do not stand in the way of the seeker who looks for him there. His final resting place is not the columbarium, but rather in the pages of an open book.

Notes

1. In *Mushrooms, Russia and History* (New York: Pantheon Books, 1957, p. 387), Gordon lavishly credits what seems to have been a professional contact with helping him succeed in Mexico:

   > Above all we wish to express our deepest appreciation to Don Agustín Legorreta, President of the Banco Nacional de Mexico, and his colleague Don Ladislao López Negrete, as well as to the Bank itself, for the way in which they furthered our plans by introducing us to the appropriate persons and by placing at our disposal their private plane, piloted by that excellent airman, Captain Carlos Borja. Without such hospitable cooperation we could have accomplished little. In Mexico we were pressed for time, and Mexico is not a country that takes time into account. Don Agustín and his Bank knew how to gear our necessities with Mexican ways.

   Gordon's professional contacts who helped in his research included more than bank executives, however. In *The Wondrous Mushroom: Mycolatry in Mesoamerica* (New York, St. Louis and San Francisco: McGraw-Hill Book Co., 1980, p. xxiv) he gratefully acknowledges "a cleaning man in my bank, who opened the doors for me to the Slovakian mushroom vocabulary, my excellent friend Nicholas Kazenchak."

2. Some background on this feud is provided in my article "The Poet and the Dreamer: A Perspective on R. Gordon Wasson and Timothy Leary," published in the occasional journal *Psychedelic Monographs & Essays* (Summer 1989).

3. This may have been true despite the fact that a paper by Gordon on *ololiuqui*, the psychoactive morning glory seeds of Mexico, appeared in the same issue. Gordon's paper was originally published as one of the Harvard *Botanical Museum Leaflets*. Conceivably, it may have been reprinted in *The Psychedelic Review* without Gordon's knowledge, though more likely he simply forgot having seen it there 25 years previously.

4. These are: *The Search for the "Manchurian Candidate,"* by John Marks (New York: Dell, 1979); *Acid Dreams: The CIA, LSD, and the Sixties Rebellion,* by Martin A. Lee and Bruce Shlain (New York: Grove Press, 1985); and *Storming Heaven: LSD and the American Dream,* by Jay Stevens (New York: Atlantic Monthly Press, 1987).

5. In a letter to P. Margot of the University of Strathclyde dated March 9, 1978, of which a copy is on file in the Tina and Gordon Wasson Ethnomycological Collection at the Harvard Botanical Museum, Gordon wrote:

   > You ask for my opinion on legislation to control the use of hallucinogenic mushrooms. I think such legislation is futile and counterproductive. These growths are everywhere. They are not addictive. They do no harm, or at any rate minimal harm to the few who abuse them. The harm that they do cannot begin to compare to the harm done by tobacco or alcohol. The enforcement of legislation would call for heavy outlays of taxpayers' money and a diversion of manpower from productive occupations. The net result would be costly failure.
   >
   > This does not mean that the hallucinogenic mushrooms cannot be abused: I deplore the abuse of them but we can do nothing about it. There were abuses in pre-Conquest times in Mesoamerica, rightly condemned by the community. Most good things in life can be abused and are.

6. The basic facts and words quoted in this paragraph are from Gordon's autobiographical "memoir postscriptum." It appears in his father's posthumous book, *That Gettysburg Address,* which Gordon arranged to have privately printed in 1965.

7. As a memorial to Thomas, Gordon dedicated to him the revised third edition of *The Hall Carbine Affair* (1971), which was also the final, and only deluxe, edition. This replaced the book's original dedication to J. P. Morgan, which appeared in the first edition published in 1941. The book's subtitle also was changed, from *A Study in Contemporary Folklore* in the first edition to *An Essay in Historiography* in the third.

Mr. Wasson and the Greeks

CARL A. P. RUCK

From the beginning, Gordon Wasson's main interest was mushrooms. His investigations of other entheogens,[1] such as *ololiuqui*, the seeds of morning glories, was incidental by comparison. In particular he focused on the fly-agaric mushroom, *Amanita muscaria*, and its surrogates, whose special sanctity he suspected lay at the roots of his own identity.

The culmination and clearest expression of this interest is Wasson's last book, *Persephone's Quest: Entheogens and the Origins of Religion* (1986), on which I, together with Jonathan Ott and Stella Kramrisch, collaborated. In it he returned to the oft-told tale of his wife Valentina Pavlovna and the first time he discovered, on their honeymoon, the difference between her great fondness for mushrooms and his initial fear of them. By the time of this last telling, however, the story had achieved the paradigmatic dimension of myth, with his bride's mushroom gathering during a forest walk linked to the ancient Greek tale of Persephone and her quest for the sacred plants of Nysa. It was their first marital crisis, as he liked to say: the first intimation of some profound difference between them. The investigation of that difference was to occupy them both for the rest of their lives, and brought him at last to the Greeks.

He did not know the classical languages; his foreign tongues were French and Spanish, although he always had a lively interest in words, all languages and their etymologies. He had traveled in Greece as a young man, walking from Athens to Sparta with the guidance of a Swedish archaeologist. Thereafter he never returned. No doubt the Greece of today could never recall that more numinous landscape of ruins amidst primitive hospitality (and bedbugs) that Gordon had found there in his youth. As an amateur academician, he always sought out guides, people whose minds he could pick, and this became the mode for all his subsequent Greek explorations. But Greece does not easily give up its mysteries, not even to those who know the language, and Wasson knew the classics only from translation (though well enough to call my attention to a curious passage of Herodotus about the Hyperboreans). It was obvious that he felt more "at home" with Vedic translations or Mesoamerican languages than he did with classic Greek.

There was something about the Greeks, however, some secret perhaps in his own family, that captured and even commanded his attention. When his father lay dying, he called for a glass of retsina. "Already far away in Antiquity," Wasson wrote,[2] "his soul was craving the bouquet of that pungent rosin-tinctured table-wine which conjures up for hellenophiles all the morning radiance that was ancient Greece." It was his father from whom he initially heard about Soma and the Eleusinian Mysteries.

Even before we worked together, Wasson had suggested that the potion imbibed by initiates at Eleusis was the essence of the Mysteries. Perhaps it included some sort of psychoactive botanical substance among its ingredients, he surmised. And when, much later, he asked me to help him, he already had a drug in hand. He had reasoned that the substance he was looking for was a higher entheogen, and also that it must have been related to barley, the symbol of the mystery revelation. Albert Hofmann suggested ergot, a fungal growth whose powerful entheogen can easily be released from the grasp of the attendant toxins by the simple intervention of man, a secret process known generally only to the priesthood of the initiation. Though he knew in advance that a fungus, of course, is a mushroom, Wasson felt a boundless joy when he first saw the fruiting bodies—not microscopic, but visible to the naked eye—rising out of the purple sclerotia of an infested grain of barley.

The Greeks also would have recognized the ergot fungus as a mushroom. This is part of the reasoning used in advancing the theory, expounded at length in *Persephone's Quest,* that ergot was a surrogate replacing an earlier fungal intoxicant used by the ancient Greeks' Indo-European ancestors. Thus, we concluded, the Eleusinian Mysteries—especially the Greater Mystery, during which the potion including the ergot was imbibed[3]—provided parallel proof of Wasson's surmise about Soma:[4] here amongst another branch of the Indo-European migration was a fungal surrogate for *Amanita muscaria* that perpetuated its sanctity.

In *Persephone's Quest,* we were able to show that the Greeks had an evolutionary view of civilization and culture, including the agrarian world. The potion and its two main ingredients—barley and mint—were symbolic of that evolving progress. The barley represented culture, the staff of life, as well as the male-dominant marital traditions of Hellenic times, while the mint commemorated the illicit polyandrous sexuality of a deposed age. Mediating between the two was the secret fungus—like all fungi in such traditions, the perfect mediator. It is a common growth on *Lolium temulentum,* the grassy weed the Greeks identified as the pre-hybridized avatar of barley itself. As the wild weed is related to the cultivated crop, so the fungal mediator likewise is related to the commemorated entheogen it serves as surrogate. Thus *Amanita muscaria,* a wild plant having no seed, defies all attempts at cultivation; but the ergot fungus *Claviceps purpurea* appears to have yielded to cultivation when the infested kernel of barley produces its crop of fruiting bodies, and the sclerotia of the host barley appears to be its seed.

In assimilating the Mother Goddess religions of the earlier inhabitants of the Greek lands, the ancient Indo-European immigrants espoused the ways of agriculture as superior to their previous nomadic existence. But they imposed on the indigenous population their ethic of male dominance and displaced the primacy of the existing local entheogen, the narcotic poppy, with their fungal entheogen. It

Gordon, in the foreground, chats with (*left to right*) Wendy Doniger, Carl Ruck and Danny Staples at the dedication of the Wasson Collection at Harvard in 1983. *Masha Wasson Britten.*

is probable that the ergot displaced *Amanita muscaria* as a sacred surrogate somewhere in the more northern, grain-producing regions along the Indo-European migration routes, and subsequently, at Eleusis, it was the ergot that replaced the poppy. Thus, they replaced the narcosis and the sanctity of the poppy's death-like sleep with a mushroom-induced sleepless night of intense mystical enlightenment.

Classicists have not been eager to accept our scenario for the Eleusinian initiation. Some have insinuated that Wasson and I could conceive of no religious experience devoid of a somatic component. Actually, such a prejudice against admitting any corporeal involvement in the rapture of the spirit may go back to the Indo-Europeans themselves, perhaps because their entheogen was undependable and difficult to obtain in their new environments (though *Amanita muscaria* has been found in appropriate locations in Greece, including Mount Olympus). More likely, however, the prejudice is rooted in an ancient deep-seated aversion to physicality and its female connotations handed down from these same, male-dominant peoples, which today is implicitly part of the Western tradition and explicit in the prejudice's ultimate assimilated form, Christianity.

In answer to our critics, I point out what should be obvious to everybody: it was the Greeks, not we, who worshipped a god of intoxication; it was they, not we, who saw something divine in what we would call drunkenness. Wasson himself had no gods, at least none he would admit to, and he was a very sober man.

Demeter and her daughter Persephone presided over the dry food of mankind, as contrasted to Dionysus/Bacchus and his wet gift of wine. Like him, Demeter balanced the subtle hybrids with commemorative honor for their wild and poisonous avatars. Wine was symbolically a triumph of cultivation, or, as we would say, of culture, a sacred hybridization that culminated and commemorated the entire history of its supposed evolution. It was also, as the Greeks recognized, a tamed and cultivated surrogate for the original entheogen of the Indo-European homeland, and thus associatively a fungal intoxicant.

The Greek word for wine appears to be a root word that the Indo-Europeans originally applied to their fungal entheogen. When these immigrants to the Greek lands first encountered wine amongst the viticultural peoples to their south, they applied their own term to this new sacred drink. Once again, the food was a symbolic commemoration of its sacred evolution. The wild fungus had succumbed to the ways of science and cultivation. It now could be grown to manufacture a drink—not of wildness, but one with all the connotations of higher culture, inspired knowledge and the refined manners of civilized society. We still perpetuate these connotations in our ritual of the cocktail party, which is most refined at higher social levels where people, including the leaders of our society, drink, but supposedly do not get drunk, and there strike deals affecting the course of our civilization.

Like the Eleusinian potion, Greek wine—the ancient retsina—incorporated commemorative acknowledgement of its deposed avatars. It was flavored with herbal components that were largely responsible for its inordinately intoxicating properties. This "bouquet" (a term still used today regarding wine) comprised the herbs added in the wine's manufacture and in the preparation of the drink for a symposium or drinking ceremony. The wine's wild avatars were also given their due commemoration in the ritually regressive orgies of the maenads—mountain revels in which women of the citizenry, who ordinarily led a sheltered and subordinate life, reenacted the gathering rituals of the pre-viticultural entheogens. The latter were symbolized during these revels by what was thought to be the cultivated vine's own biological avatar: its reputedly maddening sinister cousin, the ivy.

When we first began our investigation, I focused attention upon the available plants having known mind-altering properties. This search was complicated somewhat by the fact that most of the Old World's herbalist lore has been lost, or suppressed,[5] whereas the New World seems overwhelmingly profuse in botanical drugs. In the course of my studies I found many reputedly sacred plants that are not psychoactive, including ivy, the laurel of Apollo, and the olive of Athena and Zeus. Yet mythological traditions about these plants repeatedly suggest that they, too, served as surrogates for the fungal original of the Indo-European homeland, and therefore corroborated Wasson's surmise about the sacred mushroom. The material concerning them, however, was perhaps the most difficult for me to decode, because of the intensely rationalistic prejudice of classical scholarship. It was also, as I could judge from conferences I attended with Wasson, of less interest to our audiences, which often were composed, to his embarrassment, of people who were interested only in what they could take to "get off" locally.

But Wasson recognized, in this Greek evidence, the proof that finally vindicated his and Valentina's original quest: our last collaboration had confirmed that what he and his bride had encountered on their enigmatic walk was nothing less than the survival, into modern times, of an ancient botanical prejudice.

Notes

1. *Entheogen,* translatable as "god generated within," is the term proposed by Jeremy Bigwood, Jonathan Ott, Carl A. P. Ruck, Danny Staples and R. Gordon Wasson as a replacement for words such as *hallucinogen* to describe psychoactive botanical substances and their psychoactive chemical derivatives.
2. In a "memoir postscriptum" he wrote for a book by his father E. A. Wasson, *That Gettysburg Address* (1965), which R. G. Wasson had privately printed.
3. The identification of ergot as the likely key ingredient used in the potion of Eleusis was initially set forth by R. Gordon Wasson, Albert Hofmann and Carl A. P. Ruck in their book *The Road to Eleusis: Unveiling the Secret of the Mysteries* (New York and London: Harcourt Brace Jovanovich, 1978).
4. In his book *Soma: Divine Mushroom of Immortality* (New York: Harcourt, Brace & World, 1968, pp. 172–204), in a section titled "Europe and the Fly-Agaric," Wasson proposes that the prehistoric Indo-Europeans once venerated a mushroom, probably *Amanita muscaria,* that "must have been hedged about with all the sanctions that attend sacred things in primitive societies." Of this mushroom, which he identifies with the Soma that figures so prominently in the ancient Rig Veda, Wasson goes on to observe: "Judging by what we considered vestigial survivals in our own folkways, it must have been instinct with *mana,* an object of awe, of terror, of adoration." This, he felt, was probably the basis for the contrasting love and fear of mushrooms demonstrated today by different peoples descended from the early Indo-Europeans.
5. The attributes of the deposed pagan gods had been assigned to the Satanic antithesis of the newer and triumphant Christian traditions and were kept alive in the European mind in the subculture of witchcraft. Suppression of the latter by the Church and its civil defenders made the knowledge preserved by this subculture less accessible, though it was never completely eradicated. The fungal entheogen still is reputedly worshipped in certain Christian heresies amongst the Greeks.

Celebrating Gordon Wasson

ALEXANDER T. SHULGIN

I mentioned to Ann, my wife, that I had been asked to write a short essay recalling my impressions of and contacts with R. Gordon Wasson, to be published as part of a volume to commemorate his life and contributions to our commonweal of knowledge and perhaps give renewed life to his words and thoughts so that they will live beyond him.

Ann, who had not known him personally, said: "Wasn't he the scientist who did all those marvelous studies on mushrooms?"

I almost said no. She was certainly correct about the mushroom studies. However, I had always thought of Gordon as a scholar, not a scientist.

Ann's question was therefore provocative. It made me consider, before answering, my different definitions for a "scholar" as distinguished from a "scientist."

The latter is what I consider myself: a seeker of knowledge who uses the Baconian system of logic and challenge to sharpen my observations and temper my conclusions. I always try to disbelieve new "facts" before attempting to establish their validity by experimenting with them and failing to make them fail. Should they fail, I rejoice; for then a different, maybe better "fact" replaces it. But even should they not, I leave the quotes in place, aware that there could always be alternative explanations I haven't yet considered.

Scholars, by comparison, prefer their facts straight up, without quote marks. They build on the acceptance of a given fact as truth. Their energies are then expended on gathering together observations and arguments that can affirm the correctness of their position. Experimentation normally has no role in the process. Scholars collect and explain, whereas scientists experiment and learn from observation.

R. Gordon Wasson was therefore a consummate scholar, one whose academic research skills were legendary. However, his famous experiments in Mexico, where Gordon often ate the sacred mushrooms to observe what might happen and learn from the experience, was something else again.

So I answered Ann's question: "I guess he was also a scientist, in his way."

Gordon and I got together at several conferences and meetings held in various parts of the country, from San Francisco in the west to Washington, D.C. in the east. And we kept a modest but regular correspondence during the last few years of his life. But, as this book is intended to be a collection of more intimate thoughts, let me share what I believe Gordon himself thought to be his most valuable contribution to society.

We once had the opportunity to spend a week together at a retreat called Esalen, located near Big Sur on the Pacific about 175 miles south of San Francisco. A seminar was being held there on the subject of guruism, shamanism or New Ageism—I can't recall the exact title—and the two of us, together with several other writers and scientists and scholars, were present as part of the invited faculty.

One of my students, who had heard I was going to be there with the great Gordon Wasson, was ecstatic to learn that in fact I would be rooming with him. He gave me a copy of *Life* magazine dated May 13, 1957—the issue that included Gordon's personal account, with compelling photographs, of his experiences with the Mazatec Indians—and asked me to have Gordon autograph it for him.

When I gave it to him, Gordon said: "My heavens! A student of yours wants *me* to sign a copy of this *Life* article?"

I answered that the student both admired and acknowledged Gordon's many contributions to ethnobotany.

"And he has read this article in *Life*?," Gordon wanted to know.

"Not only has he read it," I replied. "He told me it provided him the rationale he needed to stay in college."

Gordon signed the copy, graciously including a personal note addressed to my student. As he did so, he expressed surprise that someone so young would have placed any value on events that were effectively part of the past. And then he started musing, not on past events but rather on our present circumstances that have resulted from the past. His exact words I cannot recall, but I clearly remember the spirit of the thoughts he then expressed to me:

According to Gordon, the search that eventually led to his decision to journey to Mexico was itself a thrilling story, matched only by the actual discovery of the magic mushroom ceremony. And he felt that the subsequent flow of events, including introduction of psilocybin into modern medicine and the preservation of a culture that might otherwise soon have been lost, will someday be the stuff of textbooks.

But he thought his most lasting contribution was allowing that article in *Life* to appear, and to appear in the form it took. It was, for many devout and curious readers of the magazine, their first exposure to the concept of a union between nature and God. And that there are many different ways to be in the presence of God. And that a lowly mushroom, like ordinary bread and wine, can allow, can insist, that you identify with and acknowledge the divine. He quoted Blake:

> He who does not imagine . . . in stronger and better light than his perishing eye
> can see, does not imagine at all.

It was this article that caught the attention of the populace. It was this article that served as the single most important "trigger" to initiate the psychedelic revolution of the '60s. It was this article, he felt, that resulted in what proved to be an

228

irreversible change in human awareness. And so this article, I feel certain, was the achievement of which Gordon was most proud.

Therefore this volume of essays should be dedicated, not to lamenting the death of R. Gordon Wasson, but to the celebration of his life. The world will forever be richer in knowledge and spirit because of him.

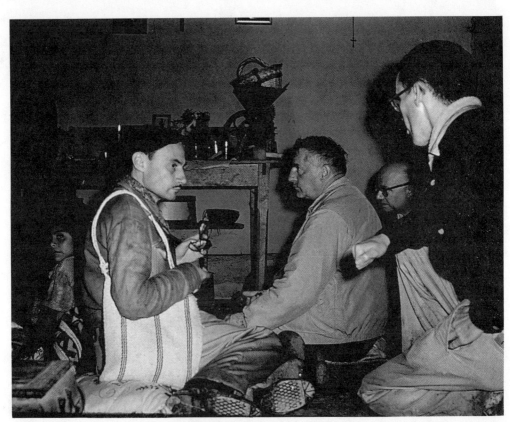
Left to right: Guy Stresser-Péan, Gordon, Roger Heim (*background*) and James Moore prepare to take sacred mushrooms during their 1956 expedition to Huautla de Jiménez. *Allan B. Richardson. Courtesy Wasson Collection.*

Travels with R. Gordon Wasson in Mexico, 1956–1962

GUY STRESSER-PÉAN

My friendly ties with Gordon Wasson spanned a period of more than 30 years. I first met him in Mexico City in July 1956, through his old friend Dr. Roger Heim, director of the French Museum of Natural History. Dr. Heim had just arrived from Paris in order to study, with Mr. Wasson, the hallucinogenic mushrooms of the Mazatec Indians of Huautla de Jiménez.

Mr. Wasson had made contact with the Indians of Huautla on two previous occasions. First, in 1953, he had witnessed a divinatory ritual led by a medicine man named Aurelio Carreras. On that occasion, Mr. Wasson had been accompanied by Robert J. Weitlaner, the distinguished ethnologist and linguist from Austria, who had personal experience with the Indians of northern Oaxaca. Two years later, in June 1955, Mr. Wasson returned to Huautla accompanied only by the photographer Allan Richardson. It was during this second visit that he met María Sabina, the native female shaman who henceforth became his principal informant.

By 1956, Mr. Wasson was eager to get better scientific data as a basis for his study of hallucinogenic mushrooms. That is why he had asked Dr. Heim, an eminent mycologist, to accompany him. Mr. Weitlaner was not available on that occasion, so Mr. Wasson invited me to join the party as a "free" ethnologist, though I had no previous experience with the Indians of northern Oaxaca or their tonal language. The two other members of our group were Mr. James Moore, a chemist from the United States, and Mr. Richardson.

Starting from Mexico City we went to Tehuacán by car. From there we had to reach Huautla—at the time a somewhat isolated Indian village not yet connected to the road system. It had previously taken Mr. Wasson 11 hours to get there from Teotitlán on horseback. But since then, a small landing field for planes had been created at San Andrés Hidalgo, not far from Huautla, so we hired two small planes. The landing field was narrow and had been cut through the ridge of a mountain at a height of 8,000 feet. It ended with a sheer rock cliff. On one side were the remains of a plane that had not been lucky enough in its landing.

The first plane dropped off four of us—Mr. Wasson, Dr. Heim, Mr.

Richardson and me—in San Andrés without problems. We waited for nearly two hours at the landing strip until the other plane arrived. It carried Mr. Moore, who was quite slim, and our luggage, which was quite heavy. The pilot had felt his plane was overloaded and feared it would be dangerous to land on the small San Andrés field. He had decided he had better put down first on a neighboring landing field located at a lower altitude, there to leave a portion of his freight before attempting to land at San Andrés. The slope of the neighboring field, however, was somewhat steep, so Mr. Moore had to jump out and gather some stones to wedge under the wheels as soon as the plane landed there. This posed linguistic problems for the pilot, who did not speak English, and Mr. Moore, who did not know a single word of Spanish. Happily, we all got together at last with our luggage in San Andrés, and set out from there to Huautla on foot—a distance of about six miles.

We stayed about 14 days in Huautla and participated in four shamanic ceremonies. The first, held during the night following our arrival, was led by María Sabina. It revealed to me the ecstasy created by the ingestion of the divine mushrooms. The next two ceremonies were led by medicine men and were not so spectacular; nevertheless, they gave rise to some interesting observations, particularly on the ritual use of the macaw feathers. The last one, again led by María Sabina, was tape-recorded in its entirety by Mr. Wasson.[1] This turned out to be a kind of first attempt preceding the second, more famous recording he made two years later on 11–12 July 1958, which was published in a splendid book.[2] Since these fascinating *veladas* have been fully described by Mr. Wasson I will not report on them here.

Every day of our stay in Huautla we were busy looking for mushrooms in their natural environment. Dr. Heim, who actively participated in this field work, had then to prepare and study all the samples—a time-consuming task. But 1956 was somewhat dry, so the mushrooms were not abundant. We could not gather enough of them for all of us to eat in the last shamanic ceremony, during which I ate instead the green leaves of *hoja de la Pastora*. It had similar hallucinogenic properties, but a flavor that was truly nauseating.

The relations both among ourselves and with the Indians were always friendly. Mr. Wasson, who already knew Huautla and its inhabitants, led our group's activities and contacts with native informants easily. The mycological erudition of Dr. Heim was held in much respect, but his ignorance of the Spanish language kept him apart from some discussions. He did, however, succeed in attracting the Indians' attention and amazement when he complained, through an interpreter, that the chili peppers he had been eating were not sufficiently hot for his taste. He then was offered, upon his insistence, yellow peppers that the Indians explained were so extremely hot they "must never be given to foreigners." He ate them without ever batting an eyelash, but later privately admitted that this prestigious performance had caused him some hard moments.

Having a personal knowledge of other Indian regions in Mexico, I was impressed with the wealth of native traditions preserved by the Mazatecs. Above all, I wondered at the frankness and candor with which they discussed the most pagan of these traditions. It seemed to me that, in the past, these Indians must have been largely protected from Spanish oppression because they were located so remotely in rough mountains and because their four-tone language is so difficult

232

to learn. Huautla remains a native village today, ruled by native local authorities and not by a *mestizo* (half-breed) middle class. Following publication of Wasson's article in *Life* magazine,[3] the rush of tourists and hippies to Huautla led not to hostility toward foreigners, but only to scorn and their commercial exploitation.

After our return, without problems, to Mexico City by way of Tehuacán, Mr. Wasson, Dr. Heim and I left by car for Oaxaca on 26 July 1956, taking with us Mr. William Upson, a linguist and Methodist missionary among the Chatino Indians. En route to Oaxaca, we stopped near Puebla to visit the fine Santa María Tonantzintla church. Its exuberant and popular baroque decoration strongly impressed Mr. Wasson, who believed that it was possibly inspired by the former ritual use of hallucinogenic mushrooms.

From Oaxaca, we traveled by plane to Juquila in the Sierra Madre del Sur. Then we went on horseback to the native Chatino village of Yaitepec, where Mr. Upson gave us accommodations in his home. The eight days spent there were interesting and agreeable. But the divinatory rituals of the Chatinos do not have the same quality of communion as those of the Mazatecs, nor are they so note-worthy. The medicine man is the only one who eats the mushrooms, in order to induce a trance. The ceremony we witnessed, during an afternoon, left Mr. Wasson disappointed. This does not mean that the native traditions of the Chatinos are not interesting. They were studied, some years later, by an Ameri-can ethnologist, Mr. Greenberg, who wrote about them in his fine book *Santiago's Sword* (1981).

In Yaitepec, we happened to be present during the feast of Saint James, the village's patron saint, which gave us the opportunity to witness the annual meeting of the "bank" of Yaitepec, cash for which was provided by the saint's funds. Each borrower arrived to stand, with a purse in each hand, before the native local authorities. One purse held the capital loaned to the borrower; the other, one year's interest on the loan. We were told that the interest rate was 100%, which Mr. Wasson, a vice president at the Morgan Guaranty Trust Company, thought amazing indeed. But we were assured that the mestizo dealers of Juquila lent money at what were perhaps still higher rates of interest. All things considered, Mr. Wasson opined that the Morgan bank was not yet prepared for lending money to the native peasants of Yaitepec. (Later, I learned from Mr. Greenberg's book that the interest rate had been possibly exaggerated by our informants—in fact, it was most likely not higher than 50%—and that the wealthier Indians were also obliged to borrow from the saint's loan fund.)

Our mycological researches led us to the top of a high mountain, where we found Indians busy with some pagan rites. We also visited a deep valley said to be haunted by a huge mythical serpent whose tail ended in a flower.

In a period of just one week, however, two murders were committed in the Yaitepec marketplace, which gave us some idea of the region's extremely high level of criminality.

Our arrival at Juquila was not easy. The landing field was located in a narrow valley that made flying in and out extremely dangerous. The pilot of the plane picking us up at the end of our stay dared not land because the wind was not propi-tious, so he returned to Oaxaca. On the following day, a smaller plane arrived and successfully landed. I boarded with the heavy luggage. The pilot was obviously very nervous. He told me frankly that we had to risk our lives, because, with an overweight cargo and a downwind, we would have to turn sharply just after take-off in order not to crash into the opposite hillside. So he kissed his lucky sheep-bone charm and off we went. The turn was barely successful. The pilot, relieved, set me down near the coast and took the luggage to Oaxaca. I was welcomed by the Negroes and mulattos of Río Grande, a neighboring village, who offered me good, sugar-cane brandy. They showed me a splendid Zapotec stele of the classic period that was lying in the bush; it now is exhibited in the Mexican National Museum of Anthropology.

On the following day, we were all reunited in Oaxaca. Mr. Wasson proceeded from there to Mexico City and New York. Dr. Heim and I pushed on to Chiapas, where we failed to find the mushrooms we wanted because it was so dry that summer.

Three years later, in July 1959, another trip brought me to Huautla with Mr. Wasson, his daughter Masha, Mrs. Irmgard Weitlaner Johnson, and Dr. Heim and his assistant, Mr. Cailleux. A new road had been created, so we enjoyed much easier travel by car. For 12 days we worked hard at collecting mushrooms. Long excursions in the direction of Río Santiago, Rancho del Cura and San Andrés Huautla found us climbing up slippery slopes and through maize fields, grasslands and, above all, damp mountain rain forests. Every evening, Dr. Heim studied and prepared the numerous specimens we had picked during the day. We attended only one divination ceremony. On this occasion all of us, including María Sabina, ingested synthetic psilocybin prepared by Dr. Albert Hofmann in his Basel laboratory.

From Huautla we went to Oaxaca and then to Mitla, which is the operational center of the Protestant linguist-missionaries of the Summer Institute of Linguistics. There we met Mr. Walter Miller, who had lived for many years among the Mixe Indians and knew their language well. Mrs. Irmgard Weitlaner Johnson, who had discovered many years before the Mixe hallucinatory mushrooms, and a Zapotec guide of Tehuántepec, Chico Ortega, who had accompanied her on her travels, joined us.

In spite of heavy rains, and though clouds covered the mountains, we were able to reach the Mixe village of Zacatepec by plane. There, a great noisy feast was being celebrated by a famous local *cacique*. Mr. Miller was leading us.

Mr. Wasson had previously visited part of the Mixe country in 1954. On our trip in 1959, Dr. Heim and his assistant collected some new hallucinogenic mushrooms. An excursion was organized, including a visit to the village of Cotzocón, but for want of a mount I could not be part of it.

In August, we were again in Mexico City. Mr. Wasson returned to New York, leaving me with Dr. Heim and his assistant. I took them to the mountains of southern Huasteca. Ancient documents mentioned that long ago, hallucinogenic mushrooms had been used there. And indeed, we found some *Psilocybes* near Zacualtipan. They were growing in a fine, birch forest covering the remains of ancient, Indian, obsidian quarries. Then we made a brief trip to Huauchinango and Necaxa, where other interesting mushrooms were collected in deep *barrancas*.

My last field trip with Gordon Wasson lasted eight days. This was in the second half of September 1962—not in 1972, the year reported incorrectly on page 44 of *The Wondrous Mushroom*.[4] Mrs. Irmgard Weitlaner Johnson came with us. In August, I had been in Pahuatlán, a small Spanish-speaking town near Tulancingo, where I had been told that a native shaman in the neighboring village of Xolotla still knew and perhaps used the hallucinogenic mushrooms. This had to be verified. So, on 15 September, my assistant, Mr. Guidon, took us by car to Pahuatlán, where we hoped to rent horses or mules. When we discovered no such animals were available we went on foot to Xolotla, some miles away. There, we made contact laboriously with Bernardo Cruz, the medicine man, a Nahuatl-speaking Indian. He knew the magic powers of the divine mushrooms and gave us their local name. But he was reticent and told us, perhaps untruthfully, that he had never eaten them. He asserted that their use only barely survived among a few very old shamans who lived somewhere else, in a place hard to reach.

We then proceeded, by foot, to a nearby village called Atla, there to speak with a young Mexican ethnologist, Mr. Montoya, who had been studying the local customs for more than six months. He had not obtained any information about the hallucinogenic mushrooms. Thus, it appeared that the Indians of this region, whose forefathers lived near an Augustinian mission center, had lost much of their native lore. It was evident, moreover, that they did not speak about their religious traditions as freely as did the Mazatecs of Huautla. Mr. Wasson was quite disappointed. We found only some *Stropharia* mushrooms and obtained ethnological data about textiles and the Volador dance.

When we returned to Mexico City, my assistant took Mr. Wasson and Mrs. Weitlaner Johnson on a trip to Huautla. But my duties as director of the French Archaeological and Ethnological Mission, then recently founded, prevented me from accompanying them.

Meanwhile, I made several individual trips in search of hallucinogenic mushrooms, especially to Tenango del Valle near the volcano Nevado de Toluca (1957) and to the Misantla region of Veracruz (1959 and 1960). The results were communicated to Mr. Wasson and Dr. Heim.

During the next 24 years, Gordon Wasson and I had no further opportunities to travel together, but we always maintained friendly contact. I was received by him and his wife at their home in New York, in January 1957, and we met many times in Mexico City and Paris. He sent me copies of all his books. Our correspondence continued until the end; his last letter reached me a few days after his death.

Thinking back on Mr. Wasson's extraordinary achievements in ethnomycology, I note that he pursued it *as a hobby* for more than half a century. He was already a middle-aged man when he entered the field of Mexican studies—a man who had worked as a journalist, who had visited many countries and studied many languages, who had reached a high position in the banking business and who had acquired remarkable organizational skills and American efficiency. Initially inspired and assisted by his Russian wife, he had long been engaged in the study of sacred hallucinatory mushrooms. But, for obvious reasons, he had never been in touch with Siberian adepts of the fly-agaric mushroom cult. It was only after years of scientific frustration that he finally discovered, in the Indians of Mexico, the chance that he might realize his dream of directly observing and perhaps taking part in a sacred mushroom ceremony.

However, ethnological field work on religious problems is a lengthy process, and the Morgan bank vice president could leave his New York job for no more than a few weeks each year. The fact that Gordon Wasson successfully managed, nonetheless, to carry out his scientific project is a testament to his intelligence and energy. He carefully prepared for his short trips by first reading extensively, then corresponding with and counting on the aid of such good friends as Irmgard Weitlaner Johnson and her father. Through the Morgan bank he easily obtained all the necessary official documents. When preparing to enter new fields, he arranged for introductions by specialized ethnologists or by Protestant missionaries.

His first, major success was achieved when, accompanied by Robert J. Weitlaner, he quickly managed to secure the friendly cooperation of the native authorities in Huautla de Jiménez. Two years later he repeated this success when he returned, without the advantage of a Spanish-speaking collaborator, and the Indian authorities ordered María Sabina to work with him. Mr. Wasson was clever in human relations and possessed great personal charm. He treated the Indians with courtesy and generosity. Understandably, he never attempted to learn the difficult Mazatec language, but he memorized the necessary words and was able to reproduce their tones correctly. He observed and inquired only about the sacred mushrooms and their ritual use, without going into more general investigations of the native religion. He focused on the Mazatecs, but also took interest in other tribes.

His main achievement was based on the use of a portable tape-recording device. He used it to make the recording of María Sabina's chants in 1958. Mr. Wasson then obtained the collaboration of Mr. and Mrs. Cowan of the Summer Institute of Linguistics, who transcribed the tapes and translated the text into Spanish and English. Almost 15 years after he made the recording he published it under the title *María Sabina and her Mazatec Mushroom Velada* (1974). This magnificent scientific document includes: four long-playing records (or, in the trade edition, cassette tapes) of the night-long chants; an ethnological exposition by Mr. Wasson; the Cowans' text; and a musical analysis by Mr. Willard Rhodes.

Mr. Wasson never forgot his early background as a journalist. He therefore took care in enlisting the help of a very good photographer, Mr. Richardson, to illustrate his writings. Mr. Richardson's beautiful images showing the inspired Mazatec priestess María Sabina in her native dress are preserved in Mr. Wasson's books. Some of these photographs also accompanied Mr. Wasson's article in *Life*. This famous article, frequently criticized for "letting out the secret" of the sacred mushrooms, was actually conceived by Mr. Wasson as a means of attracting the general public's attention to them. For he also understood the art of publicity—perhaps because his specialties at Morgan bank included "public relations" and communications. The article in *Life* was an essential piece of advertising. It paved the way for the publication and sale of his magnificent books in the field of ethnomycology.

These recollections and remarks of Mr. Wasson's one-time traveling companion have not addressed his extensive research, during free time when employed at the Morgan bank and then after retirement, in the United States, where he had access to some of the world's best libraries, universities, museums and even art dealers. He was able to enlist the assistance of top-ranking naturalists at Harvard and in Paris. But despite their tremendous importance and value to Mr. Wasson's academic research, these resources should not overshadow his field work in Mexico. For here it was that Mr. Wasson, at the cost of great physical strain and personal financial outlay, finally achieved his dream and opened a new chapter in the sciences.

Notes

1. Folkway Records of New York released this recording in 1957 as a long-playing record album titled *Mushroom Ceremony of the Mazatec Indians of Mexico*. It included an eight-page booklet containing the text of the ceremony, printed both in English and Mazatec, and an extract from Valentina Pavlovna Wasson's and R. Gordon Wasson's book *Mushrooms, Russia and History* (New York: Pantheon Books, 1957).
2. Wasson, R. G., G. Cowan, F. Cowan, and W. Rhodes. 1974. *María Sabina and her Mazatec Mushroom Velada*. New York: Harcourt Brace Jovanovich.
3. Wasson, R. G. 1957. Seeking the magic mushroom. *Life* May 13, 1957; 42(19):100–120.
4. Wasson, R. G. 1980. *The Wondrous Mushroom: Mycolatry in Mesoamerica*. New York: McGraw-Hill.

Gordon's father Edmund A. Wasson and mother Mary (1927). *Courtesy Masha Wasson Britten.*

Appendix I

Gordon Wasson's Account of his Childhood[1]

R. GORDON WASSON

My father was born in Napanee, Ontario, on March 13, 1864. His parents had come over from Ireland in the wake of the hard times of the 1840's. They must have been of the English Pale, for they and their kin were all stalwart members of the Church of Ireland. The family moved to New York when my father was only four. For reasons that remain unclear to me, the burden of raising a large family of children was borne solely by my grandmother Wasson, and as a small child I gained a deep impression of her strength of character and valor. My father almost never spoke of her, but on those occasions when he did so it was with the deepest reverence. Our mother, who had known her only at second hand, took pains to tell us of the love and awe that her mother-in-law had inspired in all her sons. Of her numerous brood, there survived to manhood only three, James, Edmund, and William; and all three became ministers of the Episcopal Church.

After attending the parish school of Trinity Church and, later, St. Stephen's College at Annandale-on-Hudson (now called Bard College), my father entered Columbia College. In the spring of 1885 he was elected to Phi Beta Kappa. I learned this from my mother some twenty-five years later: he refused to discuss the matter. He never acquired the golden Key to which he was entitled. Frugality would have been a sufficient explanation for this: there were always more reference books to buy, and many other pressing needs. But even if his friends had presented the Key to him, I am certain he would not have worn it. He always shrank from ostentation, and the display of the Key would have been for him repugnant, running counter to the very qualities that it purports to represent. Moreover, the Key is conferred for excellence in *under*graduate studies, and represents therefore only a limited intellectual achievement, certainly not one to flaunt throughout life.

My father learned Latin thoroughly, and Greek only less well. He retained his command of these languages throughout his long life, handling texts with ease and enjoyment. In his years of graduate study he also dipped into Icelandic, Hebrew, and Sanskrit; and he wrote his doctoral dissertation on 'Case Uses in Anglo-Saxon'. This thesis lies in a handwritten manuscript in the Library of Columbia University. In October 1964 I thumbed its pages: there was my father's familiar hand flooding me with a wealth of memories, the scholar's hand, full of character, but considerably more legible than I had ever known it. Since the day when it was deposited in the Library in 1888, I was apparently the only reader to have asked to see it.

Columbia University conferred a doctorate in philosophy on my father in 1888. He continued his studies at the general Theological Seminary in New York, and was duly ordained a minister in the Episcopal Church. His first charge was a mission at Eltingville, Staten Island. On July 11, 1890, he married my mother, Mary Matilda DeVeny.

My mother had been born on July 30, 1860, the youngest of four children, in Plymouth, Ohio, a market town some distance west of Columbus. She came of mixed ancestry, all Protestant, of which I recall only the Dutch and Scottish strains. The family had been in America since long before the Revolution. She alone of the children was sent to college: in 1879 she entered Wellesley. In the spring of 1883, when she was about to be graduated, Melville Dewey turned up on the campus to recruit bright young ladies of the graduating class as his helpers in the task he had just agreed to assume: the reorganization of the Library of Columbia University according to the principles of librarianship that were destined to make him world famous. He chose six girls—the 'Wellesley half-dozen', as they were to be called— and my mother was one of them. And so it came about that she was working in the Library during my father's studies there, and they met. All accounts agree on the impression that my mother made at the time: she was pretty and vivacious, but more than anything else the ripple of her gay and infectious laughter seems to have captivated all who came into contact with her. When after some time she changed her employment for a more remunerative teaching post in a girls' school—Melville Dewey was notoriously parsimonious and her annual salary had been only $500 a year—, she would return occasionally to visit her old friends. The letter that President F. A. P. Barnard of Columbia University, 76 years old at the time, wrote to the 25 year old Miss after one of these visits gives the very flavour of her presence:

> *My dear Miss Mary:*
>
> *I learned, yesterday, after you had gone away, how nearly I came to having the pleasure of seeing you. Why did not you send in your name? I would have choked off the Professor without ceremony, if I had known that you were at hand; and so I would have done with the whole Faculty, if they had been here.*
>
> *When you come again, you must not be so unobtrusive. Rather than miss you I would send away even so important a person as a trustee of the College.*
>
> *It is not often I see you now, but I miss you every time I go to the Library, and that is daily.*
>
> *Sincerely yours,*
> *F. A. P. Barnard.*

Miss DeVeny became engaged to my father in 1885, but his financial circumstances did not permit them to marry for five years. They began life together in Eltingville, and our eldest brother, Edmund, known as 'Eddie', was born there on June 14, 1891. Our parents did not stay long on Staten Island. Our father accepted an appointment to a mission in Great Falls, Montana, the Church of the Incarnation, and there the family remained until the turn of the century. Those were the final days of the old Wild West. Being a big, vital, passionate man, our father threw

Mary DeVeny Wasson (1923).
Courtesy Masha Wasson Britten.

himself into his new charge with all the enthusiasm and vigor of his youth and powerful physique. His 'parish' extended some fifty miles in every direction. Rather than on horseback, our father covered it on a bicycle. Everyone knew our parents, and I gather from the testimony of those I have met that they contributed much to the cultural life, such as it was, of the community: the Shakespeare Society, the Browning Club, and similar activities. He tutored the boys whose parents wished to send them to Eastern colleges. Years later, in the East, one of my earliest recollections is of the letters our father would receive from Charlie Russell, the cowboy artist, with little drawings enlivening every page. I also remember his visits on the occasions when he came to New York: a big, happy-go-lucky man, wearing a big cowboy hat and high-heeled cowboy boots, he would come striding down to the Rectory on Milford Avenue in Newark, New Jersey, his little wife Nancy walking resolutely by his side. She was his dedicated wife, his resourceful business manager, and his 'keeper', determined that he should hold to the 'straight and narrow'. (I know that *'strait* and narrow' is the Biblical text, but my emendation avoids the pleonasm and adds to the sense.)

My brother Thomas Campbell was born in Great Falls on February 8, 1896, and I followed on September 22, 1898. But we arrived on the scene only after a terrible tragedy had shaken the Wasson family. Our mother had gone East in 1895

Gordon's brother Eddie (date unknown).
Courtesy Wasson Collection.

to visit her family in Cleveland and had taken Eddie with her. There he had come down with diphtheria, rapidly worsened, and on September 27 died. Our father was far away. She was the embodiment of selfless motherlove. Her grief was overwhelming. Many years later, when we were tots and when the initial pangs of anguish must have lost something of their poignancy, I remember how she would call us to her knees, and tell us, in her calm, grave, well-modulated voice, about little Eddie, about his endearing ways, and how we should never see him, for he was dead. She would often hold us spellbound with her stories, but this was different. Her voice was unutterably sad, and the story ended with death. The idea of death was new and strange to us, as it must be to every child at first. We knew not what had happened to Eddie, but something mysterious and terrible that left us awestruck. She showed us the four photographs of Eddie that hung on the wall in a single frame. In the one taken a few months before his fatal illness we saw him in all his pride wearing his first pair of breeches. It was as though our mother was determined that this little member of our family should not be forgotten, and she accomplished this purpose beautifully. With her words she brought him to life for us, so that he was as real to us as we were to each other, only he was not present. There was another difference: we knew him only through the lenses of our mother's grief. Though we often heard our mother laugh, it was no longer the carefree ripple that had captivated every one in her youth: Eddie's death had stilled that laughter forever.

Our parents moved East shortly after the turn of the century when our father became the Rector of St. Stephen's parish in Newark, New Jersey. Thomas and I spent most of our boyhood in the Rectory at 77 Milford Avenue. In 1925, ten years after we had embarked on our travels, our father gave up his parish and thereafter took on only *locum tenens* assignments. Our parents then moved to the other side of town, the Branch Brook neighborhood, at 357 Parker Street, where they bought a small, solidly built, light, airy house with a garden, just suitable for their needs. One winter they passed in Jamaica, and again they went for a season to England, where they found congenial accommodation in Canterbury and absorbed the atmosphere of that ancient Cathedral town. A year or so later they also spent some months exploring Rome and its environs.

But let us return to St. Stephen's. The parish was far from wealthy; it drew its congregation mostly from the lower middle and working class. Our parents, precisely because they watched their pennies, were able to live well, although my earliest memories are of the need of what seemed to me painful economies. When I was a small boy I remember how I was pestering my father to buy the *Newark Daily Advertiser* for its 'funnies', which it carried in quantity. One day my father and I were passing in front of the newspaper dealer and I was especially insistent. My father stopped. He leaned over and whispered confidentially into my ear that we were too poor to buy the paper. I was stunned. I had always known we were not rich like the Aaron Wards and the Disses, but I had never known we were so poor as that. In the face of the appalling truth, I had nothing more to say. I was resolved not to let truth out, as it would make me ashamed. Long after, when I was older and when I perceived how well we were fed and clothed, and how there was always money to buy reference books, I saw that my father had used the pretext of poverty to divert me from unworthy occupations. Although we were far from rich, thanks to our father's sound business instinct, the Wassons were always comfortably off. It was characteristic of our father's spirit of vigorous independence that when Mr. J. P. Morgan and Bishop Lawrence of Massachusetts organized the Church pension Fund during the first World War, he told his vestry to refuse to join in the scheme, as he would provide for his own old age.

There was, however, a leak in our family's economy: the Rectory was a regular port of call for panhandlers, and they were seldom turned away empty-handed. I remember one in particular. As sure as spring would come around, there would arrive at our door a hobo, an Englishman, by name Mr. Satchel. He had had an education: he said he was an 'Oxford' man. He would captivate our father by quoting from Homer and the poets of the Greek Anthology. Sometimes for a couple of hours the two of them would remain closeted together. Our father would then rejoin the family circle looking sheepish, and anticipating our mother's reprimand that inevitably followed, 'Now, Daddy . . .' Finally there came a spring when Mr. Satchel failed to turn up, and we saw him no more.

Our mother clothed us boys and tucked us in at night, but where she excelled was in feeding us. She was always trying new dishes, and when she hit on one that we liked, it gained a lasting place in her repertory. Her dishes as prepared by her remain with me more vividly than any other sight or sound of those days: at will I conjure up their distinctive flavours, their consistency on the tongue, and my saliva still runs after fifty years and more have passed, and after my palate has been tempted and sated with the delicacies of all the Continents and all the Seas. Even her simplest dishes bore her mark. Her lima beans were young and green, served steaming hot with big lumps of melting butter, properly seasoned with salt and pepper. She would wait to serve us corn on the cob until old Mr. Drummond the farmer, as courtly a gentleman as ever walked this earth, would come to our door and announce that the corn was now ripe. Then, on a prearranged day, he would bring us the ears, either Golden Bantam or Country Gentleman, straight from his fields in Lyons Farms[2] to our kitchen, where they would be quickly husked and their silk stripped off, and then plunged into a big pot of boiling water that was waiting for them. In winter, on freezing Sunday evenings, we would gather around the coal stove in the kitchen for supper, and feast on cold flat-ribs, a subclass of spare-ribs that we considered superior and that our family butcher would set aside for us.

I know that outsiders sometimes regarded our mother's rice pudding as flat and uninteresting, but for me it has remained the touchstone. Not for me the heavy glutinous mass, usually flecked with raisins looking like flies, that ordinarily passes muster for rice pudding. The genuine article should be a smooth, creamy dish, each grain of rice suspended in the cream, with a skin lightly mottled in the oven and sprinkled over with cinnamon. Our mother did not corrupt her egg custard, as do the chefs of this world, with caramel, nor did she dole it out in tiny cupforms. Hers were rich, luscious concoctions, filling generous bowls, the surface speckled with a heavenly harmony of spices. Possibly she was best as a pastry cook. At Christmas the aroma of her cookies filled the house, and their variety and crumbly richness always outpaced our hopes. Her sponge cake would send us into a delirium of happiness: it was served fresh from the oven, its top crusty with crunchy convolutions. But we all agreed that her triumph was a cake spelled I think *allegretti*, a chocolate layer cake with a smooth chocolate filling and a white frosting covered over with a thin layer of brittle chocolate. As this was Tom's special favorite, it would always appear on his birthday. Where shall I find the like of her strawberry shortcake? Two big slabs of biscuit dough, one slab on top of the other, spread with butter when they came hot from the oven, the whole drowned in mashed strawberries, with cream to taste and more strawberries by the pitcher-full for those so inclined. Then there were the pies—cherry, apple, peach, lemon meringue—the pastry light and crispy, blending in your mouth with the juicy filling. In winter I still long for her corn-meal mush, sliced thin and fried until sizzling and almost black, served with butter. Her Christmas goose was famous for its stuffing, not the leaden sodden uneatable lump that often passes by that name and the sight of which makes my gorge rise, but an airy, steamy, crumby filling, made up of dried crumbs prepared on the back of the stove for days, seasoned with sage and interspersed with giblets and an occasional tasty little onion and (Oh, delight of delights!) just a touch of garlic to give cutting edge and zest to the whole business. Our mother did not really approve of garlic, but once we boys tasted it, we were converted, and she bowed gently in acquiescence. Of course we baked our own bread. (No 'baker's bread' for us! With scorn our mother would utter these words, sneering as she did so, which would make us all laugh, for she was not the sneering kind, and the curl of her lip, though it was intended to sear the bakers of the world, was singularly unsearing.) On Thursdays mother would knead the dough, and set it to rise overnight, covering the big bowl with old woolen garments to keep out the cold, and then bake the bread next day. She 'baked' the bread in a steamer, which gave it a thin, delicious, iridescent skin that I have never seen on bread elsewhere, and will never see again . . . Only years later, after we were launched on our travels, did I learn that toast is not everywhere cut thin and served piping hot from the stove, a crispy dark brown, *and buttered on both sides.*

From time to time our father would bring home an exotic dish. There was always a dramatic build-up for such an event. He would tell us in advance about the delicacy—where it was eaten and its history. But in our opinion these dishes did not always live up to the billing. We welcomed them with lively curiosity, albeit tinctured with skepticism. Thus liverwurst failed to kindle our appetites, nor did the tiny pot of caviar that made its appearance one day on our table. Even Shakespeare's use of the term did not whet our taste for it. Our father took pains to point out that we were 'the general' to whom Shakespeare referred: we had only ourselves to thank for our own down-rating.[3] Once our father brought home a grapefruit, not the big, juicy fruit it has since become. (How many remember that 'shaddock', after an Admiral Shaddock in the West Indies, was originally a name for it that rivaled the equally unsuitable 'grapefruit'?) It proved to be small, its rind thick, and its taste sour. Only the cheeses appealed to us, the occasional Stilton, the Roquefort, the Camembert, and many others.

When I am asked about my education, it has always been my wont to say, with a straight face but tongue in cheek, that I was brought up on the sidewalks of Newark and am a product of the public schools of that city. This is true, but how far from the whole truth! We did pass through the public schools, my brother Tom even being graduated from Barringer High. True, we did play 'shinney' in the street and baseball in the vacant lot with the Milford Avenue 'gang', with Ray Strobell and Jack Vliet and Malcolm Ross and the Gardner boys and many others. True, we followed avidly the baseball scores of the Yankees and Giants. Once we were permitted, so great was our enthusiasm, to go to new York to see the great Christie Matthewson pitch his famed fade-away. (He lost the game, alas, on that occasion.) True, we followed the ups and downs of the Newark baseball team led by Iron Man Joe McGinnity, he who in his heyday was said to have once pitched a triple header and won all three! When we were with the gang, we were part of the gang. But there came a moment in each day when we crossed the threshold and entered into the magic circle of our family life.

It was exciting to live in the Wasson family. How else am I to explain that we boys were able to take, and take eagerly, the education that our father was intent on giving us? My copy of the Bible carries at the end, in our father's hand-writing, this formidable series of statements:

245

The reading of the bible was finished by the boys on Friday, May 20, 1910. It was begun by them about Jan. 1, 1907.

Begun a second time Tuesday, May 31, 1910

Finished the second time, Wednesday, Aug. 9, 1911

Finished the third time, Thursday, March 21, 1912

Courtesy Wasson Collection.

The first time through, we read every word, Tom and I taking turns. On the later readings, we were selective, and we paused longer for explanations. Our father would expound the text according to the latest interpretations of scholars and critics—the Higher Criticism, as this body of Biblical exegesis was called. How extraordinary must have been the spectacle of us little boys, in short trousers, prattling about the Vulgate, the Septuagint, the Pentateuch and Hexateuch, the use of Yahweh and Elohim, and of Adonai, interpolations of Priestly Redactors, the various Recensions, the reconstructed Sayings of Jesus, and all the other terms then current in Biblical study! Moreover, we each learned a verse of the Bible by heart every day, and it was our practice to repeat the nine verses learned on the nine previous days as well. Our father would choose the verses to learn and as the Book of Genesis was filled with memorable passages, in our memorizing activities we had hardly finished Genesis before our three readings of the Bible reached an end. It remains true that at one point in my life I knew almost all of Genesis by heart.

How did we boys stand up under the stern Biblical indoctrination? It was rugged but it was thrilling. It was like mountain climbing, intellectual mountain climbing: the joy of reaching successive crests, the supreme satisfaction of scaling the summit, after hours of weary slogging up the slopes. Our father must have been endowed with the inspired school master's faculty of tailoring his demands on us to our capacities. Just as he would dramatize the exotic foods that he would bring home to us, so he dramatized all our education and so with the Bible. I remember that when at one stage we were under the spell of Sherlock Holmes— our father approved of Sherlock Holmes for boys—, he made us see the unbridgeable gulf separating the mysteries of Sherlock Holmes from the Mystery of the Bible, how in the *Sign of the Four* there was a readymade answer to be found in the final pages of the book, whereas in the Bible the questions were of a different order, a great variety of questions each of which posed problems of its own. How much more exciting it was to be tackling an adventure ourselves, where we were Sherlock Holmes looking for clues with our father as our guide! Thus our reading of the Bible was only the beginning. We learned how, according to the best authorities, the books of the Bible were assembled, and about the problems of authorship, and when and why the various books were written, and how scholars had arrived at all this, and the varying degrees of doubt that hung over the answers, and the *cruces* that have been argued down the ages. We learned another lesson that applies, as I have since discovered, to every field of knowledge: how, when once you probe a subject, you quickly reach the frontiers of knowledge about it; and how, if you insist on going further, you find yourself voyaging on uncharted seas where the perils of the unknown beset you on every hand. Small wonder that it became our boyish ambition to grow up to be explorers, and to push back the frontiers of knowledge, be it ever so little.

Since man first discovered how to write, less than 5,000 years ago, the mastery of this extraordinary art has slowly spread from people to people until now perhaps five per cent of the languages of the world are recorded. Of those peoples who have acquired this art, a few have taken pains, when their mastery of writing was new, to put down the content of their minds for posterity to know what manner of men they were at that turning point in their history. These initial documents in the cultural history of the race stand in a class by themselves, in which one can catch a glimpse, more or less clear, of the way of thinking and feeling and living of countless generations of our unlettered ancestors, reaching back toward the beginning of *homo sapiens*. Many of them, such as Homer and Herodotus, are of a superlative value. Over the past fifty years this class of writings has held my particular interest, and, while I am far from belittling any of them, it seems to me incontrovertible that the Bible stands in a class by itself, head and shoulders above all the rest. It is the autobiography of a gifted people, an autobiography in which every facet of that people's inner life is exposed eloquently to view, every weakness shown with naked candour, its manner of living and dying, its beliefs as to its own past, its history, the tabus and rules governing the *minutiae* of its daily existence, its aspirations, a wealth of illuminating episodes both fictional and veridical, the poetry and romance of its supreme moments, all expressed in words of matchless splendour. In many intelligent and cultivated circles the Bible suffers today from the people who keep it company, from the proprietary claims on it of extreme sectarians with whom no civilized person cares to consort. The Bible for centuries has been read chiefly for its religious teachings, and now that religion is losing its hold on us, it is dismissed more and more as (odious phase!) a 'work of literature'—a way of shelving it into a pigeon-hole along with Beowulf and the Song of Roland.

Our father did not limit himself to the Bible in educating us boys. We were constantly drilled in sound English. The distinction between 'like' and 'as', and between 'shall' and 'will' and 'should' and 'would', we knew from the first. How could we go wrong, since both our parents by ingrained habit used them correctly? The very heavens would fall on us boys if by chance we let slip the wrong pronoun after a preposition, as in 'between you and I'. But there were more subtle matters of style. I remember when for weeks Tom and I were not allowed to use the adverb 'very', this suppression being (as our father would say) the quickest and easiest step toward strength of style. ('Oh, but Daddy, surely you do not wish us, in Church on Sunday, to alter the wording of the Creed!' We were sure we had him. 'But boys! I thought you knew the difference between adverbs and adjectives, and in "very God of very God" you are dealing with an adjective, of which the adverb is "verily"!') Then there was the spell when we had to avoid the verb 'to get', which we (like everyone else) were using as an all-purpose word. Our father would insist on our preserving the distinction between 'wish' and 'want', and he deplored the way the 'half-educated' (as he would call them) used an adverb with the verbs of sense. The untutored, he would say, never make this mistake, nor do the masters of the language, but those pretending to an education that they will never attain think it nice to say, 'Tom feels badly today, but don't you think I look finely?' This was the sure mark, our father would insist, of the truly uneducated and uneducatable. Then there was the distinction between the relative pronouns 'which' and 'that'. He would tell us that they were nowhere interchangeable, and that our lawyers were responsible for the current abuse of 'which', out of a false notion of clarity and emphasis, with their clients the business men meekly following in their steps.

When Halley's comet put in its appearance in 1910, for months we had been preparing for the event. We were going to buy a telescope. This meant knowing about the sun, the moon, the planets, the stars, and the occasional visitors like our comet. It meant reading the catalogues about telescopes. Finally the telescope arrived, with a 2¼ inch objective. (We boys had hoped for a larger one, of course.) We assembled it, and on a cold clear winter night, all bundled up, we climbed up to our roof with it and viewed the comet. It disappointed us a little, as its tail did not sweep across the sky, but we felt nonetheless honored to join the company of those who, every seventy-five years, from the beginning of our race, had contemplated with wonder this phenomenon.

Though our public school teachers were perhaps not equal to their task, our father always spoke of them, and taught us to treat them, with the greatest respect. In only one small instance was it otherwise. There came a year when my teacher in Miller Street School asked us to memorize each week a verse of our own selection from the Bible, and then recite it in class. Our father felt contempt for this homeopathic approach to the vast subject of Bible study, and he conspired with me to find the most absurd, the most embarrassing, verses for me to take to class. Imagine my glee! It was known of course that my father was a clergyman, and the confusion of my teacher was all the more extreme. After she had called on me two weeks running, she thenceforth ignored my presence.

When I think back across fifty years to that little Wasson family circle, I sense a warm glow, the glow of our mother's love and of the coal stove in the kitchen, next to the kitchen table where we ate. I remember how our mother would call us to the table, and how we boys would rush, give our hands a ceremonial dip under the faucet in the sink, and slip into our seats, eagerly looking to see what dishes were coming. Then there was the intellectual glow of the conversation, fanned by my father. Not a meal went by when one or the other of us did not go scampering up the stairs to consult a dictionary; sometimes this happened repeatedly. The rule was well settled: for etymologies, consult Skeat or the big Oxford, at that time coming out in fascicles; for the history of words and their meaning, the Oxford, or in default of the Oxford, Webster's International; for pronunciations, where there was a disagreement among the authorities, turn to Webster § 277, which gave all the authorities in a most useful compendium, and then we would weigh the authorities, both as to number and quality. Funk and Wagnalls was to be ignored.

248

All my life has been irradiated by the memory of the glow of that tightly knit, somewhat isolated family circle, united by the bonds of understanding and love.

Little did we boys know what lay in store for us. Before we were ten years old our father would send us perhaps once a month, on a Saturday, to New York, lunch baskets under our arms. Business men, regular commuters, were sometimes startled to see two little urchins—the Wasson boys!—going in on the train with them. We would visit the Aquarium and St. Paul's Chapel one day, the Zoo in the Bronx on another day, the Metropolitan Museum on yet another, but our favorite was the Natural History Museum, as this institution was commonly called. How its rich collections excited our imagination![4] And afterwards there were always the inviting grounds on which to eat our lunch, in the shade of the lofty trees with the Ninth Avenue El roaring by. It is not only in retrospect that those days seem pleasant to me: at the time we knew we were having fun, fun that surely could not last. A little later our father sent us further afield. When Christmas or Easter vacation came 'round, we would visit Providence for a week, and Boston and Harvard University (to see the glass flowers), and Washington, and finally Gettysburg. We would prepare for these visits for weeks, so that we knew exactly what we had to see. In Providence we stayed with relatives, the Kelsos, but on later visits we stopped in boarding houses according to arrangements our father had made in advance. These were trial runs for bigger things.

In February 1914 my brother, just turned 18, sailed for England. I, not yet 16, joined him in June, after completing my third year in high school. By the time I had arrived my brother had toured all of Great Britain by himself on a bicycle, from London to Land's End and thence to Tain, in the extreme North of Scotland. After a week together in London we crossed to Paris. We had embarked on our travels and also on life.

Our subsequent adventures have no place here: suffice it to say that we remained abroad for most of the next seven years. The first World War broke out almost at once. I remember the announcement on shipboard of the Grand Duke Ferdinand's murder. At the time its meaning escaped me, and I could not understand why people stood in front of the bulletin shaking their heads. We were in Paris, in a boarding house, when war broke out: we lived there through the mobilization and then the Battle of the Marne. In January 1915 we set out for Spain, where my brother left me and journeyed to Greece *via* North Africa. In April 1917, as soon as our country entered the war, we both enlisted.[5]

What the separation cost our parents, and especially our mother, I can only imagine: she was too courageous to show her anxiety and pain. They both wrote us weekly letters. Our father's were lengthy questionnaires about the country we were living in—its government, politics, religious life, history, customs. We innocently thought he needed the answers and we scurried around to find them, but of course he was only training us to be intelligent travelers. Years later, when I returned to New York to live, he would send me a postcard every day. The receipt of the postcard meant that all was well at home. The message on it was some arresting fact of history, a distinction in theology or law, a point of grammar, a quotation of poetry of surpassing beauty that he had come across in his reading or that he dug up from his inexhaustible memory. I still possess these cards by the thousands [see examples on pp. 250–253]. Throughout his life our father was always reading poetry. In his younger days *Atalanta in Calydon* and *In Memoriam* and Wordsworth absorbed his attention. Later poet succeeded poet: I know not how he selected them. James Hogg, Newman's *Gerontius*, John Clare, and Chapman's *Homer* are those I remember. He did not read Chapman as a translation but as an independent Elizabethan creation. In his last years he read and re-read the *Faerie Queene.* Spenser, he would say, was beyond comparison the purest poet in the English language, such was the melody of his line, the subtlety of his verse.

After our father had launched us on the world, we returned home at intervals, sometimes staying for several months. But it was never the same as in the old days. Our mother's cooking had not changed, nor our father's character and interests. But we had changed. From nestlings looking over the edge of the nest on an exciting world, we had grown to adults returning to visit the nest. Our mother would cook the same dishes and we would enjoy them. But the pleasure of a good dish was mingled with an overtone of reminiscence, the memory of departed days. Now that mother and father and brother are all gone, I alone remain to conjure up those memories, and in a few years, when I too will be gone, there will be but the pallid evocation of a memory among those few whose eyes chance then to fall on these lines. Here in a nutshell lies the philosophic problem of reality: nothing could have been more real to us than our family circle, yet this reality is on its way to become unreality.

I know that the portrait I have given of my father is lopsided: I picture him as a small boy saw him from within the family circle. Here was the dedicated scholar and teacher whom only we were privileged to know. But there was also the pastor ministering to his parishioners. I remember the long absences of my father when he went out 'making calls' on the sick, the aged, the infirm, taking Holy

Courtesy Wasson Collection.

To ward off the evil eys, a slave rode just at the back a triumphator and amid the shouts a multitude kept repeating, "Look behind you, a remember that you are mortal",

Communion to some. From time to time there would be someone 'in trouble': then there would ensue endless sessions, often held in the Rectory behind closed doors: we were told not to ask questions. These sessions would be followed perhaps by visits to a police magistrate or judge. Our father was tireless in his parish work, and I recall vividly many manifestations of the love his parishioners bore him.

For more than forty years our father was a leading citizen of Newark, and it was our pride to walk with him through the streets, and see him greeted right and left by persons unknown to us. On the issues that interested him, he took a forthright stand, and he must have had enemies, but we did not come to know them. He owned and edited a monthly journal of opinion called *The Crown,* which, surprisingly, turned out to be a moderately profitable venture. At the time of the controversy over the discovery of the North Pole, I remember how strongly our father championed Admiral Peary and denounced Frederick A. Cook as an imposter and ridiculed his claims. There ensued some correspondence with Peary, and when Tom and I went to Washington, the highlight of our trip was a call on the great man, when he gave us some walrus tusks that I still have.

Our father maintained a lively correspondence with scholars, experts, and government officials over the whole country, all of which in the end he destroyed.

Of the issues of the day, none stirred my father more deeply than Prohibition. From the beginning he fought it root and branch. The issue, he believed, was basically religious. Prohibition was, he said, a heresy that sprang up in the 19th century among the Evangelical sects of Protestantism, or I should rather say Fundamentalists. The extreme sects of backward Christians professing to believe that the Bible was the word of God spearheaded a drive for legislation making alcoholic beverages illegal. Our father combatted them from the pulpit and lec-

Courtesy Wasson Collection.

(12m)

Further reflection leaves me still unable to decide what clothes Jesus wore on + after his Resurrection. His own clothes had been divided among the soldiers, + his grave-clothes were left in the grave.

ture platform, in *The Crown* and later in his book, *Religion and Drink,* in which he demonstrated that Prohibition was opposed to the teachings and practice of Christ and the Gospels, opposed to the Old Testament and the Church Fathers, and opposed to sound morals. He never tired of pointing out that Christ's first miracle, at Cana of Galilee, was the conversion of water into wine (not wine into grape juice), and the last act of his ministry was to invite his apostles to drink wine in remembrance of him. He would dwell on the political skullduggery to which the Fundamentalists had recourse to put over their aims, and in which they seemed to delight; skullduggery the *sequelae* to which still bedevil our body politic. He deplored the attitude of his own Church on the issue; not that its leaders favored Prohibition—Heaven forbid!—but that they maintained a pusillanimous silence toward what was a clear-cut issue. I remember when our father's Bishop wrote him a letter expressing the hope that he would be less vocal on the question. By return post our father shot back a ferocious rejoinder, whereafter no more noises issued from the timorous Bishop.

The appearance of *Religion and Drink* early in 1914 was a big event in the lives of the Wasson family. As our father knew his market better than any publisher, he brought out the book himself, putting the printer's name on the title page. It sold well, and when we boys left for Europe our home seemed to have been converted into a shipping depot. Since our father's object was to convert Fundamentalists, he accepted their premises—faith in the literal reading of the Holy Scriptures—in writing his book, although his own point of view was of course far removed from theirs.

It is a matter of history that Prohibition slowly swept the country, local option giving way to statewide Prohibition, and these in turn yielding to the Volstead Act and the 18th Amendment. When the law of the land finally outlawed alcohol in

Courtesy Wasson Collection.

drinks, our father's tactics changed. He said nothing more. Instead in his own cellar he brewed beer, made wine, and distilled strong drink. He did not try to conceal what he was doing. He flouted the law openly and proudly. All his friends and neighbors knew of his activities, and those who cared to do so shared in the products. The City officials knew of them. The policeman on the beat could count on a nip whenever he felt in need of refreshment. He flouted the law but did not flaunt his defiance: of course he did not sell liquor. He would not embarrass his numerous colleagues in the Police Department by obliging them, against their will, to prefer charges. After all, for a generation had he not been, was he not still, a high official in that same Department, the Protestant Chaplain to the Newark Police Force?

Perhaps I have not distorted the picture of our father by my emphasis on his role in our intimate family circle. There he was himself and completely relaxed. In the course of decades he came to know thousands of Newarkers, calling hundreds of them by their first names. But for all of them he remained to the end 'Dr. Wasson'. He attended their birthday parties, their outings, their reunions. Yet always a gap separated them from him, and they tacitly recognized the difference by calling him 'Dr. Wasson'. He did not encourage intimacy. (It was the same with our mother: to the end she remained 'Mrs. Wasson' to the parishioners. She was a warm, handsome woman of dignity and presence, and not one to invite familiarity.) For some of his friends he felt the highest respect: our next door neighbors in Milford Avenue, the Boehringers, enjoyed his admiration.[6] On the other hand, he did not frequent the society of his fellow-clergy; perhaps it would not be too strong to say that he shunned them. In his adult life I believe our father had only two intimate friends—one was his brother, our Uncle Will, and the other was Dr. R. P. R. Gordon of Great Falls, a diagnostician of uncanny intuition, a Scot

Courtesy Wasson Collection.

cob = hard, small, round lump.
Hence a chunky horse; also a head;
cob-ble stone; hence, as verb, = to
strike, to raise such a lump; &
hence cop = police.

Gordon's brother Thomas C. Wasson (date unknown). *Courtesy Masha Wasson Britten.*

born in Tain, whose kin, the Macdonalds of Tain and London, furnished Tom a home-base from which to operate throughout his life in Europe. I was named after Dr. Gordon, and he and his wife, my 'Aunt' Isabel, were my godparents. Both Uncle Will and Dr. Gordon died prematurely, and their deaths were heavy blows for our father.

In 1907, when Dr. Gordon developed grave cancer, our father journeyed West to take final leave of him, and he took me with him. On that pilgrimage I remember (I blush to confess it) most distinctly my first glass of beer. I had been begging for beer for a long time. My father had been putting me off until 'a suitable occasion' would come up. True to his instinct for drama, he found the occasion on our Western trip. He told me in advance about 'the beer that made Milwaukee famous', and when our train stopped in that city for twenty minutes, we hied ourselves to the nearest saloon and I was regaled with a glass of draft Schlitz. I did not like the bitter taste. Since then, at every opportunity, I have been trying to erase that error of judgement from the record, but the damnèd spot will not out, and I must ever try again.

In the middle '20's Tom joined the Foreign Service of our Government as a career officer. He was stationed in Athens in the spring of 1948 when he was posted to Jerusalem as Consul General. This was the moment of the war between the Arabs and the Jews, and Jerusalem was a trouble-spot, dangerous, but offering opportunities for conspicuous service. Tom welcomed the assignment. After he assumed his post in April, he was named to a three-man Truce Commission representing the United Nations. On May 22, when he was returning from a meeting of the Commission, he was shot in the street. He died in the early hours of the 23rd. At dawn on that day—a Sunday—it fell to my lot to break the news to my parents, who were staying with us at that moment.

Our father lived a little more than a year after that: he died on June 11, 1949, on the 59th anniversary of his wedding. The day before, as he lay in the hospital *in extremis*, he gave utterance to his last wish. He called for a glass of *rezina*. Already far away in Antiquity, his soul was craving the bouquet of that pungent rosin-tinctured table-wine which conjures up for hellenophiles all the morning radiance that was ancient Greece.

Our mother lingered a few years more: she died on June 3, 1953, when she was nearing 93.

My brother's ashes lie in the columbarium of the Cathedral on Mount St. Alban in Washington. Our parents are buried in the family plot that our father had acquired some sixty years before for his mother. The plot is marked by a Celtic cross of granite, perhaps eight feet high, on which there is no inscription, none at all. The individual graves are marked by no headstones. Our father expressed a wish that the plot be left so always.

R. Gordon Wasson

The Yosun Inn on the Isle of Marmara
August 1964;
Ardpatrick, Argyllshire
September 1964

Courtesy Wasson Collection.

How goodly are thy tents, O Jacob, and thy tabernacles, O Israel!
As the Valleys are they spread forth, as gardens by the rivers side,
as the trees of lign aloes which the Lord hath planted,
and as cedar trees beside the waters.

Notes

1. *Editor's note:* This essay comprises the bulk of "Postscriptum: A Memoir," by R. Gordon Wasson, from his father's posthumous book *That Gettysburg Address*, which Gordon arranged to have privately printed in Verona by Officina Bodoni in 1965. The original punctuation and spelling, though not always in accordance with current American style, are preserved.

2. Lyons Farms! Every Saturday afternoon for years my father would go walking in those softly rolling hills and the woods beyond, and often take me with him. How proud I was to keep up with his fast pace, and to listen as he expounded a thousand and one things to me! The 'realtors', I am told, have taken hold of the area and 'improved' it, spawning thousands of cheap houses on it. With that perfection in vulgarity which distinguishes the breed, they thought it well to change the name from 'Lyons Farms' to 'Hillside'! Until I came to write these lines, I had not thought of Mr. Drummond for more than fifty years. At first I could not recall his name. For three days at intervals I lowered the bucket into my well of memories. Each time I came closer: each time my mind's eye saw the tall, elderly gentleman more clearly, my mind's ear the timbre of his soft, cultivated voice more distinctly. The vowels and consonants of his name were struggling to fall into place. At last in a flash I had it.

3. *Editor's note:* The reference is to *Hamlet,* act 2, scene 2, lines 454–461, in which Hamlet recalls a failed play that "pleased not the million; 'twas caviare to the general." In other words, like caviar, the play could be appreciated only by those with cultivated tastes; it did not appeal to "the general [public]."

4. *Editor's note:* An anecdote about the Wasson boys' interest in the museum appears in Geoffrey Hellman's book *Bankers, Bones & Beetles: The First Century of the American Museum of Natural History* (Garden City, New York: The Natural History Press, 1968, pp. 45–47). It hinges on a letter, dated March 15, 1908, preserved in the museum's central files:

   > Dear Sir: I am a little boy of Newark New Jersey, and am interested in rocks and stones and making a collection of my own. I wonder whether at the Mus. you have duplicates or inferior specimens that you will dispose of at reasonable prices. If so, will you please name the specimens and pieces and I will arrange to come over soon to examine them.
   > Yours Truly,
   > R. Gordon Wasson

   Hellman reports that Gordon, years later, recalled for "a student of the Museum" what transpired:

   > "I remember the sequel well," he said. "Tom and I planned the solicitation, but he, perhaps fearing a brush-off, put his younger brother up to initiating the correspondence. We turned up on the appointed day, and Mr. Gratacap showed us over the mineral collections and explained things to us. He gave us, of course without charge, from the cellar, as much in the way of diversified specimens as we could carry, which, considering our appetite, was plenty. I remember going out of my mind with impatience as he meticulously labeled each stone with its Latin name. He also admonished us, with an emphasis that is still graven on my memory, how *not* to make a collection of stones. One does not solicit stones from a museum. One makes field trips and discovers the stones in their natural habitat."

5. *Editor's note:* In Gordon's case, it seems to have been a delayed enlistment. He did not formally withdraw from school to enter the service until 25 June 1917.

6. Later the Boehringers moved to Muri-Bern, in Switzerland. Dr. Rudolph Boehringer, an eminent chemist, died in 1952. His widow still lives, and I never fail to visit her and her descendants when I am within reach of Bern. Their house is for me my Continental home.

Appendix II

Bibliography:
R. Gordon Wasson and
Valentina Pavlovna Wasson

J. CHRISTOPHER BROWN

Wasson, V. P. 1939. *The Chosen Baby.* First edition. With illustrations by H. Woodward. New York: Carrick and Evans, Inc.

Wasson, R. G. 1941. *The Hall Carbine Affair: A Study in Contemporary Folklore.* First edition. Privately published. Printed in New York by Pandick Press. Limited to 100 numbered copies.

Wasson, R. G. 1945. An analysis of Beveridge's 'Full employment in a free society.' *The Harvard Business Review* Summer 1945. Later reprinted in a book; see separate listing, same title, 1960.

Hughes, J. P., and R. G. Wasson. 1947. The etymology of Botargo. *American Journal of Philology* October 1947; 68(4).

Teale, E. W., and R. G. Wasson. 1947. The lost years of W. H. Hudson. The *Saturday Review of Literature* 12 April 1947. Later reprinted in a book; see separate listing, same title, 1959.

Wasson, R. G. 1948. *The Hall Carbine Affair: A Study in Contemporary Folklore.* Second edition. Slightly revised version of the first (1941) with a new preface. Privately published. Printed in New York by Pandick Press. Limited to 750 unnumbered copies.

Wasson, V. P. 1950. *The Chosen Baby.* Second edition. Philadelphia and New York: J. B. Lippincott Co.

Wasson, R. G. 1951. *Toward a Russian Policy: A Second Look at Some Popular Beliefs about Russia and the Soviet Regime.* A 25-page book. Stamford, Connecticut: The Overbrook Press.

Wasson, R. G. 1956. Lightning bolt and mushrooms: an essay in early cultural exploration. In *For Roman Jakobson: Essays on the Occasion of his Sixtieth Birthday.* RGW's first published work on ethnomycology. The Hague: Mouton & Co., 1956, pp. 605–612. Republished in Dutch, with revisions, amplifications and illustrations; see separate listing, same title, 1960.

Wasson, R. G. 1957. The hallucinogenic mushrooms of Mexico: an adventure in ethnomycological exploration. *Transactions* of the New York Academy of Sciences, Series II, February 1957; 21(4):325–339.

Wasson, R. G. 1957. Seeking the magic mushroom. With watercolors by R. Heim and photographs by A. Richardson. *Life* May 13, 1957; 42(19). First public announcement of the Wassons' discoveries in Mexico. A Spanish language version was published soon after; see separate listing for RGW's "En busca de los hongos mágicos" (1957).

Wasson, R. G. 1957. En busca de los hongos mágicos. *Life en Español* 3 June 1957. Spanish language version of RGW's "Seeking the magic mushroom" (1957); see separate listing.

Wasson, R. G. 1957. [Untitled introductory paragraph dated October 11, 1956.] In booklet accompanying *Mushroom Ceremony of the Mazatec Indians of Mexico,* an LP (long-playing) phonograph record of a *velada* performed by María Sabina on the evening of 21–22 July 1956. New York: Folkways Records and Service Corp., Album No. FR8975. The booklet also includes excerpts from *Mushrooms, Russia and History* (see next entry).

Wasson, V. P., and R. G. Wasson. 1957. *Mushrooms, Russia and History.* Deluxe first (and only) edition. Two volumes with 85 plates, of which 26 are water color illustrations by H. Fabre reproduced by D. Jacomet, Paris, using the *pochoir* process. New York: Pantheon Books. Printed in Verona by Stamperia Valdonega. Limited to 512 numbered copies, of which two are designated A and B, the others from 1 to 510.

Wasson, V. P. 1957. I ate the sacred mushrooms. *This Week* May 19, 1957. Later reprinted in a book; see separate listing, same title, 1982.

Heim, R., and R. G. Wasson. 1958 (1959). *Les Champignons Hallucinogènes du Mexique.* With 37 full-page plates, of which 17 are in color; 83 illustrations in the text, of which 14 are in color; and three maps. Paris: Editions du Muséum National d'Histoire Naturelle. Printed in 1958 by A. Lahure, Paris. Released in 1959.

Wasson, R. G. 1958. The divine mushroom: primitive religion and hallucinatory agents. *Proceedings* of the American Philosophical Society, June 24, 1958; 102(3):221–223.

Wasson, V. P., and R. G. Wasson. 1958. The hallucinogenic mushrooms. *The Garden Journal* January–February 1958:1–6.

Teale, R. G., and Wasson, R. G. 1959. The lost years of W. H. Hudson. In *The Saturday Review Gallery.* New York: Simon and Schuster. Article reprinted from earlier publication; see separate listing, same title, 1947.

Wasson, R. G. 1959. Wild mushrooms: a world of wonder and adventure. *The Herbarist,* Boston, 1959; 24:13–28.

Wasson, R. G. 1960. An analysis of Beveridge's 'Full employment in a free society.' In *The Critics of Keynesian Economics,* edited by H. Hazlitt. New York: D. Van Nostrand Co., Inc. Article reprinted from earlier publication; see separate listing, same title, 1945.

Wasson, R. G. 1960. Lightning bolt and mushrooms: an essay in early cultural exploration. *Antiquity and Survival,* The Hague, 1960; 3(1):59–73. Reprinted from 1956 (see separate listing, same title) in Dutch, with revisions, amplifications and illustrations.

Heim, R., and R. G. Wasson. 1961. Une investigation sur les champignons sacrés des Mixtèques. In the *Comptes rendus* of the Académie des Sciences, Paris, session of 29 January 1961.

Wasson, R. G. 1961. The hallucinogenic fungi of Mexico: an inquiry into the origins of the religious idea among primitive peoples. (Annual lecture of the Mycological Society of America, Stillwater, Oklahoma, August 30, 1960.) Harvard *Botanical Museum Leaflets* 1961; 19(7):137–162. Later reprinted in *The Psychedelic Review;* see separate listing, same title, 1963. Two abbreviated versions, also listed separately, were later published: "Hallucinogenic fungi of Mexico" (1962) and "Mushroom rites of Mexico" (1963).

Wasson, R. G., and S. Pau. 1962. The hallucinogenic mushrooms of Mexico and psilocybin: a bibliography. Harvard *Botanical Museum Leaflets* September 1962; 20(2):25–73. Second printing, with corrections, was published the following year; see separate listing, same title, 1963.

Wasson, R. G. 1962. Hallucinogenic fungi of Mexico. *International Journal of Parapsychology* Autumn 1962; 4(4). First abbreviated version of "The hallucinogenic fungi of Mexico: an inquiry into the origins of the religious idea among primitive peoples" (1961).

Wasson, R. G. 1962. A new Mexican psychotropic drug from the mint family. Harvard *Botanical Museum Leaflets* 28 December 1962; 10(3):73–84.

Wasson, R. G., and S. Pau. 1963. The hallucinogenic mushrooms of Mexico and psilocybin: a bibliography. Harvard *Botanical Museum Leaflets* 10 March 1963; 20(2a):25–73c. Corrected version of 20(2); see separate listing, same title, 1962.

Wasson, R. G. 1963. The hallucinogenic fungi of Mexico: an inquiry into the origins of the religious idea among primitive peoples. *The Psychedelic Review* June 1963; 1(1). Reprinted from earlier publication; see separate listing, same title, 1961.

Wasson, R. G. 1963. Mushroom rites of Mexico. *The Harvard Review* Summer 1963; 1(4). Second abbreviated version of "The hallucinogenic fungi of Mexico: an inquiry into the origins of the religious idea among primitive peoples" (1961).

Wasson, R. G. 1963. Notes on the present status of *ololiuhqui* and the other hallucinogens of Mexico. Harvard *Botanical Museum Leaflets* 1963; 20(6):161–193. A slightly corrected version was later published; see separate listing, same title, 1964. The original version was later reprinted under a different title in a book; see separate listing for "*Ololiuhqui* and the other hallucinogens of Mexico" (1966).

Heim, R., and R. G. Wasson. 1964. Note préliminaire sur la folie fongique des Kuma. In the *Comptes rendus* of the Académie des Sciences, Paris, session of 3 February 1964.

Heim, R., and R. G. Wasson. 1964. La folie des Kuma. *Cahiers du Pacifique* June 1964; 6:3–27. Later reprinted in English as "The 'mushroom madness' of the Kuma;" see separate listing under that title, 1965.

Wasson, R. G. 1964. Notes on the present status of *ololiuhqui* and the other hallucinogens of Mexico. *The Psychedelic Review* 1964; 1(3):275–301. Slightly corrected version of the paper published under this title in 1963.

Heim, R., and R. G. Wasson. 1965. The 'mushroom madness' of the Kuma. Harvard *Botanical Museum Leaflets* 11 June 1965; 21(1):2–36. English language version of "La folie des Kuma" (1964); see separate listing.

Wasson, R. G. 1965. Postscriptum: a memoir. In *That Gettysburg Address,* by E. A. Wasson. An autobiographical account of RGW's childhood, included in his father's book. Privately published. Printed in Verona by Officina Bodoni. Limited to 225 unnumbered copies.

Wasson, R. G. 1965. Rite of the magic mushroom. In *The Drug Takers,* a special Time-Life Books Report. New York: Time, Inc.

Wasson, R. G. 1966. *Ololiuhqui* and the other hallucinogens of Mexico. In *Summa Antropológica en Homenaje a Roberto J. Weitlaner.* Mexico: Instituto Nacional de Antropología e Historia, Secretaría de Educación Pública, 1966, pp. 329–348. Previously published as "Notes on the present status of *ololiuhqui* and the other hallucinogens of Mexico;" see separate listing under that title, 1963.

Heim, R., R. Cailleaux, R. G. Wasson and P. Thevenard. 1967. *Nouvelles Investigations sur les Champignons Hallucinogènes.* With 12 full-page plates, of which five are reproduced by the *pochoir* process and one in four colors; 34 illustrations in the text, of which five are in color; and one map. Paris: Editions du Muséum National d'Histoire Naturelle. Printed in Paris by A. Lahure.

Wasson, R. G. 1967. Fly-agaric and man. In *Ethnopharmacologic Search for Psychoactive Drugs,* edited by D. H. Efron, B. Holmstedt, and N. S. Kline. NIMH Workshop Series No. 2. Dept. of Health, Education and Welfare. Washington, D.C.: U.S. Government Printing Office, 1967, pp. 405–414.

Wasson, R. G. 1967. Soma: the divine mushroom of immortality. *Discovery* Fall 1967; 3(1).

Wasson, R. G. 1968. *Soma: Divine Mushroom of Immortality.* Ethnomycological Studies No. 1. Deluxe first edition. With two full-page color plates printed by D. Jacomet, Paris, using the *pochoir* process. A Helen and Kurt Wolff Book. Published by Harcourt, Brace & World, Inc., New York, and simultaneously by Mouton & Co., The Hague. Printed in Verona by Stamperia Valdonega. Limited to 680 numbered copies, of which two are designated A and B, the others from 1 to 678.

Wasson, R. G. [1968] *Soma: Divine Mushroom of Immortality.* Ethnomycological Studies No. 1. First hardcover trade edition. Does not include the two *pochoir* prints found in the deluxe first edition (1968). A Helen and Kurt Wolff Book. Published by Harcourt

Brace Jovanovich, Inc., New York. Neither the publisher's location, the date of publication nor the printer's name appear in the book. Printed in Italy, presumably in Verona by the Stamperia Valdonega.

Wasson, R. G. [1968] *Soma: Divine Mushroom of Immortality.* Ethnomycological Studies No. 1. First paperbound trade edition. Same information applies as for the first hardbound trade edition (see preceding entry), except the paperbound edition was apparently not printed by the Stamperia Valdonega.

Wasson, R. G. 1968. Primitive lighters from India. *Lore* Spring 1968; 18(2).

Wasson, R. G. 1969. [Review of Carlos Castaneda's book *The Teachings of Don Juan: A Yaqui Way of Knowledge.*] In *Economic Botany* April-June 1969; 23(2):197.

Heim, R., and R. G. Wasson. 1970. Les putka des Santals, champignons doués d'une âme. *Cahiers du Pacifique* September 1970; 14.

Wasson, R. G. 1970. Soma of the Aryans: an ancient hallucinogen? *Bulletin on Narcotics* July–September 1970; 22(3). Later reprinted in *The Psychedelic Review;* see separate listing, same title, 1971. Published in French as "Le *soma* des aryans: un ancien hallucinogène?" (1970); see separate listing.

Wasson, R. G. 1970. Le *Soma* des aryans: un ancien hallucinogène? *Bulletin des Stupéfiants* July–September 1970; 23(3). Published in English as "Soma of the Aryans: an ancient hallucinogen?" (1970); see separate listing.

Wasson, R. G. 1970. Letter to the editor. *Times Literary Supplement* 21 August 1970.

Wasson, R. G. 1970. Drugs: the sacred mushroom. Guest editorial in *The New York Times* September 26, 1970.

Wasson, R. G. 1970. Letter to the editor. *Times Literary Supplement* 29 September 1970.

Wasson, R. G. 1970. Comments inspired by Professor Kuiper's review. *Indo-Iranian Journal* 1970; 12(4):286–298. (Kuiper's review immediately precedes the reply in the same issue.)

Ingalls, D. H.H., and R. G. Wasson. 1971. *R. Gordon Wasson on Soma, and Daniel H. H. Ingalls' Response: An Essay of the American Oriental Society, No. 7, 1971.* New Haven: American Oriental Society. Reprinted from an earlier publication; see separate listing, below, for RGW's "The Soma of the Rig Veda: what was it?" (1971).

Wasson, R. G. 1971. *The Hall Carbine Affair: An Essay in Historiography.* Deluxe third edition, with new subtitle and preface. Privately published. Printed in Verona by Stamperia Valdonega. Limited to 276 numbered copies, of which 26 are designated from A to Z, the others from 1 to 250.

Wasson, R. G. 1971. The Soma of the Rig Veda: what was it? *Journal of the American Oriental Society* April–June 1971; 91(2):169–187. Published along with "Remarks on Mr. Wasson's Soma," by D. H. H. Ingalls, in the same issue. Both papers were later reprinted together; see separate listing, above, for *R. Gordon Wasson on Soma, and Daniel H. H. Ingalls' Response: An Essay of the American Oriental Society, No. 7, 1971,* by Ingalls and Wasson (1971). RGW's paper was later republished in Spanish; see separate listing, English title, 1973.

Wasson, R. G. 1972. The divine mushroom of immortality. In *Flesh of the Gods,* edited by P. Furst. New York: Praeger Publishers, pp. 185–200. In addition to this paper, which is designated Chapter 6, includes a second paper by RGW as Chapter 7; see separate listing, below: "What was the Soma of the Aryans?" (1972).

Wasson, R. G. 1972. What was the Soma of the Aryans? In *Flesh of the Gods,* edited by P. Furst. New York: Praeger Publishers, pp. 201–213. In addition to this paper, which was designated Chapter 7, a second paper by RGW was included as Chapter 6; see separate listing, above, for "The divine mushroom of immortality" (1972).

Wasson, R. G. 1972. [Review of Carlos Castaneda's book *A Separate Reality: Further Conversations with Don Juan.*] In *Economic Botany* January–March 1972; 26(1):98–99.

Wasson, R. G. 1972. The death of Claudius, or mushrooms for murderers. Harvard *Botanical Museum Leaflets* April 1972; 23(3):101–128.

Wasson, R. G. 1972. *Soma and the Fly-agaric: Mr. Wasson's Rejoinder to Professor Brough.* Ethnomycological Studies No. 2. Cambridge, Massachusetts: Botanical Museum of Harvard University, November 1972.

Imazeki, R., and R. G. Wasson. 1973. *Kinpu,* mushroom books of the Tokukawa period. *The Transactions of the Asiatic Society of Japan,* Third Series, December 1973 (April 1975); 11:81–92.

Wasson, R. G. 1973. [Review of Carlos Castaneda's book *Journey to Ixtlan: The Lessons of Don Juan.*] In *Economic Botany* January–March 1973; 27(1):151–152.

Wasson, R. G. 1973. The rôle of 'flowers' in Nahuatl culture: a suggested interpretation. Harvard *Botanical Museum Leaflets* 1973; 23(8):305–324.

Wasson, R. G. 1973. The Soma of the Rig Veda: what was it? *Plural,* Mexico City, January-February 1976. Reprinted (in Spanish) from an earlier publication; see separate listing, same title, 1971.

Wasson, R. G. 1973. Mushrooms and Japanese culture. *The Transactions of the Asiatic Society of Japan,* Third Series, December 1973 (April 1975); 11:5–25.

Wasson, R. G., G. Cowan, F. Cowan, and W. Rhodes. 1974. *María Sabina and her Mazatec Mushroom Velada.* Ethnomycological Studies No. 3. Deluxe first edition. With four LP (long-playing) phonograph records and a musical score. A Helen and Kurt Wolff Book. Published by Harcourt Brace Jovanovich, New York and London. Printed in Verona by Stamperia Valdonega. Limited to 250 numbered copies, of which 50 are designated with Roman numerals from I to L, the others from 1 to 200.

Wasson, R. G., G. Cowan, F. Cowan, and W. Rhodes. 1974. *María Sabina and her Mazatec Mushroom Velada.* Ethnomycological Studies No. 3. First hardcover trade edition. With four audio cassette tapes and a musical score. Published by Harcourt Brace Jovanovich, New York and London. Printed in Verona by Stamperia Valdonega.

Wasson, R. G. 1974. [Review of Carlos Castaneda's books *Tales of Power* and *Las Enseñanzas de Don Juan.*] In *Economic Botany* July–September 1974; 28(3):245–246.

Guzmán, G., R. G. Wasson, and T. Herrera. 1975. Una iglesia dedicada al culto de un hongo, 'Nuestro Señor del Honguito' en Chignahuapan, Puebla. *Boletin de Sociedad Mexicana de Micologia* 1975; 9:137–147.

Wasson, R. G. 1977. Presentación. In *Vida de María Sabina, la Sabia de los Hongos,* by A. Estrada, Mexico: Siglo XXI Editores, pp. 9–22. Also published separately (see next entry). Later published in a Brazilian edition of Estrada's book; see RGW's "Apresentaçao" (1984).

Wasson, R. G. 1977. María Sabina y los hongos. In *Vuelta* July 1977; 1(8):24–27. A separately printed version of RGW's "Presentación" in Estrada's *Vida de María Sabina;* see preceding entry.

Wasson, V. G. 1977. *The Chosen Baby.* Third edition. Revised, with new illustrations by G. Coalson. Philadelphia and New York: J. B. Lippincott Co.

Wasson, R. G., A. Hofmann, and C. A. P. Ruck. 1978. *The Road to Eleusis: Unveiling the Secret of the Mysteries.* With a new translation of the *Homeric Hymn to Demeter* by Danny Staples. Ethnomycological Studies No. 4. First edition. A Helen and Kurt Wolff Book. Published by Harcourt Brace Jovanovich, Inc., New York and London. Printed in Verona by Stamperia Valdonega.

Wasson, R. G., A. Hofmann, and C. A. P. Ruck. 1978. *The Road to Eleusis: Unveiling the Secret of the Mysteries.* With a new translation of the *Homeric Hymn to Demeter* by Danny Staples. Ethnomycological Studies No. 4. First paperbound trade edition. A Helen and Kurt Wolff Book. New York and London: Harcourt Brace Jovanovich, Inc. A Harvest/HBJ Book. Printed in the United States.

Wasson, R. G., A. Hofmann, and C. A. P. Ruck. 1978. *LSD: Ho Dromos Gia ten Eleusina: He Apokalypse tou Mystikou ton Mysterion.* Modern Greek edition of *The Road to Eleusis* (1978). Athens: Synergatikes Ekdoseis Koinoteta.

Wasson, R. G. 1978. Wasson's velada with María Sabina. *Head* February 1978; 2(7):52–56.

Wasson, R. G. 1978. Presenting Keewaydinoquay. In *Puhpohwee for the People: A Narrative Account of Some Uses of Fungi among the Ahnishinaubeg,* by Keewaydinoquay. Ethnomycological Studies No. 5. Introductory paragraph dated November 19, 1977. Cambridge, Massachusetts: Botanical Museum of Harvard University, p. v.

Wasson, R. G. 1978. Introducing Mr. Ott. In *Mr. Jonathan Ott's rejoinder to Dr. Alexander H.*

Smith, by J. Ott. Ethnomycological Studies No. 6. Introductory paragraph dated 30 October 1978. Cambridge, Massachusetts: Botanical Museum of Harvard University, p. 1.

Wasson, R. G. 1978 (1979). Soma brought up to date. Harvard *Botanical Museum Leaflets* 1978; 26(6):211–223. Also published in the *Journal of the American Oriental Society;* see separate listing, same title, 1979.

Wasson, R. G. 1979. Foreword. In *Phantastica: Rare and Important Psychoactive Drug Literature from 1700 to the Present.* An antiquarian book seller's catalog (No. 13). Privately published by William and Victoria Dailey Antiquarian Books and Fine Prints, 8216 Melrose Avenue, Los Angeles, California 90046.

Wasson, R. G. 1979. Soma brought up to date. *Journal of the American Oriental Society* 99(1). Also published in Harvard *Botanical Museum Leaflets;* see separate listing, same title, 1978 (1979).

Wasson, R. G. 1979. Traditional use in North America of *Amanita muscaria* for divinatory purposes. *Journal of Psychedelic Drugs* 1979; 11(1–2).

Wasson, R. G. 1980. *The Wondrous Mushroom: Mycolatry in Mesoamerica.* Ethnomycological Studies No. 7. Deluxe first edition. New York, St. Louis and San Francisco: McGraw-Hill Book Co. Printed in Verona by Stamperia Valdonega. Limited to 501 signed and numbered copies, of which 26 *hors commerce* are designated from A to Z, the others from 1 to 475.

Wasson, R. G. 1980. *The Wondrous Mushroom: Mycolatry in Mesoamerica.* Ethnomycological Studies No. 7. First paperbound trade edition. New York, St. Louis and San Francisco: McGraw-Hill Book Co. Printed in the United States.

Wasson, R. G., A. Hofmann, and C. A. P. Ruck. 1980. *El Camino a Eleusis: Una Solucion al Enigma de los Misterios.* Spanish translation of *The Road to Eleusis* (1978). Mexico: Fondo de Cultura Económica.

Wasson, R. G. 1981. Retrospective essay. In *María Sabina: Her Life and Chants,* by A. Estrada. Translated by H. Munn from the earlier Spanish language edition. Santa Barbara, California: Ross-Erikson Inc., pp. 13–20.

Wasson, R. G. 1982. *R. Gordon Wasson's Rejoinder to Dr. Rolf Singer.* Ethnomycological Studies No. 9. Cambridge, Massachusetts: Botanical Museum of Harvard University, February 8, 1982. RGW's response to *A Correction by Rolf Singer.* Ethnomycological Studies No. 8. Same publisher, February 1, 1982. Singer's essay concerned J. Ott's earlier contribution to this series; see separate listing under Wasson, R. G., "Introducing Mr. Ott" (1978).

Wasson, R. G. 1982. The last meal of the Buddha. With memorandum by W. Rahula and epilogue by W. Doniger O'Flaherty. *Journal of the American Oriental Society* October-December 1982; 102(4):591–603.

Wasson, V. P. 1982. I ate the sacred mushrooms. In *Shaman Woman, Mainline Lady: Women's Writings on the Drug Experience,* compiled by C. Palmer and M. Horowitz. New York: William Morrow and Co., Inc., 1982, pp. 182–186. Reprinted from earlier publication; see separate listing, same title, 1956.

Ott, J., and R. G. Wasson. 1983. Carved 'disembodied eyes' of Teotihuacán. Harvard *Botanical Museum Leaflets* 1983; 29(4):387–400.

Wasson, R. G. 1983. *El Hongo Maravilloso Teonanácatl: Micolatria en Mesoamerica.* Spanish edition of *The Wondrous Mushroom* (1980). Mexico: Fondo de Cultura Económica.

Wasson, R. G., A. Hofmann, and C. A. P. Ruck. 1984. *Der Weg nach Eleusis: Das Geheimnis der Mysterien.* German edition of *The Road to Eleusis* (1978). Frankfurt am Main: Insel Verlag.

Wasson, R. G. 1984. Apresentaçao. In *A Vida de María Sabina, a Sábia dos Cogumelos,* by A. Estrada. Translated from earlier Spanish edition (1977). Sao Paulo, Brasil: Livraria Martins Fontes Editora Ltda., pp. 13–23.

Wasson, R. G. 1985. In pursuit of mushrooms. *Discovery* 1985; 18(2):9–15.

Kramrisch, S., J. Ott, C. A. P. Ruck., and R. G. Wasson. 1986. *Persephone's Quest: Entheogens and the Origins of Religion.* Ethnomycological Studies No. 10. Deluxe first edition.

Published by Yale University Press (New Haven and London, though these cities are not named in this edition). Printed in Verona by Stamperia Valdonega. Limited to 300 numbered copies, of which 26 *hors commerce* are designated from A to Z, the others from 1 to 274.

Wasson, R. G., S. Kramrisch, J. Ott, and C. A. P. Ruck. 1986. *Persephone's Quest: Entheogens and the Origins of Religion.* Ethnomycological Studies No. 10. First hardcover trade edition. New Haven and London: Yale University Press.

Appendix III

The R. Gordon Wasson Ethnomycological Studies Series

Beginning with *Soma: Divine Mushroom of Immortality* (1968), R. Gordon Wasson designated all his self-published works on sacred mushrooms and related matters the Ethnomycological Studies series. Thus *Soma* was Ethnomycological Studies No. 1, and his last book, *Persephone's Quest: Entheogens and the Origins of Religion* (1986), was No. 10. The present volume has been designated No. 11 by mutual consent of his children Peter Wasson and Masha Wasson Britten and by Richard Evans Schultes, Director (Emeritus) of the Harvard Botanical Museum, whom Gordon put in charge of the Tina and Gordon Wasson Ethnomycological Collection when it was established in 1983. In a letter to the present volume's editor dated January 24, 1989, Dr. Schultes writes: "Your plan to call the Festschrift that you have edited and will publish Ethnomycological Studies No. 11 seems to me to be fitting and appropriate, even though it follows Wasson's death in 1986. I recommend that this book be the final number in this numbered series." Gordon's children, representing his estate, concur. A list of all eleven titles in the series follows:

1. *Soma: Divine Mushroom of Immortality*
   By R. Gordon Wasson. With Wendy Doniger O'Flaherty. New York: Harcourt Brace Jovanovich, 1968.
2. *Soma and the Fly-Agaric: Mr. Wasson's Rejoinder to Professor Brough*
   By R. Gordon Wasson. Cambridge, Massachusetts: Botanical Museum of Harvard University, November 1972.
3. *María Sabina and her Mazatec Mushroom Velada*
   By R. Gordon Wasson, George and Florence Cowan, and Willard Rhodes. A Helen and Kurt Wolff Book. New York and London: Harcourt Brace Jovanovich, 1974.
4. *The Road to Eleusis: Unveiling the Secret of the Mysteries*
   By R. Gordon Wasson, Albert Hofmann, and Carl A. P. Ruck. With a new translation of the *Homeric Hymn to Demeter* by Danny Staples. A Helen and Kurt Wolff Book. New York and London: Harcourt Brace Jovanovich, 1978.
5. *Puhpohwee for the People: A Narrative Account of Some Uses of Fungi Among the Ahnishinaubeg*
   By Keewaydinoquay. Cambridge, Massachusetts: Botanical Museum of Harvard University, February 1978.

6. *Mr. Jonathan Ott's Rejoinder to Dr. Alexander H. Smith*
   By Jonathan Ott. Cambridge, Massachusetts: Botanical Museum of Harvard University, October 1978.
7. *The Wondrous Mushroom: Mycolatry in Mesoamerica*
   By R. Gordon Wasson. New York: McGraw-Hill & Co., 1980.
8. *A Correction*
   By Rolf Singer. Cambridge, Massachusetts: Botanical Museum of Harvard University, February 1982.
9. *R. Gordon Wasson's Rejoinder to Dr. Rolf Singer*
   By R. Gordon Wasson. Cambridge, Massachusetts: Botanical Museum of Harvard University, February 1982.
10. *Persephone's Quest: Entheogens and the Origins of Religion*
    By Stella Kramrisch, Jonathan Ott, Carl A. P. Ruck, and R. Gordon Wasson. New York and London: Yale University Press, 1986.
11. *The Sacred Mushroom Seeker*
    Edited by Thomas J. Riedlinger. Rochester, Vermont: Park Street Press, 1997.

Index

Abel, R. 11
Abelson, S. 11
Abies 103
Academie des Sciences 53
Acid Dreams 220
Addison Emery Verrill Medal 16, 23, 44
Adonai 246
Afghanistan 25
agapé 14, 62
Agaricales 100
agarics 100
Ahnishinaabeg (var. Ahnishinaubeg) 144, 145
Ahnishinaabem 142, 145
Alarcón, H. Ruíz de 44, 76
Alaska 101
Albers, J. 45
alcohol 8, 25, 132, 153, 157, 185, 220, 251–253
Alexander, H. 209
Alice in Wonderland 166, *167*, 172, 217
Allegro, J. 179, 190
Alpert, R. (see also Ram Dass) 56, 131
Alta Mixteca 99
Alotepec 139
Amanita muscaria (see also fly-agaric) 7, 10, 15, 17, 21, 26, 51, *57*, 72, 94, 100, 101, *101*, 112, 114, 132, 148, 166, 167, 172, 178, 180, 183, 221, 222, 225
Amazon 14
American Book Prices Current 49
American Museum of Natural History 37, 112, 135, 249, 256
Ames, O. 17
Ami, C. 11
"Among the Drunkards" 169
Anales 76
Andrews, L. 131
Anglo-Saxon 239
Annandale-on-Hudson 239
Annual Review of Psychology 212
Antonia, M. 139
Anungoday 141
Aphrodite 169
Aphyllophorales 99
Apollo 224

Apolonia (María Sabina's daughter) 122
Arabia Felix 25
Aragon, Father 122
Archeological and Ethnological Mission (see also Mission Arqueologique et Ethno-mycologique) 235
Ardpatrick 255
Ardrey, R. 149
Argentina 20
Argyllshire 255
Aristides 62
arrieros (mule guides) 37
Arroyo, C. 88, *89*
Arroyo, M. 88, *89*
The Artificially Steered Mind 123
Aryans 21, 62, 148
Asclepias 25
Ascomycetes 99
Asia 14, 15, 33, 100, 136, 147, 162, 178
Asia Minor 25
Aspen Mushroom Conference 184
Athena 224
Athens 254
Atiisokanek 144
Atla 235
Atlanta in Calydon 250
Aurelio (María Sabina's son) 199; *plate 7*
Aurelio, Don: see Carreras
Aurora (María Sabina's daughter) 122
Avestan 178
axis mundi 68–69
ayahuasca 78
Ayautla 117–118
Aztec 7, 13, 15, 44, 68, 70, 74, 75, 76, 115, 148

Bacchae 25, 26
Bacchus 224
Bachrach studio 193
backstrap loom 38
Bactria 25
Badham 103
badoh 15
badoh negro 75
bagasse 197

267

Bahamas 132
Baltimore, Maryland 208, 216
Banco National du Mexico 194, 220
Bandala Muñoz, V. 106
Banisteriopsis 178
B. caapi 78
Bank of Mexico 194
Bankers, Bones & Beetles 256
barbasco 84
Bard College 239
barley 148, 222
Barnard, F. 240
Barneby, R. 158
Barrera, A. 90
Barringer High School 245
Basel, Switzerland 72, 74, 75, 127, 234
basket weaving 37
Bass ale 214
Battaglia, Y. 40, 147, 209, 214, 216, 218
Battle of the Marne 249
Beals, R. 136
beatniks 131
Beaver Island 205–207
Beckett, S. 131
beer 157, 254
Belize 41, 216
bemushroomed 65
Bemitez, F. 92
Beowulf 247
Bernal, I. 92
beta blocker 126
Bethlehem Chapel 24, 42
Bevan, B. 135
bhang 25
Bible 245–248
Big Sur, California 228
Bigwood, J. 16, 22, 67, 185, 225
bimadisiwin 145
Bingham, H. 43
Binghamton, New York 24, 40
birch forest 234
Black, P. 23
Blake, W. 62, 228
Boehringer, R. 256
Boehringers 253, 256
Boethius, A. 19
boland gomba (crazy mushroom) 163
Bookman's Price Index 49
Borhegyi, S. 43
Borja, C. 194, 220
Boston 208, 216, 249
Boston University 15
Botanical Institute, Harvard University 126
Bourke, J. 172
Brack, A. 99
Brahmin 56
The Brearley School 193, 200
British Museum 207
Britten, M. Wasson 11, 20, *30*, 120, 123, 136, 193, 197, 198, 199, 234, 265
Bronx Zoo 249

Brough, J. 51, 131
Brown, C. 11, *23*
Brown, H. 45
Browning Club 241
Brugger, J. 210
Brugmansia candida 152, 154, 158
Buddha 114, 163, 188
Bulletin of the Mexican Mycological Society 104
Butler, S. 57, 58

caballero 197
Cage, J. 56
Cailleux, R. 99, 136, *138*, 234
Calderón, E. 78
Calegio Nacional 92
calendar, Mixe ritual 139, *140*
Caleti, A. 68
Cambridge, Massachusetts 14, 64, 183
Cambridge University 51
Campbell, J. 62, 113
Campillo, C. 92
Cana of Galilee 252
candles, ceremonial *plate 6*
cannabis: see *Cannabis sativa*
Cannabis sativa 78, 166, 180
Canterbury, England 243
Carreras, Don A. 38, 122, 123, 231; *plate 15*
Carroll, L. 166
Carvajal 103
Casas Campillo 103
"Case Uses in Anglo-Saxon" 239
Cashman, J. 205–207
Cashman, W. 205–207
Caso, A. 68, 100
Castaneda, C. 78, 79, 131
Castaneda's Journey 78
Castillo 104
Castle Howard 17
caterpillar 166, *167*
Catskills 111, 161
cauiqua terequa 101
caviar 34, 245, 256
Celtic cross 255, *255*
Central Intelligence Agency: see CIA
Centro de Estudios Avanzades 104
Centro Dermatológico Pascua 105
Century Club 188, 199
Cerletti, A. 99
Chapman, G. 250
Chatino 98, 233
Chaucer, G. 163
Chávez de la Mora, A. 103
Chekhov, A. 34
Chiapas 104, 234
Chiapas, State of 98
Chicago Tribune 209
Chico: see Pedro
Chignahuapan 88–91, 184
Chignahuapan, church of 88–91, *89*, 184
Child, J. 45
chili pepper 154, 157, 232

268

Chiltepec 157
Chinantla 99
Choapan 136
chocolate 188
chocolatl 122
Chol Maya 44
Church of Ireland 239
Church of the Incarnation 240
CIA (U.S. Central Intelligence Agency) 53, 73, 199, 203, 212
Cid, H.: see Martínez Cid
Cifuentes 105
Ciudad Victoria 104
Civil War 210
Clare, J. 250
Clathrus crispus 99
Clavaria truncata 98
Clavariadelphus truncatus 98
Claviceps 76
C. purpurea 7, 168, 222
Cleland, T. 214
Clerk of Oxford 163
Cocteau, J. 61
codices, Maya 10, 63, 100
Coe, M. 76
Coe, P. 45
Coe, S. 43
colador del brujo 99
Colegio de Postgraduados de Chapingo 104
Colima, State of 98
Colombia 132
Columbia College 239
Columbia College, Library of 239, 240
Columbia School of Journalism 19, 24
Columbia University 240
Columbus, Ohio 240
Comptes rendus 53
Concord, Massachusetts 212
Conocybe 74
C. siligenoides 98; *plate 11*
Consejo Nacional de Ciencia y Tecnología (CONACYT) 105, 106
A Contribution to Our Knowledge of Rivea corymbosa, the Narcotic Ololiuqui of the Aztecs 74
"A Conversation with R. Gordon Wasson" 59
Convolvulaceae 76
Cordyceps 99
C. capitata 98, 99
C. ophioglossoides 98, 99
Cook, F. 251
Cooke, M. 166, 172
Coombes 167
copal incense 120, 122, 154, 155; *plates 6, 13*
A Correction 52, 266
Cortés Hotel 194
Cotzocón: see San Juan Cotzocón
Cowan, F. 14, 22, 130, 236, 265
Cowan, G. 14, 22, 130, 236, 265
Crane clan 142
The Crown 251

Cruz, B. 235
Cruz, J. 88
Cthulhu mythos 175
Culture in Context 149
curandera, curandero (healer) 71, *87*, 112, 118, 119–120, 121–124, 131, 135, 139, 163, 197, 199
Current Opinion 20

d-lysergic acid diethylamide (see also LSD) 72
Dailey, V. 132
Dailey, W. 132
D'Alembert 48
Dallas, Texas 194
Danbury, Connecticut 10, 21, 24, 40, 43, 44, 62, 147, 183, 184, 185, 186, 188, 201, 207
Dartmouth College 78
Darwin, C. 190
The Darwinian Theory 190
Datura 74, 76, 152, 154, 157
D. stramonium 152, 154
De Avila 101
De Mille, R. 78, 79
Dead Sea Scrolls 179
Delay, J. 99
Demeter 224
Department of Near Eastern Studies, University of California 178
Desierto de los Leones 96
DeVeny, M.: see Wasson, M. DeVeny
Dewey, M. 240
Dictyophora indusiata 98, 99
D. phalloides 98
Diederot 48
Dionysian religion 180
Dionysus 25, 26, 224
Dioscorea 84
disembodied eyes 69, 188
Disses 243
"The Divine Mushroom of Immortality" 66
Dobkin de Rios, M. 78
Dolmatoff, G. 78
Don Juan 78, 79
The Don Juan Papers 78, 79
Don Luis: see Rodríguez
Doniger (O'Flaherty), W. 9, 15, 20, 58, 113, 129, 223, 265
The Doors of Perception 211
Dorantes, J. 135
Drugs and Medicines of North America 169
"Drugs and the Mind" 131
Drummond, Mr. 244, 256
Dubovoy 103
Duke Forest 147
Duke University 78
Dylan, B. 209

Eckholm, G. 37, 135
Eduardo 118
Eerde, E. 93
Efron, D. 178

269

18th Amendment 252
Eje Neovolcanico 106
El Salvador 96
Elaphomyces 99
E. variegatus 98, 99
Eleusinian Mysteries 7, 15, 22, 28, 61, 65, 66, 114, 132, 148, 163, 168, 222, 224, 225
Eleusis: see Eleusinian Mysteries
Eliade, M. 180
Elohim 246
Eltingville, Staten Island 240
Emboden, W. 78
Encyclopédie 48
Endless Caverns 175
England, English 39, 106, 142, 145, 189, 217, 219, 231, 243, 249
English Pale 239
Ensenada 104
entheogen 9, 16, 22, 55, 67, 147, 163, 186, 211, 215, 221, 225
Ephedra 25
Episcopal Church 240
Epling, C. 126
ergot 15, 22, 72, 75, 98, 148, 222–223
Erickson 161
Esalen Institute 179, 228
Escalante, R. 101
Escuela de Agricultura, Chapingo 104
Escuela Nacional de Antropología 76
Escuela Nacional de Ciencias Biológicas 104, 158
Esparza García 103
Espíritu Santo Tamazulapan 137
Estrada, A. 113
Ethnomycological Studies series 52, 61, 187, 265–266
ethnomycology 8, 13, 72, 83, 105, 111, 124, 161, 165
"Ethnopharmacologic Search for Psychoactive Drugs" 177
Ethnopharmacology Newsletter 64
ethyl alcohol: see alcohol
Etidorhpa 132, 168–172, *170*, 175
Euripides 25, 26
Europe, European 13, 14, 15, 21, 63, 72, 73, 106, 132, 148, 162, 168, 185, 216, 222–225, 252, 254
"Europe and the Fly-Agaric" 225
extrasensory perception attributed to mushroom intoxication 39-40, 123, 197–198

Facultad de Ciencias, University of Mexico 104
Faculty of Medicine, University of Mexico 105
The Faerie Queen 250
Far East 20, 21
The Father of Ethnomycology 8
Ferdinand, Grand Duke 249
Fernandez, J. 78
Fitz Hugh Ludlow Library 28
Fitzgerald, E. 206

Flashback Books 175
Flattery, D. 178
Flesh of the Gods 62, 66, 78
flesh of the gods (teonanácatl) 148
flies 100
Flor de cacao 136
Flora Neotropica 106
Florrie: see James, F.
fly-agaric (see also *Amanita muscaria*) 17, 26, 51, 56, *57*, 67, 72, 78, 100, 145, 206, 221
Folkway Records 237
Forte, R. 59, 179, 211, 212
4-hydroxy-indol 126
4-phosphoryloxy-dimethyltryptamine 126
Framboises 142, 143
France, French 14, 19, 20, 24, 34, 48, 53, 61, 106, 142, 145, 147, 189, 221, 231, 235
Freedom of Information Act 212
Fuchicovsky 104
Fuente, B. de la 92
Fuente, R. de la 92
Fundamentalists 251, 252
Fungi Mexicana 100
Funk & Wagnells Dictionary 248
Furst, D. 80
Furst, P. 62, 66

Galindo 104
Gandára 104
Ganoderma lobatum 90, 91
García (Mendoza), Cayetano 197
García, Consuela 119–120, 122, 123
García (Mendoza), home of Cayetano and Guadalupe *plate 3*
García, J. 117
García (Mendoza), G. 197
The Garden Journal 37
Garden of Eden 179
Gardner boys 245
Gasteromycetes 99
Gelpke, L. 123
Gerontius 250
Geschickter Fund 203, 212
Gettysburg 249
Giants (baseball team) 245
Gilberti, M. 101
glass flowers 249
Glückspilz 67, 72
God 228
Gomphus floccossum 98
goose eggs *plate 6*
Gordon, Dr. R. 253–254
Gordon, I. 254
Gospels 252
Gothenburg Ethnographical Museum 71
Graves, R. 20, 26, 32, 37, 56, 112, 165, 180
Grail: see Holy Grail
grapefruit 245
Gratacap, Mr. 256
Great Britain 249
Great Falls 19, 240, 253

270

Great Lakes 145
Greece, Greek 7, 15, 22, 25, 62, 63, 65, 114, 148,
    151, 179, 180, 211, 221–225, 239, 243, 249,
    255
Greek cross 219
Grof, S. 111, 215
Guadalajara 105, 106
Guadelupe 116, 121
Guaranty Co. of New York 20
Guatemala 14, 15, 43, 65, 100, 101, 162
Guerrero, State of 76, 105
Guggenheim Memorial Foundation 88, 106
Guzmán, G. 20, 52, 53, *88*, 105, 158, 183
Guzmán Davalos, L. 106

Haight Ashbury (district of San Francisco) 114
Hainuwele 158
Halffter, G. 84, 96
Halifax, J. 28, *28*
The Hall Carbine Affair 23, 47, 130, 210, 220
Halley's Comet 217, 248
"The Hallucinogenic Fungi of Mexico" 189
"The Hallucinogenic Mushrooms" 37
Hallucinogenic Plants of North America 52, 186,
    187
"Hallucinogens and the Shamanic Origins of
    Religion" 78
Hamlet 256
haoma 178
harmala alkaloids 178
harmaline 178
harmine 178
Hartford, Connecticut 208
Harvard Botanical Museum 16, 19, 22, 37, 48,
    49, 52, 63, 65, 66, 69, 74, 76, 131, 133, 172,
    209, 216, 220, 265
Harvard *Botanical Museum Leaflets* 13, 48, 63,
    131, 188, 220
The Harvard Club 113
The Harvard Review 131
Harvard University 62, 63, 65, 74, 76, 113, 126,
    140, 161, 177, 202, 218, 237, 249
Hebrew 239
Heim, R. 14, 20, 44, 48, 53, 62, 63, 74, 84, 88, 92,
    98, 99, 104, 129, 136, 138, *139*, 149, 158, 163,
    185, 186, 199, *230*, 231, 232, 233, 234; *plates*
    8, 9
Herald Tribune 20
Hernandez, R. 88
Herodotus 221, 247
heroin 73
Herrera, T. 84, 88, *88*, 91, 99, 103, 104, 105, 183
Hexateuch 246
Heyerdahl, T. 149
Hidalgo, State of 105
High Priest 56, 58
High Times 56, 58, 131
Higher Criticism 246
Hill of the Mushrooms 100
Hillside 256
Himalayas 148

Hindu 7, 21, 148
Hindu Kush 148
hippies 33, 44, 113, 124, 131, 199, 233
Hispahkwe 145
Hofmann, Albert 14, 15, 22, 28, 39, 61, 62, 72,
    73, 74, 75, 91, 99, 105, 114, *117*, 130, 132,
    146, 168, 209, 222, 225, 234, 265; *plate 19*
Hofmann, Anita 115, 116, 120, 123
Hogg, J. 250
hojas de la Pastora (see also *Salvia divinorum*)
    115, 118–121, 122, 123, 126, 232
hojas de María Pastora (see also *Salvia*
    divinorum) 115
Holmes, Sherlock 246
Holmstedt, B. 177
holograms 215
holonomic theory of consciousness 215
Holy Grail 185, 219
Homer 243, 247
Homer 250
Homeric Hymn to Demeter 22, 61, 265
Honor March of Bears 145
Hoogshagen, S. 100, 158; *plate 1*
hookah 166, *167*
Horowitz, M. 10, 133, 172, 175
Hospital de Enfermedades Tropicales 104
Huauchinango 234
Huautla de Jiménez 37, 38, 71, 75, 76, 84, *85*,
    87, 93, 103, 112, 113, 118, 121, 124, 135, 136,
    168, 197, 207, 209, 215, 231, 234, 235, 236;
    plates 2, 6
Huautla de Jiménez, trail to *plate 3*
Huaxteca 136, 234
Hudson, W. 24
Hueyapan 101
Hughes, L. 19
Huichol 67, 69, 70, 76, 78
huipile 38, 137
Hungary 163
Huxley, A. 56, 171, 211
Hyperboreans 221

"I Ate the Sacred Mushroom" 39, 131
iboga 180
Icelandic 239
In Memoriam 250
INAH (see also Instituto Nacional de
    Antropología é Historia) 70
India 13, 15, 17, 25, 58, 62, 136, 148, 151, 178,
    179, 183, 188
Indian Harbor 143
Indian Ocean 147
Indochina 147
Ingalls, D. 62
"Inquiry into the Origins of the Religious
    Idea" 190
Institute for Experimental Biology 104
Instituto de Biología 104
Instituto de Botánica, University of Guadala-
    jara 105, 106
Instituto de Ciencias, Tutla Gutíerrez 104

Instituto de Ecología 106
Instituto Mexicano del Seguro Social 105
Instituto Nacional de Antropología é Historia (INAH) 69
Instituto Nacional de Investigaciones sobre Recursos Bióticos 105
Instituto Nacional Indigenista 121
Instituto Tecnológico, Ciudad Victoria 104
Interpersonal Diagnosis of Personality 131, 212
Ipomoea violacea 15, 69, 70, 75, 117–118, 152, 158
Iran 178
Irish Ecclesiastical Record 172
Isikowa 103
Isla de Beheur 104
isopropylamino–2-hydroxy-propyl-4-*hydroxy-indol* 126
Ixtapalapa 104
Ixtepec 195

J. P. Morgan & Co.: see Morgan & Co., J. P.
Jacomet, D. 36
Jakobson, R. 149
Jagger, M. 131, 209
Jalapa de Díaz 115, 121
Jamaica 243
James, F. (Florrie) 34
James, W. 9, 175
Japan, Japanese 28, 96, 97
Játiva, C. 126
Jensen, A. 158
Jerusalem 254
Jesus Christ *90*, 152, 246, 252
Johnson, I. Weitlaner 38, 71, 75, 88, 115, 122, 123, *134*, *138*, 158, 183, 234, 235
Johnson, J. 38, 71, 115, 135
Johnson, S. 11
Joralemon, D. 158
Journal of the American Oriental Society 58
Journal of Entheogenic Drugs (proposed) 186
Journal of Ethnopharmacology 64
Journal of Psychoactive Drugs 8, 186
Journal of Transpersonal Psychology 207
Juchitán 195
Juquila 233

Kallunki, J. 158
Kamchatka Penninsula 100
Kandy, Ceylon 147
Karolinska Institute 177
Katerina Ishmailova 64
Kaum 141, 142, 143
Kazenchak, N. 220
Keats, J. 163
Keewaydinoquay 15, 141, 142, *142*, 144, 145, 265
Kelsos 249
Kenwabakesi 141
Kerouac, J. 131, 206
Klauber, L. 149
Kleicomolo 175

Knapp, J. 170
Kobayasi 96
Kobel, H. 99
Kodachrome film 194
Korjaks 166
Kramrisch, S. 15, 22, 50, 130, 216, 221, 266
Kroll, A. 72
Kubla Khan 197
künstlich gesteurte Seele, Die 123

La Barre, W. 78, 114, *146*, 149
La Malinche 138
La Providencia 117
Lacaud, L. 135
lactones 158
Lagunas 195
Lahr, B. 131
Lake Michigan 205
Lamperière, T. 99
Land's End 249
Lapland 17
"The Last Meal of the Buddha" 58
The Last Supper 219
laurel 224
Laviada, I. 140
Lawrence, Bishop 243
Leary, T. 8, 44, 56, 58, 129, 131, 211–212, 216, 218
Lee, M. 220
Lennon, J. 131, 209
Lentz, B. 172
Les Champignons Hallucinogènes du Mexique 20, 48, 51, 99, 129; *plate 20*
Leví-Strauss, C. 143
Life 7, 8, 10, 49, 71, 84, 92, 130, 177, 186, 193, 197, 199, 202, 203, 209, 210, 212, 218, 228, 233, 236
Life en Español 71
Linnaeus (var. Linneus) 100
Linnean Society 16
Lipp, F. 100, *150*
Lloyd, C. 169
Lloyd, J. U. 132, 168, 175
Lloyd Library of Botany and Pharmacy 169
Lloydia 169
Logan Airport 208
Logretta, Don A. 220
Lolium temulentum 222
London 249
London School of Economics 19
López González, A. 101
López Negrete, Don L. 220
López Ramírez, M. 106
Lophophora williamsii 70
Lovecraft, H. P. 175
Lowy, B. 100, 101, *160*
LSD (see also *d*-lysergic acid diethylamide) 7, 72, 75, 76, 98, 111, 132
Luce, H. 209
Lycoperdon mixtecorum 98
L. marginatum 98

Lydia 25
Lyons Farms 244, 256

Macdonalds 253
McCarthy, J. 43
McGinnity, "Iron Man" J. 245
macaw feathers 232
Machu Picchu 43
Madrid Codex 100
Madrigal, X. 93
Mahabharata 56
maize 153
Majorca 37, 112
Mandaic 178
Mangelsdorf, P. 16
Mapes, C. 103
Margot, P. 220
Marks, J. 199, 203, 212, 220
Mardersteig, G. (Hans) 36, 37, 47, 96, 112, 129, 149
Mardersteig, M. *128, 133*
María Sabina and her Mazatec Mushroom Velada 14, 22, 72, 130, 135, 190, 200, 218, 232, 236, 265; *plate 20*
marijuana 25, 78
Marstrand Foundation 158
Martínez Cid, H. 38, 118, 119, 121, 123, 197, 199; *plate 15*
Martínez Mancilla 103
Masks of the Gods 62
Mason, R. 11; *plate 20*
Massachusetts Correctional Institute 212
Matlazinca 101
Matthewson, C. 245
Maya 7, 43, 44, 96, 100
Mayer, K.-H. 96
Mazahua 98, 151
Mazatec 13, 14, 15, 20, 51, 52, 62, 65, 71, 74, 84, 92, 98, 112, 115, 116, 130, 131, 136, 148, 151, 163, 228, 231, 232, 233, 235, 236
Mazatlán de los Mixes 71, 151, 195–196
Mazatlán de los Mixes, trail to *plate 1*
MDA (methylenedioxyamphetamine) 177
meadow sage 119
Medea 25
Mendel, G. 191
Merrill, A. 165
mescal 157
methylenedioxyamphetamine: see MDA
Metropolitan Museum 249
Metropolitan University, Ixtapalapa 104
Metzner, R. 177
Mexican Mycological Society 92, 96, 104, 105
Mexico 13, 14, 101, 188
Mexico, State of 95, 99, 103, 105, 106
Mexico City 14, 37, 40, 68, 69, 71, 83, 84, 87, 92, 94, 95, 96, 104, 105, 115, 121, 123, 183, 184, 194, 197, 200, 202, 231, 233, 234, 235
Michoacan, State of 97, 98, 101
Miller, W. 100, 136, 139, 158, 194, 195, 196, *196*, 234; *plate 1*

milpas (corn fields) 38
Milwaukee, Wisconsin 254
Milwaukee Public Museum 43
Mind-at-Large 211
Miniss Kitigan 141, 142, 145
Miniss Kitigan Press 144
mint 222
Misantla 235
Miskwedo 145
Mission Arqueologique et Ethnomycologique (see also Archeological and Ethnological Mission) 106
Mr. Jonathan Ott's Rejoinder to Dr. Rolf Singer 52, 266
Mitla 234
Mixe, Mixeria 70, 87, 98, 136, 138, 139, 140, 151–159, 194, 234
Mixtec 98
Molohovetz, V. 36
Molokhovets, E. 34
"Monsieur Among the Mushrooms" 172–174, 175
Monterrey 104
Montoya Bello, L. 106, 235
Moore, J. 199, 212, *230*, 231, 232; *plate 8*
morel 216
Mother Goddess 222
The Mother of Ethnomycology 8
Mount Olympus 223
Mount St. Alban 255
Mountbatten, Lord 147
Mora, V. 101
Morelos, State of 101, 106
Morgan, J. P. 23, 220, 243
Morgan & Co., J. P. 7, 23, 47, 149, 208, 209–211, 236
Morgan Guaranty 208, 233
morning glory (see also ololiuqui) 15, 69, 70, 72, 74, 76, 77, 117, 136, 157, 220, 221
Morris, D. 149
Muri-Bern, Switzerland 256
muscimole 17
Muséum National d'Histoire Naturelle 20, 74, 87, 129, 136, 231
Museum of Anthropology, Mexico 76
Museum of Natural History, France: see Muséum National d'Histoire Naturelle
Museum of Natural History, Mexico City 96, 106
Mushroom Ceremony of the Mazatec Indians of Mexico 237
mushroom dryer 93
mushroom dust 79
mushroom intoxication, accounts of 38, 92–95, 155–157, 165, 197–198, *198*
mushroom stones 15, *21*, 43, 77, 96–97, *97*, 100, *101,* 103, 104, 162, 177
Mushrooms, Russia and History 7, 20, 25, 34, 43, 47, 49, 62, 63, 64, 66, 67, 71, 79, 92, 112, 129, 151, 162, 166, 172, 185, 189, 220, 237; *plate 20*

musicians (at festival in San Juan Metaltepec) *138*
mycolatry 163
Mycologia 52, 53, 186, 187
Mycological Society of America 162
mycophiles 16, 20, 36, 55, 56, 67, 77, 161
mycophilia 163
mycophobes 15, 20, 36, 55, 78, 161, 177
mycophobia 163

Nahuatl 68, 95, 96, 98, 147, 148, 163, 235
Nanacamilpa 184
nanácatl 95
Naples, Florida 47
National Cathedral: see Washington Cathedral
The National Geographic 202
National Museum of Anthropology, Mexico City 68, 96, 115, 234
National Science Foundation 158
Natividad Rosa 119
Natural History 72
Natural History Museum: see American Museum of Natural History
Nava, Don I. 92, 93, 94, 95
Necaxa 96, 234
Neolinco 92
Nepanee, Ontario 239
Nevado de Toluca 84, 99, 101, 103, 235
"The New Accelerator" 168
New Guinea 15
New Haven Register 19
New York 194, 196, 234, 235, 239, 241
New York Botanical Garden 23, 158, 187
New York Times 147
Newark, New Jersey 19, 24, 217, 241, 243
Newark Daily Advertiser 243
Newark Police Force 253
Newman 250
Newman, A. 172
Nexapa 95
Nicholas-Charles, P. 99
Nicotiana rustica 70
nicotine 73
Nieto 103
nitrous oxide 168
non-specific amplifiers 211
North Africa 249
North Pole 251
Nouvelles Investigations sur les Champignons Hallucinogènes 20, 48, 99, 129; *plate 20*
Nuestro Señor del Honguito 184
Nymphaea ampla 62
Nysa 221

Oakes Ames Library 183
Oaxaca (city) 137, 139, 140, 194, 195, 233
Oaxaca, State of 13, 14, 15, 16, 20, 22, 53, 70, 74, 75, 84, 87, 92, 98, 99, 100, 112, 116, 131, 135, 136, 138, 151, 186, 234
O'Flaherty, W. Doniger: see Doniger

Officina Bodoni 256
Ojibway (see also Ahnishinaabeg) 15, 143, 145
Ojibwe: see Ojibway
Okiiyasin 141
Old Testament 252
olive 224
ololiuqui (see also morning glory) 15, 43, 70, 75, 76, 77, 115, 118, 163, 220, 221
ololuc 75
Olympia, Washington 91, 185
On the Road 131
Orinoco Delta 78
Orissa, State of 188
Orizaba, Valley of 115
Ortega, F. (Chico) 136, 138, 234
Otomi 98
Ott, J. 15, 16, 22, 50, 52, 58, 67, 88, 130, 216, 221, 225, 266
Our Lord of the Little Mushroom 184
Oxford, Maine 165
Oxford Dictionary 248
Oxford University 166

Pahlavi 178
Pahuatlán 235
Palace of Quetzalpapálotl 188
Palacios, M. 84, *86*, 93
Palenque 44
Panaeolus 13, 87
P. campanulatus 87
P. fimicola 87
P. papilionaceus 165, 168
P. sphinctrinus 87
Papua, New Guinea 15
Paris 74, 136, 231, 235, 237, 249
Pastrana, A. Lemos 158
Patzcuaro 103
Paz, O. 32, 63, 92
Peabody Museum 22, 43, 44
Peaine Airport 141
Peary, Admiral 251
Pedro (Chico) 116, 117, 121
Peganum 180
P. harmala 178
Peloponnesus 19, 144
Pentagon 73
Pentateuch 246
Pentheus 25
Perfido (boy subject of 1958 mushroom velada) *plates 16, 18*
Persephone 221, 224
Persephone's Quest 9, 15, 22, 40, 50, 130, 179, 186, 188, 190, 216, 218, 221, 222, 265, 266; *plate 20*
Persian 178
Petaluma, California 175
Peterkin, D. 210
peyote 13, 43, 67, 69, 70, 74, 76, 78
peyotl 70
Phantastica (catalog) 132

"The Pharmaceutical Alchemist" 175
Phi Betta Kappa 239
Phrygia 25
piciétl 70, 163
Pike, E. 194
Pim, H. Moore 172–174, 175
Pinchot, P. 99
pine 156, 158–159
Pinus 103
pipiltzintzíntli 70
A Plain and Easy Account of British Fungi 166
Plantillo 195
"The Plattner Story" 168
Plymouth, Ohio 240
pochoir process 36, 50
The Poet and His Muse 61
"The Poet and the Dreamer" 220
Pointe De Galle 147
"Politics of Consciousness Expansion" 131
Pollack, S. 53, 92
Polytechnic Institute, Mexico City 83, 88, 90, 103, 104, 106
Pope Pius XII 210
Popocatéptl 84, 95, 96, 115
Popoloápan 115
poppy 222–223
pork 157
Portilla, L. 92
Posada Rosaura 121
Post, E. 201
"Postscriptum: A Memoir" 256
Praeger Publishers 78
prayer to the sacred mushroom 153–154, 155; *plate 17*
Priestly Redactors 246
Prince of Flowers 148
Prohibition 132, 185, 251–253
Providence, Rhode Island 249
psilocin 14, 75, 99, 124, 158
psilocin, medical uses of 124
psilocybin 14, 38, 75, 99, 103, 122, 123, 124, *125*, 126, 131, 158, 168, 171, 174, 175, 187, 234
psilocybin, medical uses of 103, 124, 228
Psilocybe 20, 51, 70, 74, 88, 91, 92, 94, 98, 99, 103, 105, 106, 125, 151, 153, 158, 163, 167, 168, 212, 234
Psilocybe, map of species distribution in Mexico *107*
P. angustipleurocystidiata 101, *102,* 105, 106
P. aztecorum 95, 96; *plate 11*
P. barrerae 105
P. caerulescens 74, 91, 96, 101, 103, 152; *plates 4, 10, 11, 15*
P. cordispora 152
P. cubensis 93, 98, 99, 101, 103; *plate 10*
P. hoogshagensis 152
P. mexicana 74, 79, 87, 91, 96, 99, 101, 103, 123, 152; *plate 11*
P. muliercula 51, 53, 88, 92, 99, 101, 103, 105, 186

P. sanctorum 101, 105
P. wassonii 20, 51, 53, 88, 92, 103, 186
P. wassoniorum 20, 53, 92
P. yungensis 152
P. zapotecorum 103, 105; *plate 10*
psychedelic 211
psychedelic healing 78
Psychedelic Monographs & Essays 220
psychedelic movement 8, 114, 211, 212, 214, 228
The Psychedelic Review 177, 212, 216, 220
Puebla (city) 123, 197, 233
Puebla, State of 88, 91, 96, 115, 183
Puharich, A. 190
Puhpohwee for the People 146, 265
Pulitzer Traveling Scholarship 19
Purepecha (Tarascan) 98, 100, 101, 103
"The Purple Pileus" 167

Quararibea funebris 136
Quecha 78

R. Gordon Wasson's Rejoinder to Dr. Rolf Singer 52, 265
Rattlesnakes 149
Ram Dass (see also Alpert) 56
Rancho del Cura 93, 234
Rancho Lucas Martin 87
Ravicz, R. 100
Recensions 246
recipe cards 34–35
Reko, B. Pablo 13, 14, 37, 75, 96, 135
Religion and Drink 185, 251
"The Religious Experience: Its Production and Interpretation" 212
retsina (var. retzina) 222, 255
ReVision 211
Revista Mexicana de Micología 104
Revue de Mycologie 53
Reyes, L. 139
Rhine, J. B. 149
Rhodes, W. 22, 130, 236, 265
Richardson, A. 11, 38, 112, 131, 189, *192, 200, 201,* 231, 232, 236; *plate 5*
Richardson, M. 193, 196, 201
Riedlinger, Jennifer 11
Riedlinger, June 11, 205
Riedlinger, T. 31, 202, 266
Rig Veda, Vedas, Vedic 15, 25, 49, 56, 62, 72, 114, 148, 163, 190, 221, 225
Rio Grande 234
Rio Santiago 118, 234
Rio Santo Domingo 117, 118
Rivea corymbosa 75, 76, 77
The Road to Eleusis 9, 15, 22, 47, 61, 63, 113, 130, 133, 168, 190, 225, 265; *plate 20*
Roberts, T. 11
Rockefellers 193
Rodríguez, L. (Don Luis) 137–140, *139*
Rome 243
Romero 104

275

Rosetta stone 148
Rosovsky, Dean 161
Ross, M. 245
Rowlandson, T. 214
Ruck, C. 15, 16, 22, 28, 50, 61, 62, 67, 113, 130, 132, 168, 185, 186, 216, 223, 225, 265, 266
Ruíz Oronoz, M. 104
Russell, C. 241
Russell, N. 241
Russia 34
Russian cuisine 34–35
rye 22, 72, 148

Sabina, M. 14, 39, 44, 49, 71, 72, 103, 112, 113, 121, 123, 131, 148, 151, 162, 197, 207, 209, 231, 232, 236; *plates 7, 12, 13, 16, 17, 18*
Sabino 119
sacred fungus of the church of Chignahuapan 88–91, *90, 91,* 184
The Sacred Mushroom and the Cross 179
The Sacred Mushroom Seeker 266
Safford, W. 74, 75
Sahagún, B. de 38, 44, 68, 70, 76, 95, 96, 97
St. Anthony's fire 73
St. James 233
St. James, Michigan 205
St. John of Patmos 62
St. Joseph of Arimathea 219
St. Joseph of Arimathea, Chapel of 219
St. Paul's Chapel 249
St. Stephen's College 239
St. Stephen's parish 243
Salvia 127
S. divinorum (see also hierba de María Pastora, hojas de María Pastora, ska Pastora and ska María Pastora) 15, 22, 70, 119, 127, 136; *plate 18*
San Agustín Loxicha 71
San Andrés (town and landing strip) 84, *86,* 231–232, 234; *plate 2*
San Bartolo Yautepec 75
San Cristóbal Chichicaxtepec 137
San Diego Museum of Man 78
San Francisco 185, 205, 227, 228
San Isidro mushroom 93
San José Tenango 118, 120, 122
San Juan Cotzocón 138, 139, 140, 234
San Juan Metaltepec 137
San Miguel-Hidalgo 118
San Pedro cactus 78
Sánchez Marroquín 104
Sandoz, Ltd. 72, 99
Sanskrit 57, 62, 72, 148, 239
Santa María Mixistlan 137
Santa María Nativitas Coatlán, informants from *plate 1*
Santa María Tonantzintla church 233
Santiago 121, 137
Santiago's Sword 233
Santo Domingo Petapa 195, 196
Sapir, E. 149

Sapper, C. 96
Sartre, J.-P. 207
"Sartre's Rite of Passage" 207
Satchel, Mr. 243
Scatalogic Rites of All Nations 172
Schultes, R. E. 11, 20, 25, 37, 38, 44, 49, 62, 64, 65, 69, 74, 76, 77, 78, 88, 91, 95, 99, 114, *146,* 148, 149, 158, *160,* 161, 183, 219, 265
Schwartz, M. 178
Science 165
scopolamine 158
Scotland 249
The Search for the "Manchurian Candidate" 199, 203, 212, 220
Second International Congress of Mycology 105
Secretería de Agricultura y Recursos Hidráulicos 105
Secretería de Desarrollo Urbano y Ecología 105
Secretería de Educación Pública (SEP) 105
Secretería de Programación y Presupuesto 105
"Seeking the Magic Mushroom" 7, 50, 71, 84, 92, 131, 209
Seeler, M. 51
Selected Letters and Marginalia (of H. P. Lovecraft) 175
Septuagint 246
serotonin 75
The Seven Sisters of Sleep 166, 172
shaddock 245
Shaddock, Admiral 245
Shakespeare 163, 245
Shakespeare Society 241
shaman, shamanic, shamanism 9, 14, 15, 50, 56, 62, 65, 72, 76, 78, 111, 112, 131, 145, 148, 151, 153, 158, 163, 165, 180, 189–190, 231, 232, 235
Shandy, T. 50
Sharon, D. 78
Shelley, P. 163
Shi'a Islam 178
Shlain, B. 220
Shostakovitch 63
Shulgin, Alexander 177
Shulgin, Ann 227
Siberia 15, 72, 101, 112, 148, 151, 163, 166, 235
Sierra Costera 71
Sierra de Puebla 68
Sierra Madre del Sur 233
Sierra Mazateca 116, 121, 195; *plate 19*
Sign of the Four 246
Silliman College 45
Simon & Schuster 79
Simlipal Hills 188
Singapore 59
Singer, R. 51, 84, *86,* 87, 88, 93, 98, 99, 100, 103, 104, 184, 186–187, 266
"The Siren" 34
Sistema Nacional de Investigadores 105, 106

ska María Pastora (see also *Salvia divinorum*) 115, 127

ska Pastora (see also *Salvia divinorum*) 115

Skeat 248

Smith, A. 51, 100, 184, 186

Smith, B. 58

Smithsonian Institute 169

Sociedad Mexicana de Antropología 71

Society for Economic Botany 16

soil flowers 103

Soma (book) 20, 47, 56, 62, 72, 113, 129, 131, 162, 183, 184, 189, 190, 201, 206, 218, 225, 265; *plate 20*

Soma (sacred substance) 7, 10, 15, 17, 20, 25, 26, 49, 51, 56, 63, 72, 78, 148, 149, 162, 178, 222, 225

Soma and the Fly-Agaric 265

"Soma: The Three-and-One-Half Millenia Mystery" 149

"Some Comments on Hallucinogenic Agarics and the Hallucinations of Those Who Study Them" 187

Song of Roland 247

Sontag, S. 45

sorcery 156, 157

Sosa de García, D. 118

Spain, Spanish 4, 19, 22, 24, 68, 70, 71, 74, 76, 84, 93, 95, 100, 118, 120, 131, 137, 189, 197, 221, 249

Spaniards 13

Spenser, E. 250

Sputnik 131

Sri Lanka 147

Stamperia Valdonega 47, 129, 133, 149, 216

Staples, D. 16, 22, 185, *223*, 225, 265

State University of New York (SUNY), Buffalo 25

Stein, S. 186

Stevens, J. 220

Stillwater, Oklahoma 162

Stockholm Zoo 17

Storming Heaven 220

Stresser-Péan, G. 136, *138*, 199, *230*; *plate 8*

Strobell, R. 245

Stropharia 44, 74, 235

S. cubensis 98, 99, 169

Summa Antropológica en homenáje á Roberto J. Weitlaner 70

Summer Institute of Linguistics 136, 194, 234, 236

surya 25

Syntex Laboratories 84

Tabernanthe iboga 78

Tagetes 152

T. erecta 152, 154, 158

Tain, Scotland 249, 253

Talocan 68

tamale 159

Tampa, Florida 105

tape recording equipment *plates 15, 18*

Tarascan (Purepecha) 98, 100, 101

Tarascan lexicon 101

Tarbay, G. 11

The Teachings of Don Juan 79

Tehuacán 37, 197, 231, 233

Tehuántepec 136, 234

Teodosio 116, 117, 121

teonanácatl 13, 70, 74, 95, 96, 100, 115, 121, 135, 147, 148, 163, 184, 186

Teopancaxco mural *64*

Teotihuacán 68, 76, 81, 188

Teotitlán del Camino 95, 231

teotlaquilnanácatl 91, 96

Tenango del Valle 99, 103, 235

Tenochtítlan 68, 96

Tepantitla mural 63, 68, *69*, 76, 77

termites 153

terpenes 158

Terrés, H. 63, 92

That Gettysburg Address 126, 130, 220, 225, 256

Thebes 25

Theological Seminary, New York 240

Thévenard, P. 99

Third International Mycological Conference 97

This Week 39, 71, 131

Thomas, K. 123

Thompson, E. 199

thunderbolt 100

The Times Literary Supplement 9

Times Square 189

Tina and Gordon Wasson Ethnomycological Collection 11, 16, 22, *23*, 34, 48, 65, 66, 133, 140, 161, 172, 202, 216, 218, 220, 265; *plate 20*

Tlaloc 68

Tlalócan 68

Tlaxcala University 104

tlitliltzin 75

tobacco 8, 70, 78, 220

Tobacco and Shamanism in South America 78

tonalpohualli 139

Topkis, G. 78

Townshend, P. 131

"Transcultural Use of Narcotic Water Lilies in Ancient Egyptian and Maya Drug Ritual" 64

transmitter-receiver concept of reality 126

transpersonal psychology 205

The Treatises on Superstitions 44

tree of life 68–69

Trejo Leal, A. 106

Trichocereus pachanoi 78

Trinity Church 239

Troncoso, F. 76

tropane alkaloids 158

Troxler, F. 125

Tukanoan Desana 78

Tulancingo 235

Tulane University 106

Turbina corymbosa 15, 69, 136, 151, 153, 158

turkey 159
Turkish 178
Tuxtepec 115
The Twenty Mountains 136

UCLA (University of California, Los Angeles) 77, 78, 79
Ulloa 103, 105
United Nations Truce Commission 255
United States Air Force 194
U.S. Army 210
U.S. Central Intelligence Agency: see CIA
U.S. Department of Health, Education and Welfare 178
University of Baja California 104
U. of Bridgeport 16, 23
U. of California, Los Angeles: see UCLA
U. of California Press 78, 178
U. of Chile 177
U. of Delaware 212
U. of Guadalajara 105
U. of Guanajuato 104
U. of Mexico 84, 88, 103, 104
U. of Morelos 104
U. of Nuevo León 104
U. of Strathclyde 220
U. of Veracruz 104
Unknown Mortals—In the Northern City of Success 172
Upson, W. 233

"Vale" 189
The Varieties of Religious Experience 9, 175
Vatican 210
Venezuela 132
Veracruz, State of 87, 92, 98, 235
Verona, Italy 37, 47, 112, 133, 136, 148, 216, 256
Viet Nam 147
Vindobonensis Codex 100
Virgin de la Soledad 195
Vision 203
Visken 125, 126
Vliet, J. 245
vodka 34
Volador dance 235
Volstead Act 252
Vuelta 63
Vulgate 246

Wabakesi 141
Waiting for Godot 131
Wall Street 7, 55, 72, 211
Wallace, B. 206–207
Warao 78
Ward, A. 243
Washington, D.C. 203, 219, 227, 249
Washington (National) Cathedral 24, 42, 217, 219, 255
Washington, State of 185
Wasson Collection: see Tina and Gordon

Wasson Ethnomycological Collection
Wasson, E. (brother) 240, 242, *242*
Wasson, E. (father) 19, 126, 130, 185, 217, 225, *238*, 239–255, 256
Wasson, J. (uncle) 239
Wasson, M. (daughter): see Britten
Wasson, M. DeVeny (mother) *238*, 239–255, *241*
Wasson, P. (son) 20, 31, 265
Wasson, R. Gordon . . . accepting of imminent death 216 . . . accused of betrayal for publicizing sacred mushrooms 8 . . . accustomed to command 55, 142 . . . acknowledges debt to Roberto Weitlaner 71 . . . acknowledges Wendy Doniger as co-author of *Soma* 56, 71 . . . aged gracefully 44 . . . alerted to connection between Soma and urine 56 . . . alerted to existence of mushroom cults 13–14, 20, 37, 96, 112 . . . aloofness from youth counterculture 131 . . . alleged to be aloof and snobbish 17 . . . allegedly drank urine to test Soma hypothesis 59 . . . almost knifed in Mazatlán 195 . . . amateur status 9, 18, 23, 56, 83, 149, 190, 221, 235 . . . ambivalence toward children 31 . . . appointed Harvard Botanical Museum Research Associate 16, 22 . . . arranges for *Life* to publish his article 199 . . . asks colleague to gather reindeer urine 17 . . . avoided computers, radio and television 215 . . . banking contacts utilized in his research 194, 220, 236 . . . a baseball aficionado in childhood 245 . . . believed psychoactive drugs should be legalized 8, 132, 220 . . . Biblical training 245–247, 248 . . . a bibliophile 47–48 . . . birth 18, 241 . . . bookplates 51–54, *52* . . . both a scientist and scholar 227 . . . breaks news of brother's death to parents 255 . . . carefully prepared for field trips 136, 236 . . . and Carlos Castaneda 79 . . . childhood 217, 239-255 . . . childhood jaunts to New York City 249 . . . children 20, 31–32 . . . chronicles false rumor J.P. Morgan sold defective carbine rifles to U.S. Army 23–24, 210 . . . close bond with older brother Thomas 216–217 . . . co-discovers Mexican Indian use of *Salvia divinorum* 119 . . . commissions G. Mardersteig to print his books 36 . . . compared to Darwin 190 . . . contents of CIA file 212 . . . a cultivated man 148 . . . curiosity 19, 32, 83, 210 . . . death 15, 17, 24, 40-42, 50, 130, 188 . . . derides abuses of Bible study 247 . . . description of home in Danbury 207-208, 218 . . . devastated by older brother's death 217 . . . devotion to wife 56 . . . diary entry when contacted by CIA *213* . . . disdain for CIA experiment with LSD 212 . . . disdain for hippies 33, 44, 131 . . . disdain for sloppy scholarship 33 . . . dispute with Rolf Singer

278

51, 186–187 . . . dissatisfied with existing terms for hallucinogenic substances 16, 55, 67, 147, 185, 211, 225 . . . diversity of his friends 32, 147 . . . diversity of his professional contacts 136 . . . documentary film about 23 . . . drilled in correct grammar by father 247–248 . . . early career as journalist 19-20, 43, 49, 236 . . . eats cobra's liver 59 . . . eats live beetles 33 . . . eats morning glory seeds 75 . . . eats synthetic psilocybin during a velada 122 . . . economy of expression 189 . . . education 7, 19, 55, 245 . . . efficiency 235 . . . elected Fellow of Linnean Society 16, 22 . . . elitist about hallucinogens 8 . . . eloquence/erudition 24, 62 . . . embarrasses grammar school teacher by reading select Bible verses 248 . . . energy 236 . . . enthusiasm 92 . . . excellent memory 185, 256 . . . exceptional writing ability 61, 62, 147, 189, 198 . . . fastidious appearance 17, 55, 114, 142 . . . Father of Ethnomycology 9 . . . financial arrangement with Cleland for house purchase 214 . . . financial arrangement with photographer Richardson 193, 202–203 . . . first taste of beer 254 . . . first outsider (with Richardson) to eat sacred mushrooms in a velada 7, 14, 20, 38, 112, 113, 121, 131, 189, 191 . . . first to widely publicize hallucinogenic mushrooms 7, 84, 165 . . . fondly recalls mother's cooking 243–245 . . . friendliness 55, 56, 147, 183, 209 . . . frugality taught in childhood 243 . . . a gentleman 13, 16, 19, 43, 177, 187 . . . given Walrus tusks by Admiral Peary 251 . . . a good publicist 236 . . . heart murmur caused by childhood rheumatic fever 216 . . . helpful to students 26, 28 . . . helps trigger "psychedelic movement" with *Life* article 212 . . . hikes through Greece as a young man with Axel Boethius 19, 144, 221 . . . his minister father fought Prohibition 217, 251–253 . . . his father's formality 253 . . . honeymoon and start of mushroom quest 33, 77, 111, 161, 221 . . . "horror" of those with "pseudo-religious" approach to telepathy 40 . . . ignored or slighted by some scholarly and scientific peers 190, 217 . . . impressed as a child by publication of father's book 252 . . . innovative thinker 65, 162, 189, 191 . . . insightful 148 . . . intellectual glow of childhood family circle 248 . . . internment at Washington Cathedral 24, 217, 219 . . . interviewed on audio tape by Jonathan Ott 185, 189 . . . intimacy of childhood family circle 248-249 . . . intolerant of foolish conversation 194 . . . introduces idea of union between God and nature to general public 228 . . . joins staff of J. P. Morgan & Co. 20 . . . kindness, generosity, hospitality 28, 62, 114, 147, 191, 219, 236 . . . knowledge

and interests 16, 32, 43, 62, 113, 206–207, 210, 211, 219, 236 . . . laments repressive drug control laws 212 . . . last wishes 42 . . . lasting significance of his discoveries 126 . . . learns in childhood to make his own field trips 256 . . . lectures on his discoveries 200 . . . lived in Europe as an adolescent 19, 217, 249 . . . love of exploring learned in childhood 247 . . . love of scholarship learned in childhood 246 . . . loved nature 24 . . . marriage 20 . . . meets Rolf Singer in the field 84 . . . member of Harvard's Visiting Committee to the Slavic Department 16 . . . memory training in childhood 246 . . . mentally alert in old age 44 . . . meticulous scholarship 17, 26-27, 32, 49, 55, 68, 148, 149, 162, 184 . . . military service 19, 217, 249, 256 . . . mischievous 114, 218 . . . mistakenly approves as safe a potentially toxic mushroom 80 . . . monetary value of his limited editions 7, 47, 49, 55, 62, 129–130, 185 . . . multidisciplinary approach to ethnomycology 20, 23, 33, 62, 68, 111, 148, 161-162, 189 . . . multilingual abilities 24, 32, 48, 84, 142, 189, 221, 236 . . . mushroom messianism 179 . . . named for Dr. R. P. R. Gordon 253–254 . . . named Honorary Life Manager, New York Botanical Garden 23, 187 . . . named honorary member, Mexican Mycological Society 92 . . . named Honorary Research Associate, New York Botanical Garden 22 . . . on relevance of hallucinogens to modern society 9, 124 . . . opinion of Timothy Leary 211–212, 216, 218 . . . a patient teacher 191 . . . persistent in defending his theories 10, 57, 58 . . . personal charm 236 . . . photographs of 2, 12, 21, 28, 30, 32, 41, 86, 88, 89, 117, 128, 134, 142, 146, 150, 160, 196, 200, 223, 230; plates 1, 4, 7, 8, 13, 14, 15, 19 . . . physical characteristics 209 . . . pioneer ethnomycologist 20, 72, 83, 111, 161 . . . plans for a second edition of *Mushrooms, Russia and History* 20, 62–63, 66, 172 . . . predicts resurgence of interest in hallucinogens 214 . . . president, Society for Economic Botany 22 . . . pressured his children to "follow in his footsteps" 31 . . . proudest of *María Sabina and her Mazatec Mushroom Velada* 22, 190, 217 . . . quality of his limited editions 17, 47, 49, 112, 129, 133, 144, 149, 162, 184, 216 . . . rates Bible above all other classical works 247 . . . recalls father's first book as "a big event" 252 . . . receives Addison Emery Verrill Medal 16, 23 . . . receives Distinguished Economic Botany award 16, 22 . . . receives first Pulitzer Traveling Scholarship 19 . . . receives honorary doctorate 16, 23 . . . relations with bank associates 209-211 . . . relied on help from specialists 25, 55, 92, 162, 221, 236 . . . religious back-

ground and beliefs 9, 190, 223, 245–247 . . . respect for Harvard University 49 . . . responsibilities at J. P. Morgan & Co. 210 . . . said to be a connoisseur of food and drink 43 . . . said to be "spartan" about food and drink 55 . . . said to have regretted *Life* article 50 . . . said to have been pleased with results of Life article 228, 236 . . . a scholarly teamworker 92 . . . a scrupulous grammarian 147, 207 . . . self-financed his research 190, 237 . . . sends Hofmann morning glory seeds for analysis 75 . . . sense of humor 17, 19, 114, 218 . . . sensitivity to others' values and feelings 33, 59, 143 . . . shrewd businessman 185, 214 . . . shunned adulation and publicity 23, 44, 219 . . . slept outdoors on studio porch 44, 48, 114, 208, 217 . . . sobriety 223 . . . teaches English at Columbia University 19 . . . tends bar at party in Mazatlán 195 . . . theory Buddha died of mushroom poisoning 114 . . . theory *Claviceps purpurea* was sacramental basis of Eleusinian Mysteries 7, 15, 61, 222 . . . theory decorations on Xochipilli statue represent hallucinogenic plants 44, 148 . . . theory fruit of the Tree of the Knowledge of Good and Evil in the Garden of Eden was a psychoactive mushroom 179 . . . theory hallucinogens inspired origins of religion 9, 15, 51, 179, 189, 216 . . . theory prehistoric Europeans once venerated a mushroom 225 . . . theory Rig Veda's Soma was *Amanita muscaria* 7, 15, 20, 25, 51, 56, 72, 148 . . . theory words engender physiological responses 189 . . . took ancient writers at their word 49 . . . took hallucinogenic mushrooms about 30 times 215 . . . treated Indians as equals 71, 184 . . . treated Indians with respect 236 . . . a trustee of Fitz Hugh Ludlow Library 28 . . . turns down CIA request for his cooperation 212 . . . unconventional approach to scholarship 113 . . . unpretentious 184, 188, 208 . . . welcomed controversy and debate 57, 61–62 . . . willingness to try new approaches 33 . . . withdraws from college to enlist in the U.S. Army 24 . . . writings influenced Timothy Leary to try hallucinogens 8, 56 . . . writings preserve information that might otherwise have been lost 8, 228 . . . zest for life 32, 145
Wasson, T. (brother) 217, 220, 241–255, *254*
Wasson, V. Pavlovna (wife) 7, 8, 20, 25, *30,* 31, *33,* 34, 38, 40, 56, 71, 77, 111, 112, 161, 165, 189, 199; *plate 7*
Wasson, W. (uncle) 239, 253, 254
The Wasson Theory 189
water lilies 10, 63, 65
Watts, A. 129
Watts, J. 113
Wausseine 143

WaussungNaabe 143, 144
The Way of the Animal Powers 113
Webster, J. 168
Webster's International Dictionary 248
Weil, A. 8
Weitlaner, R. 69–71, 72, 75, 115, 135, 195, *196,* 202, 231, 236; *plate 1*
Weitlaner Johnson, I.: see Johnson, I. Weitlaner
Wellesley College 240
Wellesley half-dozen 240
Wells, H. G. 167–168
Wenner-Gren Foundation for Anthropological Research 158
West Bloomfield, New York 168
Where the Gods Reign 219
whistling language 37
Whittaker, G. 44, 76
Whorf, B. 149
Wilbert, J. 78
Wilgas, N. 168, 175
Williams, T. 24
Wilson, E. 189
Winchester, Virginia 201, 202
wine 224
witchcraft 225
Wolff, H. 265
Wolff, K. 265
The Wondrous Mushroom 8, 22, 42, 77, 92, 105, 113, 130, 145, 162, 190, 218, 220, 235, 266; *plate 20*
Wordsworth, W. 250
World War I 19, 217, 249
World Wildlife Fund 219

Xalapa 87, 104, 105
Xochipilli 44, 148
Xolotla 235

Yahweh 246
Yaitepec 233
yajé 78
Yale University 16
Yale University Press 45, 130
Yankees (baseball team) 245
Yaqui 78
yoga 148, 178
Yorkshire, England 17
Yosun Inn 255
Yucatan 99
Yvonne: see Battaglia

Zacatecas 100
Zacatepec 136, 137, 138, 139, 140, 234
Zacualtipan 234
Zapotec 15, 98, 152, 194, 234
Zeiss Contaflex camera 194, *201*
Zempoaltepetl 136
Zenteno 103, 105
Zeus 224
Zoroastrianism 178
Zurich, Switzerland 96

Contributors

Michael R. Aldrich is Curator of the Fitz Hugh Ludlow Memorial Library, San Francisco, the nation's largest private collection of psychoactive drug literature. He is the author of *The Dope Chronicles 1850–1950* and numerous papers on drug history and literature.

Masha Wasson Britten is Associate Dean and Associate Professor, The Decker School of Nursing, State University of New York, Binghamton. She is the author of numerous papers on topics relating to medical/surgical nursing.

J. Christopher Brown is presently a graduate student in Botany at the University of Massachusetts, Amhurst. When employed as a student assistant at Harvard Botanical Museum he helped R. Gordon Wasson catalog and organize the Tina and Gordon Wasson Ethnomycological Collection.

Michael D. Coe is Professor of Anthropology at Yale University and Curator of Anthropology at Yale's Peabody Museum of Natural History. He is the author of *The Jaguar's Children: Pre-Classic Central Mexico, America's First Civilization: Discovering the Olmec* and several other books.

Robert C. Demarest is Director of Libraries for Collier County, Florida. He also owns and operates Mycophile Books, Naples, Florida, a retail firm specializing in books about mushrooms and particularly the works of R. Gordon Wasson.

Wendy Doniger is Mircea Eliade Professor of the History of Religions at the University of Chicago. Her books include *Hindu Myths: A Sourcebook Translated from the Sanskrit* and several others published under the name Wendy Doniger O'Flaherty; and *Mythologies,* an English-language version of Yves Bonnefoy's *Dictionnaire des Mythologies,* now in press under the name Wendy Doniger.

William A. Emboden is Professor of Biology at California State University, Northridge. His books include *Narcotic Plants, Leonardo da Vinci on Plants and Gardens* and *Albrecht Dürer on Plants.*

Peter T. Furst is Professor Emeritus of Anthropology at the State University of Albany and a member of the American Department of the University Museum, University of Pennsylvania. Among his books are *Flesh of the Gods, Hallucinogens and Culture,* and (with Jill L. Furst) *Precolumbian Art of Mexico* and *North American Indian Art.*

Gastón Guzmán, of the Institute of Ecology in Xalapa, Mexico, is a co-founder and past President of the Mexican Mycological Society. His many published works include a monumental book on psychoactive mushrooms, *The Genus Psilocybe: A Systematic Revision of the Known Species Including the History, Distribution and Chemistry of the Hallucinogenic Species.*

Joan Halifax, anthropologist and teacher, is founder of The Ojai Foundation and The Foundation School, an interdisciplinary, cross-cultural educational center in Ojai, California. Her books include *Shamanic Voices, Shaman: The Wounded Healer* and the forthcoming *River of Sorrow, Mountains of Joy.* She also co-authored (with Stanislav Grof) *The Human Encounter with Death.*

Albert Hofmann, the discover of LSD and psilocybin, is the retired Director of the Pharmaceutical-Chemical Research Laboratories of Sandoz Ltd., Basel, Switzerland. His books include *LSD: My Problem Child, Insight-Outlook* and (with Richard Evans Schultes) *Plants of the Gods.*

Michael Horowitz is founder and Director of the Fitz Hugh Ludlow Memorial Library. He also is proprietor of Flashback Books, Petaluma, California, a mail-order bookshop specializing in rare and out-of-print books of the 1960s counterculture. Among his own publications are *Moksha: Aldous Huxley's Writings on Psychedelics and Visionary Experience* (co-edited vith Cynthia Palmer) and *An Annotated Bibliography of Timothy Leary* (co-authored with Karen Walls and Billy Smith).

Irmgard Weitlaner Johnson is a specialist in Prehispanic and contemporary Mesoamerican textiles. Among other affiliations in Mexico, she is past Curator of the *Bodego de Etnografía* at the *Museo Nacional de Antropolgía.* Her numerous published works include *Design Motifs on Mexican Indian Textiles.*

Keewaydinoquay is Mashkikikwe of Miniss Kitigan, Michigan. An erstwhile lecturer on Great Lakes native ethnology, she is the author of *Puhpohwee for the People,* numerous botanical treatises and a series of booklets and cassette tapes recounting Legends of the Ahnishinaabeg.

Weston La Barre is James B. Duke Professor Emeritus of Anthropology, Duke University. He is the author of *The Peyote Cult, The Human Animal, The Ghost Dance: Origins of Religion* and *Muelos: A Stone Age Superstition about Sexuality.*

Frank J. Lipp is Professor, Department of Anthropology, Universidad de las Américas-Puebla, Mexico, and Research Associate, Department of Anthropology, New School for Social Research. He is co-author (with Siri von Reis) of *New Plant Sources for Drugs and Foods* and author of the forthcoming book *People of the Mountains: Religion, Ritual and Healing among the Mixe of Oaxaca.*

Bernard Lowy, Professor Emeritus, Botany Department, Louisiana State University, founded and is Curator of that institution's mycological herbarium. Among other accomplishments during his long and distinguished career in mycology, he served 15 years as a member of the editorial board of *Mycologia.* He is the author of numerous papers on mycology and ethnomycology.

Terence McKenna, author, explorer and lecturer, specializes in shamanism and ethnopharmacology of the Amazon Basin. He founded and is Secretary-Treasurer of Botanical Dimensions, a non-profit botanical research garden in

282

Hawaii for collecting and propagating plants of ethnopharmacological interest. He co-authored (with Dennis McKenna) *The Invisible Landscape* and is the author of *True Hallucinations.*

Claudio Naranjo, a psychiatrist and lecturer, pioneered the use of ibogaine, harmala alkaloids, MDA and MMDA in psychotherapy. He co-authored (with Robert Ornstein) *On the Psychology of Meditation,* and authored *The One Quest* and *The Healing Journey.*

Jonathan Ott is a writer and founder of Natural Products Co., a small chemical manufacturing business. His books include *Hallucinogenic Plants of North America, Teonanácatl: Hallucinogenic Mushrooms of North America* and *The Cacahuatl Eater: Ruminations of an Unabashed Chocolate Addict.*

Allan B. Richardson is a retired commercial photographer now living in Winchester, Virginia. He writes an occasional column for the alumni newsletter of Princeton University, his alma mater, and is working on an article that will provide a firsthand look behind the scenes at the magazine *Vanity Fair* during its heyday in the 1930s and '40s.

Thomas J. Riedlinger is a trade magazine editor for Delta Communications, an Elsevier Business Press company. His freelance writings on psychology have been published in *The Journal of Transpersonal Psychology, Medical Hypotheses* and *Psychedelic Monographs & Essays.*

Carl A. P. Ruck is Professor of Classical Studies at Boston University. In addition to collaborative books with R. Gordon Wasson (*The Road to Eleusis* and *Persephone's Quest*), he is the author of works in epigraphy, translations from the Greek, studies in classical literature, and basic textbooks in both the Greek and Latin languages.

Richard Evans Schultes is Edward Charles Jeffrey Professor of Biology, Harvard University, and Director (Emeritus) of Harvard Botanical Museum. A specialist in the flora and ethnobotany of the Northwestern Amazon, his books include *Where the Gods Reign: Plants and Peoples of the Colomban Amazon,* two collaborative efforts with Albert Hofmann (*The Botany and Chemistry of Hallucinogens* and *Plants of the Gods*) and one co-authored with Robert F. Raffauf (*The Healing Forest: Medicinal and Toxic Plants of the Northwest Amazon*).

Alexander T. Shulgin is a specialist on the chemistry of hallucinogenic substances who lectures on toxicology for the School of Public Health, University of California, Berkeley. In addition to over 150 scientific papers, he has authored a book, *The Controlled Substances Act: A Resource Manual of the Current Status of the Federal Drug Laws,* and with his wife Ann is completing another on the centrally active phenethylamines.

Guy Stresser Péan is a Professor with Ecole des Hautes Etudes, Paris, a researcher with the Centre National de la Recherche Scientifique, and a founder and first Director of Mission Archéologique et Ethnologique Française au Mexique. His many published works on the archeology, history, ethnology and linguistics of Northeastern Mexico include *El arado criollo en México y América Central.*